SYSTEMATIC BOTANY MONOGRAPHS

VOLUME 45

Phylogenetic Systematics and Biogeography of the Tribe Robinieae (Leguminosae)

Matt Lavin
Mario Sousa S.

THE AMERICAN SOCIETY OF PLANT TAXONOMISTS
21 August 1995

SYSTEMATIC BOTANY MONOGRAPHS
ISSN 0373-8211

ISBN 0-912861-45-2

Printed in the United States of America

PHYLOGENETIC SYSTEMATICS AND BIOGEOGRAPHY OF THE TRIBE ROBINIEAE (LEGUMINOSAE)

Matt Lavin
Department of Biology
Montana State University
Bozeman, Montana 59717

Mario Sousa S.
Herbario Nacional, Instituto de Biología
Universidad Nacional Autónoma de México
Apartado Postal 70–367
04510 México, D.F., Mexico

ABSTRACT. A phylogenetic analysis of the predominantly tropical North American tribe Robinieae is presented and based primarily on morphological and phytochemical data. Recent well-supported findings from molecular systematic investigations are integrated by enforcing appropriate topological constraints during parsimony analysis. Clade robustness is evaluated by Bremer support analysis and this reveals two well-supported generic groupings: the *Robinia*-group (*Robinia*, *Coursetia*, *Olneya*, *Peteria*, *Genistidium*, and *Sphinctospermum*) and the *Gliricidia*-group (*Hybosema*, *Poitea*, and *Gliricidia*). The classification of the 12 constituent genera of the tribe is built around these two groups. The *Sesbania*-group (*Sesbania*, including *Glottidium*) is the putative sister group to the rest of the tribe and has the only native occurrence in the Paleotropics. The *Hebestigma*-group comprises the monotypic Cuban endemic, *Hebestigma*. The *Gliricidia*-group tentatively includes also the genus *Lennea*, and is mainly from tropical Mexico and Central America, but with *Poitea* principally of the Greater Antilles. The *Robinia*-group is centered in the drier regions of northern Mexico and southwestern U.S.A., but with *Robinia* restricted to temperate deciduous forests and woodlands of the southern U.S.A. and adjacent Mexico, and *Coursetia* mainly from the seasonally dry tropical forests of North and South America. An ancestral area analysis is performed to explore the feasibility of determining the relative age of Robinieae on the North versus South American continent. This ancestral area analysis is performed explicitly outside the realm of vicariance biogeography and is not akin to centers of origin concepts. The phylogenetic and biogeographic patterns of the genera of Robinieae, if found to be common for Neotropical taxa, will provide important insights into the relationships of Tertiary North American tropical biotas. For example, the closest relatives of Robinieae include the Old World genera *Millettia*, of the tribe Millettieae, and *Wisteria* and *Caragana*, of the tribe Galegeae. There is no evidence to suggest that the sister group of Robinieae comes from South America. These phylogenetic and biogeographic patterns are more consistent with Wolfe's Boreotropical flora hypothesis than Chaney's Neotropical Tertiary Geoflora concept. Formal species-level taxonomic treatments are provided for each genus, except *Poitea*, *Robinia*, and *Coursetia*, for which recent treatments have been published, and *Sesbania*, which is in the process of systematic study. One new variety, *Lennea viridiflora* var. *novogaliciensis*, and one new section, *Coursetia* sect. *Madrenses*, are described. Six new combinations, *Sesbania* sect. *Daubentoniopsis*, *Sesbania* sect. *Glottidium*, *Poitea* sect. *Sabinea*, *Poitea* sect. *Sauvallella*, *Poitea* sect. *Corynella*, and *Gliricidia brenningii*, are proposed.

INTRODUCTION

Tribal- and subtribal-level taxonomies of the tropical woody Papilionoideae are confounded by poorly circumscribed genera. Geesink (1981, 1984) and Polhill (e.g., 1971) detail the complexity of the situation in the tribes Millettieae and Dalbergieae, and Geesink (1984) considers the problem unresolvable due to the paucity of cladistically or taxonomically informative morphological characters. Part of the problem lies with the pantropical distribution of tribal and infratribal taxa. The circumscription of such taxa has been influenced by provincial taxonomies or biogeographic hypotheses that assume a Gondwanan origin of the family Leguminosae. The difficulty of conducting a comprehensive study of a pantropical taxon is exacerbated by potentially erroneous suppositions that Paleotropical genera should connect to the New World via South American taxa. The tribe Robinieae, however, provides a good starting point from which to resolve the generic relationships among these tropical papilionoid groups. In part, this is due to its concentration in the Neotropics, and there concentrated in North America. By determining the phylogenetic relationships of this primarily North American legume group, it may be possible to begin to draw preliminary conclusions about general biogeographic patterns for tropical taxa, which can be evaluated by performing similar studies on other tropical groups.

This approach to providing taxonomic resolution to the higher taxa of Robinieae, particularly genera and generic groupings, includes a species-level revision, and building from this by searching for shared derived characters from among the species. Such data are then used to circumscribe monophyletic genera and higher level taxa. With the present study, all genera of the tribe Robinieae, except *Sesbania,* have been the subject of species-level monographs. The result has been the successful circumscription of monophyletic genera and, to a large degree, higher taxa, including generic groups, as well as possibly the tribe itself. It is hoped that the phylogenetic framework provided here will facilitate other such studies of related Papilionoideae, particularly from the tribes Millettieae and the woody component of Galegeae.

TAXONOMIC HISTORY

The legume genera included in the present circumscription of Robinieae have been recognized, more or less, as a distinct element in the Papilionoideae since Rydberg's (1923a, 1923b, 1924a, 1924b) classic treatments. He classified these genera in the large heterogeneous tribe Galegeae, mostly in the subtribe Robiniinae. Other genera now included in Robinieae were variously aligned in several different subtribes of the tribe Galegeae. The genus *Sesbania* and its segregate genera, for example, were included in the subtribe Sesbaniinae, *Peteria* was allied with *Tephrosia* (then *Cracca* L.) and placed in subtribe Craccinae, and the species of *Poitea* with paripinnate leaves were placed in the subtribe Corynellinae (as the genera *Sabinea*, *Corynella*, *Notodon*, and *Bembecidium*). Thus, it is Rydberg's subtribe Robiniinae that forms the nucleus around which the present tribe Robinieae is built.

Rydberg (1924a) characterized subtribe Robiniinae as a woody native-American group closely allied to the subshrubby elements of subtribe Astragalinae, including *Homalobus* Nutt. and *Kentrophyta* Nutt. He suggested little difference between Robiniinae and these genera of Astragalinae, but he maintained the two as distinct, because Robiniinae was predominantly New World, tropical, and woody, and Astragalinae was, in contrast, predominantly Old World, temperate, and herbaceous. It is doubtful Rydberg would have maintained the distinction had more of the woody Old World members of Galegeae been present in North America.

Polhill and Sousa (1981) made the next significant contribution to the taxonomy of the tribe by ranking the entire assemblage, including the subtribes Sesbaniinae, Robiniinae, and Corynellinae, the genus *Peteria*, and additionally the monotypic subtribe Diphysinae, at the tribal rank, as Robinieae. They excluded the possibility of any potentially close relationship with the genera of Galegeae (including Astragalinae) because, in part, this latter tribe was characterized by epulvinate leaves lacking stipels, in addition to being a predominantly temperate, herbaceous Old World group. A relationship with Astragalinae was not emphasized, especially because Polhill and Sousa found stronger affinities of Robinieae to the genera of the tribes Aeschynomeneae, Dalbergieae, and Millettieae (then Tephrosieae). This newly characterized Robinieae was divided into the barbistyled and glabristyled species; that is, those species with a pollen brush on the style and those with a glabrous style. The barbistyled genera were characterized as bearing stipitate glands on the herbage and inflorescences and having spinescent stipules, whereas the glabristyled genera lacked such indumentum and stipules. The genus *Peteria* was separated from *Tephrosia* because it possessed all of the diagnostic characteristics of the barbistyled Robinieae. This latter group otherwise included the genera *Robinia*, *Olneya*, *Sphinctospermum*, *Genistidium*, and *Coursetia* (then segregated into the genera *Cracca* Benth., *Neocracca*, and *Poissonia*). The glabristyled genera included *Hebestigma*, *Lennea*, *Gliricidia* (then including *Hybosema*), *Yucaratonia* (here included in *Gliricidia*), and *Poitea* (then divided into six segregate genera). This group also included *Sesbania*, which was considered anomalous but better included in Robinieae than placed elsewhere, because, in part, the seeds accumulated the non-protein amino acid canavanine, as in the rest of Robinieae. The genus *Diphysa* was also included and allied with *Sesbania*. Polhill and Sousa did not intend to portray their treatment as definitive, but only as a preliminary step to improving the taxonomy of the woody Papilionoideae. Indeed, many aspects of their work are preserved here.

Lavin (1987) presented a cladistic analysis of the tribe Robinieae, as circumscribed by Polhill and Sousa (1981), in an effort to find appropriate outgroups for a more detailed phylogenetic systematic revision of the genus *Coursetia* (Lavin 1988). The notable outcome of this study was that the barbistyled genera of Robinieae, including *Coursetia*, were determined to be monophyletic. They were informally referred to as the *Robinia*-group in an effort to put less emphasis on the single character of the style brush, since the glabristyled genera appeared to be a paraphyletic group. Another finding from this cladistic analysis was the exclusion of three genera previously included in the tribe. The genus *Diphysa* was determined to be part of the tribe Aeschynomeneae subtribe Ormocarpinae. This taxonomic placement was based on many morphological attributes distinctive to this subtribe and has found additional support recently (Sousa & Delgado 1993). Also excluded was *Poecilanthe*, a genus referred to Robinieae only by Geesink (1984). The exclusion of *Poecilanthe* is still supported in the present analysis. The genus *Sphinctospermum* was also excluded, based primarily on the distinctive morphology of the genus, most closely approximated by *Tephrosia* subg. *Macronyx*. Yet, subsequent cladistic analyses of morphological and chloroplast DNA data strongly suggest that *Sphinctospermum* is very closely related to the *Robinia*-group of the tribe Robinieae (Lavin 1990; Lavin & Doyle 1991); hence, it is included in the present monograph of the tribe.

The taxonomy of two of the larger genera in the tribe has followed a pattern of broad generic circumscriptions (Lavin 1988, 1993). The genus *Coursetia* was circumscribed to include three previously recognized genera (*Cracca*, *Neocracca*, and *Poissonia*), because *Coursetia* and *Cracca* as then circumscribed were paraphyletic with respect to each other and to *Poissonia* and *Neocracca* (Lavin 1988). The West Indian genus *Poitea* was also broadly circumscribed to include *Sauvallella*, *Sabinea*, *Corynella*, *Notodon*, and *Bembecidium*, because these were either not monophyletic or were only weakly differentiated (Lavin 1993). Aside from this treatment of *Poitea*, the taxonomy of the glabristyled genera has not been comprehensively enumerated, excepting the preliminary phylogeny presented in Lavin and Luckow (1993). Thus, the present monograph provides the first comprehensive taxonomy based on phylogenetic relationships of all constituent genera of the tribe.

PHYLOGENY

Though the phylogenetic relationships of the genera of Robinieae have been estimated previously (Lavin 1987; Lavin & Doyle 1991; Lavin & Luckow 1993), a more thorough character analysis is presented here with a comprehensive sample of putative and potential sister groups in a cladistic analysis. Aside from more precisely defining certain character states and correcting a few mistaken character state assignments, the most significant change from previous analyses is the inclusion of both *Wisteria* and *Caragana* (both of the tribe Galegeae) among the outgroup genera. These two woody temperate genera represent a lineage of mostly temperate Old World legumes that have lost the chloroplast DNA inverted repeat (Lavin et al. 1990; Liston 1995). The reason for including representatives of this particular group is that a cladistic analysis of the entire family Leguminosae (Chappill 1995) and recent nuclear DNA evidence from the phytochrome gene family (Mathews et al. 1995; Lavin et al. unpubl.) suggest that Robinieae is either a sister group to these mostly temperate Old World legumes or is the clade from which they were derived. Closer inspection of these genera, especially of *Caragana*, has revealed some notable morphological similarities with genera of Robinieae that have heretofore gone undetected (see discussion of the outgroup genera below).

The following cladistic analysis is based on morphological and phytochemical data. Though molecular data have been and are being gathered for these legume groups, the details are not presented here. However, there are two findings from molecular investigations on these legume genera that are becoming well established and that are used here to constrain the parsimony analysis. The first is the finding from chloroplast DNA restriction site analysis on the inverted repeat region that strongly suggests that the tribe Robinieae as here defined is monophyletic (Lavin & Doyle 1991; Lavin, unpubl.). The second is that the papilionoid legumes that have lost the chloroplast DNA inverted repeat region (e.g., *Wisteria* and *Caragana*) form a monophyletic group (Lavin et al. 1990; Liston 1995).

The data set of 25 terminal taxa and 37 morphological and phytochemical characters (table 1) was subjected to parsimony analysis with the computer program PAUP version 3.1 (Swofford 1993) and Hennig86 version 1.5 (Farris 1988). All characters (bi- and multistate) were unordered (non-additive) during all analyses, except as noted in the discussion of characters. Strict consensus trees were gener-

Table 1. Data matrix of 25 terminal taxa by 37 characters for the genera of the tribe Robinieae and putative sister groups. Character states 0, 1, and 2 are described for each character in the phylogeny section; ? = missing data, A = multistate taxon, B = uncertain state taxon. See phylogeny section for a discussion of the coding of each genus and the ordering of multistate characters.

	5	10	15	20	25	30	35	
Dalbergia	01000	00000	00000	00000	00000	00000	10000	00
Ormocarpinae	01001	11000	02000	01100	00000	01000	10000	00
Poecilanthe	00000	01100	00010	00010	00000	01000	00100	00
Ostryocarpus	10001	00101	10010	01000	00100	00000	00101	00
Philenoptera	10011	00101	11110	01100	10100	00000	00101	00
Kunstleria	10001	00101	11110	00000	00100	10200	10101	00
Platycyamus	00111	01101	10101	00010	10100	00100	00101	00
Lonchocarpus	10111	10101	11110	0A000	01100	10200	00101	00
Tephrosia	12111	11101	11112	01020	01100	00100	10101	00
Schefflerodendron	10111	10101	00010	00021	??000	01000	00101	00
Millettia sect. *Podocarpae*	10111	10100	00110	01021	01000	11100	00101	00
Caragana	00001	10110	00110	01021	10000	11010	21101	10
Wisteria	12101	11101	10110	01021	10000	0000A	00101	00
Sesbania	10001	10100	00110	01A11	10000	01000	11001	00
Hebestigma	12101	10100	00110	01021	10000	01000	00111	00
Lennea	12101	12100	01111	01021	10000	01000	00111	01
Hybosema	02101	12110	01110	01121	10001	01000	10121	01
Poitea	020A1	12110	00110	01121	10001	110A0	11121	A0
Gliricidia	02101	10110	A0110	01121	10001	11000	00121	00
Robinia	12101	11100	00112	11021	10010	01001	20111	00
Coursetia	02101	11100	00112	11021	10010	01001	20121	A0
Olneya	12101	11100	00112	11021	10010	01001	20111	11
Peteria	12101	11100	00112	11021	10010	010B1	20111	01
Genistidium	12001	11100	00112	11021	10010	01011	20111	01
Sphinctospermum	12001	11100	00112	11021	10000	01011	20111	00

ated to determine which clades were resolved consistently among the most parsimonious sets generated in a particular analysis. Adams consensus were also generated in order to identify those terminal taxa that were the most variably placed among the set of most parsimonious cladograms. Bremer support analysis (Bremer 1988; Källersjö et al. 1992) and permutation tests were performed with RNA (Farris 1994; see also Källersjö et al. 1992, and Farris et al. 1994). Enforcing topological constraints to reflect the findings of molecular data were performed with PAUP. PAUP also allowed the analysis of the data matrix (table 1) that had missing data, multistate taxa, and uncertain state taxa; this data matrix was subject to the heuristic search option and enacting the various options, such as steepest descent and the random addition sequence. Character analysis, tree comparisons, and graphical output were performed with Phylip version 3.52c (Felsenstein 1993), Component version 2.0 (Page 1993), and MacClade version 3.01 (Maddison & Maddison 1992). The programs Hennig86, RNA, Component, and Phylip were run on a Zenith 486/25 with 4 MB RAM, whereas PAUP and MacClade were run on a Macintosh 950 Quadra with 8 MB RAM.

The 25 terminal taxa used in this analysis include the 12 genera of Robinieae plus 13 genera from the tribes Aeschynomeneae, Dalbergieae, Galegeae, and Millettieae. These last four tribes have been considered to be very closely related to Robinieae (Geesink 1981, 1984; Lavin 1987, unpubl.; Polhill & Sousa 1981; Sousa & de Sousa 1981), and the representative genera from these tribes were selected on evidence that they might potentially be a sister group to the tribe Robinieae. The characters that suggest this are enumerated in the discussion of each outgroup genus (see below). Also, all potential sister group genera from South America (e.g., *Platycyamus* and *Poecilanthe*) were scrutinized so as to test most thoroughly the alternative biogeographic hypothesis outlined in the section following this phylogenetic analysis.

Ingroup Terminal Taxa: a Synopsis of the Tribe Robinieae

The emphasis in this synopsis is on the autapomorphies of each genus in Robinieae, and thus, implicitly, on the justification for use as a terminal taxon in the cladistic analysis. A detailed discussion of each of the taxa listed below is provided in the taxonomic section following the phylogeny and biogeography sections.

I. The *Sesbania*-group

1. *Sesbania*: non-explosively dehiscent legumes with a well-developed, often spongy, inner epidermis; paripinnate leaves with a leaflet nyctinasty consisting of forward movements; base chromosome number x = 6.

II. The *Hebestigma*-group

2. *Hebestigma*: conspicuously penicillate stigma; opposite to subopposite branching; leaflets abaxially concave.

III. The *Gliricidia*-group

3. *Lennea*: distally incurled style; diminutive flowers and inflorescence rachis relative to the size of the legumes, which at maturity are large and woody.
4. *Hybosema*: standard callus single along the midrib in the region of the nectar-guide.
5. *Poitea*: short shoots with distichously arranged stipule-pairs; flowers with distinctly long keel petals; legumes with thinly woody legume valves, which are readily explosively dehiscent.
6. *Gliricidia*: standard petals fully reflexed (180° backward) at anthesis; the persistent calyx becoming leathery, or nearly so, at fruit maturation.

IV. The *Robinia*-group

7. *Robinia*: legume valves weakly elastically dehiscent; seeds oblique and borne on a long funiculus.
8. *Coursetia*: style with a latrorse pollen brush; keel petals rostrate; legumes with seed compartments caused in part by the inflexion of the pericarp between seeds.
9. *Olneya*: flower buds and flowers with calyx lobes overlapping, not valvate, in bud; seeds orbicular; stout, recurved spinescent stipules.
10. *Peteria*: inflorescence a terminal raceme; herbs arising from extensive underground tuber-systems; trichomes in the canaliculate portion of the leaf rachis distinctly more dense than on the other parts of the leaf.

11. *Genistidium*: trifoliolate leaves; inflorescence often a new shoot with solitary flowers borne in the axils of trifoliolate leaves basally and unifoliolate leaves distally.
12. *Sphinctospermum*: annual habit; simple linear leaves; seeds with a central constriction.

Outgroup Terminal Taxa: Putative Sister Groups of Robinieae

The following discussion of each of the outgroup genera is limited to the evidence that suggests a sister group relationship to the tribe Robinieae.

1. *Dalbergia* L. f. This pantropical genus of about 100 species is most diverse in South America and tropical Asia (Prain 1904; Pittier 1922; Polhill 1981a). Representative of the genera of Dalbergieae, the combination of a well-developed floral hypanthium and a forward type of leaflet nyctinasty in *Dalbergia* approaches Robinieae, especially *Sesbania*. Dalbergieae and Robinieae have been suggested to be closely related (Polhill 1981a).

2. Aeschynomeneae subtribe Ormocarpinae Rudd. This group, as circumscribed by Rudd (1981), and emended by Lavin (1987) and Sousa and Delgado (1993), includes a distinct woody component of the tribe including *Ormocarpum* P. Beauv. (widespread in the Paleotropics), *Ormocarpopsis* R. Vig. (endemic to Madagascar), *Belairia* A. Rich. (endemic to Cuba), *Pictetia* DC. (centered in the Caribbean), and *Diphysa* Jacq. (centered in Mexico). This subtribe also includes three herbaceous genera centered in Mexico, *Fiebrigiella* Harms, *Chaetocalyx* DC., and *Nissolia* Jacq. The subtribe Ormocarpinae is used as the outgroup, and the characteristics of the woody genera are taken as representative. It is not possible to select one of these woody genera as the outgroup, because generic circumscriptions, which are distinguished primarily by fruiting characteristics (Rudd 1981), are problematic. For example, certain species of *Diphysa* have flat, articulated fruits that do not develop vesicular valves, like those of *Ormocarpum*. The other species of *Diphysa* have non-articulated fruits that have vesicular valves, like those of *Belairia* and *Ormocarpopsis*. Like Dalbergieae, Ormocarpinae approach Robinieae, and especially *Sesbania*, by having a well-developed floral hypanthium and leaflets with a forward leaflet nyctinasty.

3. *Poecilanthe* Bentham. This genus of most likely eight species is confined to tropical South America and has been placed in Dalbergieae (see Polhill 1981a; Lavin 1987), Millettieae (Geesink 1981), and Robinieae (Geesink 1984). It distinctively approaches Robinieae in its inflorescence, which comprises axillary racemes with single flowers at each node, and calyces with well-developed lobes, a characteristic of the *Robinia*-group. See "Excluded Taxa" for further discussion of this genus.

4. *Ostryocarpus* Hooker f. A member of Millettieae, this genus of six species occurs in tropical Africa and southeast Asia, and includes in synonymy the well-known genera *Aganope* Miq., *Ostryoderris* Dunn, and *Xeroderris* Roberty (Geesink 1984). The following combination of characters in *Ostryocarpus* is suggestive of a relationship with Robinieae: inflorescence nodes bearing solitary flowers (even though the flowers are arranged in terminal and axillary panicles), wing petals that are free from the keel petals, and staminal filaments in a diadelphous arrangement (though lacking basal fenestrae).

5. *Philenoptera* A. Richard. A member of Millettieae with 15 species restricted to tropical Africa, *Philenoptera* includes the species formerly referred to *Lonchocarpus* sect. *Paniculati* Benth. (Geesink 1984). A putative relationship of this genus with Robinieae is suggested by its inflorescence nodes, which bear solitary flowers (although arranged in terminal and axillary panicles), wing petals free from keel petals, and seeds that accumulate canavanine.

6. *Kunstleria* Prain. The eight species of this genus of Millettieae occur in southeast Asia (Geesink 1984). A possible close relationship of *Kunstleria* with Robinieae is evidenced by its flowers with wing petals only loosely adherent to the keel. The style of *Kunstleria* is ciliate just at the base, and the ovarian trichomes do not trail up into the style, as in, for example, *Lennea*. Thus, this genus is coded in the phylogenetic analysis as having a glabrous style (see character #15, below). Geesink reports that the inflorescences of *Kunstleria*, like those of Robinieae, consist of solitary flowers at each node (though arranged in axillary panicles); however, all specimens observed during this analysis had pseudoracemes of the biflorous type that were described for *Lonchocarpus* by Sousa and de Sousa (1981).

7. *Platycyamus* Bentham. This is one of the few genera of Millettieae restricted to South America, with one species each in Brazil and Peru (Geesink 1984). A close relationship of *Platycyamus* with Robinieae is suggested by its flowers, which have wing petals free from the keel petals and a diadelphous staminal tube (though without basal fenestrae), and its seeds, which accumulate canavanine. Also, the calyx lobes are well developed (similar to those of members of the *Robinia*-group), the style is ciliate and the ovary densely hairy (like in *Lennea*), and the standard petals are glabrate on the abaxial surface (like in nearly all Robinieae but unlike most other Millettieae). Unlike Robinieae and most of its putative sister groups, *Platycyamus* has anthers in which the connective is very narrow and not distinctly pigmented. Also unlike Robinieae is the pseudoracemose inflorescence of *Platycyamus*. Geesink (1984) suggested that the flowers of *Platycyamus* are solitary at rachis nodes and thus like the simple raceme of Robinieae; however, the nodes of the inflorescence of *Platycyamus* actually bear a fascicle of flowers along a short shoot. Such an arrangement is common to many genera of Millettieae.

8. *Lonchocarpus* H. B. K. This genus of Millettieae, according to Sousa and Delgado (1993), is characterized by a pseudoracemose inflorescence in which two flowers are borne on a common pedicel (the "biflorous" inflorescence). *Lonchocarpus* includes about 150 species and is most diverse in tropical North America and secondarily in South America. It may be very closely related to tropical Asian *Paraderris* (Miq.) Geesink, a genus that is also characterized by an inflorescence in which each pedicel bears two flowers (Geesink 1984; Sousa unpubl.). Although few, if any, morphological data suggest a relationship with Robinieae, *Lonchocarpus* is used as an outgroup anyway, because it is often considered to be part of a lineage of Millettieae that is most divergent from Robinieae (see Sousa & de Sousa 1981). It is thus included to determine how other outgroup genera are related to it and to Robinieae.

9. *Tephrosia* Persoon. Centered primarily in Africa, Asia, and Mexico, this genus of over 400 species is a member of Millettieae. Its ebracteolate flowers with well-developed calyx lobes and its elastically dehiscent legumes often with many (e.g., 10–20) ovules suggest that it is possibly closely related to Robinieae. Sousa and de Sousa (1981) allied *Tephrosia* with Robinieae and specifically excluded it

from *Lonchocarpus* and close allies (their subtribe Lonchocarpinae) based on the dichasial cymose (determinate) inflorescences of the latter, unlike the racemose (indeterminate) inflorescences of *Tephrosia* and Robinieae.

10. *Schefflerodendron* Harms. *Schefflerodendron*, comprising four species confined to tropical Africa (Geesink 1984), is a member of Millettieae, and is potentially closely related to Robinieae by virtue of its simple axillary racemes, mature legumes with thick-walled tanniniferous valves, and flowers with a diadelphous staminal tube (but without basal fenestrae).

11. *Millettia* Wight & Arnott. This genus of about 90 species is centered in Africa and Asia (Dunn 1912), but five species referred to the Mexican genus *Hesperothamnus* Brandegee might be part of this group (Geesink 1984). Unfortunately, the genus *Millettia* traditionally has been the repository for all otherwise undistinctive species of Millettieae with elastically dehiscent legumes. Thus *Millettia* is very likely paraphyletic, as recently revealed by Liston (1995), who has shown that some species (e.g., *Millettia japonica*, *M. reticulata*, and *M. atropurpurea*, all placed at one time or another in the genus *Callerya* Endl.) lack the chloroplast DNA inverted repeat region, unlike other species of this diverse genus (Lavin et al. 1990). Also, Evans et al. (1985) found that certain sections of *Millettia* accumulate canavanine in the seed, whereas others accumulate hydroxyhomarginine. *Millettia* sect. *Podocarpae* is used as the representative of the heterogeneous *Millettia,* because, of all subgroups in the genus, it is probably most closely related to Robinieae (Geesink 1984). The inflorescence rachises bear single flowers at each node at maturity, though developmentally they are initiated with two (thus they are coded for having pseudoracemose or pseudopaniculate inflorescences). The flowers are ebracteolate at maturity, though having bracteoles very soon caducous, and have a well-developed hypanthium and no intrastaminal disk. Unlike Robinieae, *Millettia* sect. *Podocarpae* accumulates hydroxyhomarginine in the seed.

12. *Caragana* Fabricius. This mostly cool-temperate genus of about 80 species is centered in eastern Europe and Asia (Polhill 1981b). Until the advent of the phytochrome DNA data set (Mathews et al. 1995; Lavin et al. unpubl.), the remarkable similarities of this genus with those of Robinieae went undetected, and now the addition of this taxon has singularly changed the phylogenetic perspective of the tribe Robinieae relative to previous analyses (see discussion below). The combination of its generally yellowish flowers and paripinnate leaves is markedly similar to *Sesbania*, and its spinescent stipules, an uncommon trait in papilionoid legumes, are like those of the *Robinia*-group. Also, its calyx tube, which persists into fruit, is very similar to that found in *Hybosema*, *Poitea*, and *Gliricidia*. Possibly as an adaptation to cool-temperate climates, *Caragana* forms bud-scales by a fusion and lamination of the stipule-pairs on short shoots, somewhat like those found in *Sesbania* and *Poitea*. Unlike in these two genera of Robinieae, the fused stipules of *Caragana* occur only on the short shoots; the spinescent stipule-pairs at each node on the long shoots are not connected.

13. *Wisteria* Nuttall. This genus comprises eight species of mostly lianas. *Wisteria* is native to cool-temperate to subtropical regions of China, Japan, and eastern United States; the single species native to eastern North America has been included in *Wisteria* s.s., whereas the seven species native to temperate and subtropical eastern Asia have been placed in the segregate genus *Rehsonia* Stritch (Stritch 1984, 1985). Regardless, *Wisteria* and *Rehsonia* are most closely related to each

other (Stritch 1985) and are considered here together under *Wisteria*; the generic segregation of *Rehsonia* probably is unwarranted. The flowers of *Wisteria* occur singly from nodes along the inflorescence rachis (forming a simple raceme), lack bracteoles, and have wing petals free from the keel, and are thus very similar to those found among the species of Robinieae. Also, one species of *Wisteria*, the native North American *W. frutescens* (L.) Poir., occasionally forms glandular trichomes, though very distinct in morphology from those found in the *Robinia*-group; a persistent whitish "spaghetti-like" glandular exudate is exuded by a relatively robust trichome. Because of the variable presence in just this one species and distinct morphology, *Wisteria* was coded as a multistate taxon for this particular character.

Character Analysis

The following character systems, whose direction of evolution was determined by the use of the above outgroups in a cladistic analysis, were selected by two criteria. First, they displayed fixed variation (one character state occurs uniformly in a set of taxa, and the alternative state occurs uniformly in the remaining set). Second, they presumably have had independent evolutionary origins. This latter criterion was estimated by ensuring, as much as possible, that two or more of the following characters were not simply different descriptions of essentially the same attribute. Using the options provided in PAUP, polymorphic taxa were coded as either multistate taxa or as uncertain-state taxa.

I. Accessory Reproductive Characters

1. Pedicels: articulated with calyx = 0; confluent with calyx = 1. This zone of disarticulation is not so much based on whether there are textural or indumentum differences at the interface of the pedicel and calyx (as in Lavin 1987), but by whether the flower falls, after abortion, separate from, or firmly attached to, the pedicel.

2. Bracteoles: present = 0; present and distinctively persistent on the distal end of the pedicel even after floral abortion = 1; absent (presumably lost) = 2. Though bracteoles are very reduced in *Caragana* (typical of many genera of Galegeae), they are nevertheless present. Many genera of Dalbergieae and Aeschynomeneae have bracteoles that persist usually as a pair on the distal end of the pedicel after floral abortion, whereas the genera of Millettieae and many other woody papilionoid groups have flowers with early caducous bracteoles. These persistent bracteoles of Dalbergieae and Aeschynomeneae subtend the joint at the base of the calyx. A step matrix for this character was implemented during an analysis with PAUP from the following: 1 ↔ 0 ↔ 2. Thus a transformation from persistent bracteoles to the absence of bracteoles requires two steps. There is a two-fold bias against this transformation relative to one that involves a change from caducous stipules to either persistent or absent.

3. Standard claw: gradually tapering from the blade = 0; abruptly contracted and strongly reflexed from the blade = 1. Standard petals from flowers that are bird-pollinated (e.g., most species of *Poitea*), flowers with elongate-tubular calyces (e.g., *Caragana*), and very small flowers (*Genistidium* and *Sphinctospermum*) often

lack the abruptly contracted claws and inflexed auricles. Regardless, some genera and generic groups lack the standard petal with a contracted claw and inflexed auricles in spite of pollination syndromes, or flower shapes and sizes (e.g., *Dalbergia*, *Poecilanthe*, and Aeschynomeneae).

4. Wing petals: free from keel petals = 0; firmly adnate to keel petals by a boss on the wing that fits into a socket of the keel petal = 1. The boss-and-socket joint (see Lavin 1993) attaches the wings firmly to the keel petals, and this joint is held fast by the adnation of the wing petal to the keel in the region of this joint. Contrary to Geesink's report (1984), the wing and keel petals of *Platycyamus* are adnate in the region of the boss-and-socket joint. A derived subgroup of *Poïtea* (the *Poitea florida*-alliance) has flowers with wing and keel petals connected by a boss-and-socket joint, but the petals are not adnate in this region as they are in the genera of Millettieae and Galegeae. The genus *Poitea* is coded as a multistate taxon for this character, but in all analyses it was consistently grouped within the genera of Robinieae, which otherwise all lack the boss-and-socket joint. This is consistent with earlier findings that the boss-and-socket is derived within the genus (Lavin 1993). *Poitea* could be coded 0 not only because of these results, but also because the wing and keel petals are not firmly adnate in the species with the boss-and-socket joint.

5. Keel petals: free = 0; valvately connate = 1. According to Tucker (1987a), fusion of the keel petals occurs by either epidermal appression (as characterized by *Lupinus* L.) or by interlocking marginal hairs (e.g., as characterized in *Aeschynomene americana* L.). These two conditions have not been surveyed intensively among the taxa included in this analysis, because cursory inspection with the light microscope of the fused keels of various genera reveals little distinction. Thus, only valvately connate is scored when the keel petals are distally fused. Keel petals that simply overlap at their distal ends are coded as free.

6. Distal end of keel: obtuse = 0; falcate = 1. Blunt and rostrate describe the two extremes of continuous variation found in the falcate (asymmetrical) keel petals. The obtuse keel petals are ones that are essentially bilaterally symmetrical.

7. Calyx lobes: small, less than the length of the calyx tube, and evenly spaced (sometimes resulting in a truncated calyx if lobes are very small) = 0; well developed, as long or longer than the length of the calyx tube, and essentially evenly spaced = 1; unequally spaced with two vexillary and three carinal resulting in a conspicuously bilobed calyx = 2. These character states were ordered 1 ↔ 0 ↔ 2, because there is no case (at least in the genera used in this analysis) where well-developed lobes are unequally displaced to both the carinal and vexillary sides of the calyx tube. The bilabiate condition occurs when lobes are less than one-half the length of the calyx tube.

8. Carinal-most calyx lobe: markedly longer than the others in the same calyx and diverging from the calyx tube at a distinct angle = 0; similar to the adjacent lateral calyx lobes = 1. The calyces in *Dalbergia* and Aeschynomeneae have a distinctly enlarged carinal calyx lobe that diverges from the calyx tube at a distinctly different more reflexed angle than the lateral and vexillary lobes. This condition could be considered ancestral to the symmetrical arrangement of calyx lobes, because it is found also among the genera of Sophoreae (e.g., *Myrospermum* Jacq. and *Myroxylon* L. f.), which are considered as some of the earliest divergent lineages of the papilionoid legumes (Herendeen 1995).

9. Calyx tube in fruit: caducous = 0; persistent = 1. The calyx tube that

persists into fruit maturation grows in size and thickness, and thus is usually neither ruptured nor becomes withered until at or after fruit maturation. The calyx of *Gliricidia*, which becomes leathery upon fruit development and maturation, is considered to represent an additional derivation of this character.

10. Hypanthium: well developed and usually nectariferous = 0; absent or essentially so = 1. When the calyx is articulated with the pedicel, a hypanthium is distinctive as an attenuate stipitate-like base of the calyx. When no zone of articulation is present, the well-developed condition of the hypanthium is manifest as a conical-attenuated base, often differentially pigmented from the distal portion of the calyx tube. A flower lacking a hypanthium is distinguished by the calyx tube that abruptly expands from the distal end of the pedicel, and the staminal filaments and petal claws arise directly from this point of abrupt expansion.

II. Androecial Characters

11. Floral nectaries: from hypanthium = 0; from intrastaminal nectariferous disk = 1. The conditions of this character are not dependent on those of character #10. For example, flowers with a hypanthium can produce nectaries from an intrastaminal disk (e.g., *Gliricidia brenningii*) and flowers essentially lacking a hypanthium can also lack the intrastaminal disk (e.g., *Schefflerodendron*). For this character, *Gliricidia* was coded as a multistate taxon (see discussion of *G. brenningii* in the taxonomic section). It could just as well have been coded 0 for this particular trait because of its consistent grouping with the genera of Robinieae, *Poitea* and *Hybosema*, which lack the intrastaminal disk.

12. Stamens: 9+1 diadelphous arrangement = 0; pseudomonadelphous arrangement = 1; 5+5 diadelphous arrangement = 2. Pseudomonadelphous staminal tubes have a vexillary filament that is free at base but fused to the others so it forms a staminal tube in the mid-section of the filament. The 5+5 diadelphous staminal tube is an autapomorphy of the Ormocarpinae in this analysis, and it consists of a staminal column that splits along the vexillary and carinal margin. When the ovary begins to mature into fruit, the two halves of the column often become upwardly reflexed just before they fall or wither. Because it is most likely that the 5+5 arrangement has no direct link to the pseudomonadelphous staminal tube, a step matrix for this character was implemented from the following transformation: 1 ↔ 0 ↔ 2.

13. Basal fenestrae of the staminal tube: absent = 0; present = 1. In certain papilionoid legumes, a pair of fenestrae occurs at the base of the staminal tube on the vexillary side. Fenestrae are located just inside the flower from the nectar-guide on the standard petal, and presumably allow insect visitors to access the nectaries. The fenestrae are absent if the vexillary filament is adnate to at least one of the adjacent filaments at its base. If the fenestrae are present, the vexillary filament is always completely free at the very base from adjacent filaments. The presence of basal fenestrae is not dependent on the vexillary filament's being free at just the base and fused to the tube at its mid-section (i.e., the derived condition in character #12). Many genera with 9+1 diadelphous staminal tubes also form distinct pairs of basal fenestrae (e.g., all genera of Robinieae; illustrated in Lavin 1990); however, all 5+5 diadelphous staminal tubes so far observed lack basal fenestrae.

14. Anther connective: inconspicuous = 0; broad and distinctly pigmented

relative to the anther sacs = 1. Because Sophoreae and Dalbergieae lack a well-developed anther connective, the inconspicuous connective is considered the putative ancestral condition in this analysis. Crisp and Weston (1987) ultimately used the inconspicuous anther connective, in part, to group the Australian *Templetonia*-group with the American Brongniartieae, so in their study the final cladograms suggest the reverse polarization.

III. Gynoecial Characters

15. Style: glabrous = 0; ciliate = 1; with a pollen brush = 2. The pollen brush is defined specifically as trichomes of uniform size and orientation that are borne only from the style and function in secondary pollen presentation (Lavin & Delgado 1990). Species of *Kunstleria* may have uniseriate trichomes from the ovary barely extending onto the base of the style, but not in the same manner as the ciliate style where these trichomes trail up the ventral (introrse or adaxial) side of the style to near the stigma (e.g., *Platycyamus* and *Lennea*). This character was unordered during analysis, because, given the taxonomic distributions of ciliate styles and pollen brushes in Papilionoideae (Lavin & Delgado 1990), it is very likely that any one state can be readily transformed to any other.

16. Style base: thin (solid and confluent with placental margin of the maturing fruit) = 0; bulbous (hollow and often displaced toward the carinal margin of the maturing legume) = 1. This derived condition is characteristic of the *Robinia*-group and is distinctive in the mature fruit as a nubbin of swollen style base at the distal end of the legume, but in line with the carinal, not placental, margin.

17. Ovules per ovary: one to four = 0; more than four (mostly more than six) = 1. Though seemingly representing continuous variation, all species of the tribe Robinieae, with the exception of *Sesbania* sect. *Glottidium* Lavin [*S. vesicaria* (Jacq.) Ell.], have ovaries with more than four ovules. Only *Hebestigma cubense*, *Lennea melanocarpa*, and *L. modesta*, each with ovaries having 4–6 ovules, lie near the dividing line here imposed. In contrast, other genera of related tribes, such as *Dalbergia* and *Poecilanthe*, all have constituent species with fewer than five ovules per ovary. The species of *Lonchocarpus*, in particular, have mostly fewer than four ovules per ovary, but some species have more than four; *Lonchocarpus* is thus coded as a multistate taxon for this character.

18. Ovary position: sessile or on a gynoecial stipe less than half the length of the calyx tube = 0; stipitate, the stipe nearly as long as or longer than the calyx tube and often extending well beyond the calyx tube in fruit = 1. In the derived condition, as found in *Poitea* and *Gliricidia*, the stipe is pronounced in fruit. *Sesbania* sect. *Sesbania* has sessile ovaries, whereas the rest of the genus has predominantly stipitate legumes. *Sesbania* is coded here as a multistate taxon.

IV. Fruit and Seed Characters

19. Legume valves: indehiscent = 0; tardily dehiscent = 1; elastically dehiscent = 2. In the elastically dehiscent condition, the valves dehisce explosively and catapult seeds for several meters; *Gliricidia sepium* is documented to throw seeds up to 50 m (Hughes 1987). The genus *Robinia* is coded as "2" but approaches condition "1"; the valves strongly twist not immediately upon dehiscence, but slowly as they

dry; this is unlike other genera that are characterized by tardily dehiscent legumes (e.g., *Sesbania*). In *Robinia*, the elastic nature of the legume valves no longer functions in catapulting seeds. Considering legume dehiscense throughout the Leguminosae, it appears that all transformations of these character states are readily possible. Thus, this character was unordered during analysis.

20. Mesocarp: without tannins = 0; with tannins and these accumulated in relatively large cells = 1. The latter condition does not include those species with legumes in which just the exocarp accumulates tannins, as in many genera of Millettieae (the valves only light brown and tannin deposits not thick). All Robinieae, plus *Millettia*, *Wisteria*, *Caragana*, and *Schefflerodendron* have the derived condition.

21. Canavanine in seed: not accumulated = 0; accumulated = 1. In addition to recent surveys (Lavin, unpubl.), this information is reported in Bell (1981), Evans et al. (1985), and Lavin (1986, 1993).

22. Hydroxyhomarginine in seed: not accumulated = 0; accumulated = 1. The distribution of this non-protein amino acid comes from Bell (1981), Evans et al. (1985), and Lavin (unpubl.). Though the heterogeneous *Millettia* is recorded to have both canavanine and hydroxyhomarginine accumulated in the seed, *Millettia* sect. *Podocarpae* is known to accumulate only hydroxyhomarginine, and is thus coded to reflect this.

23. Hilum: apical to lateral = 0; recessed = 1. The "kidney-bean" or reniform shape of certain seeds (e.g., those of *Lonchocarpus* and *Tephrosia*) is essentially the result of the recessed hilum. The lenticular or orbicular seeds (e.g., those of the genera of Robinieae) have an apically or laterally attached hilum. The latter condition occurs when the hilum is displaced slightly off center, usually to the proximal side (that toward the side of the seed nearest the pedicel).

24. Seed testa: uniformly pigmented = 0; mottled with large purplish patches = 1. Seeds of all species analyzed have a testa that varies from red-brown to light brown, and in all cases, as the seed ages (as on older herbarium specimens) the testa becomes dark brown. But species of the *Robinia*-group have, in addition to this basic background color, a mottling of large patches of purple. This mottling is very distinct from the nearly imperceptible patches of purplish that may be found occasionally on the seed of species of *Sesbania* sect. *Sesbania* or *Lennea viridiflora*.

25. Cotyledon leaves as enclosed by the testa: thick with rounded edges = 0; thin with acute edges, thus resulting in very thin lenticular seeds = 1. *Hybosema*, *Poitea*, and *Gliricidia* have relatively thin lenticular seeds, and thus cotyledons, compared to seeds from other genera. This trait does not necessarily correspond to whether or not the cotyledons become foliar after germination (see character #37). *Hybosema* has thin cotyledons in seed, but the cotyledon leaves in the seedling do not become fully expanded and leaf-like, but rather thicken and seem to serve a function that is as much for photosynthate storage as for photosynthesis.

V. Inflorescence Characters

26. Inflorescence: on hornotinous stems (produced during the current year) and in axils of coeval (sometimes reduced) leaves = 0; on annotinous stems (produced during the previous years) and at leafless nodes (often on short shoots) = 1. In the ancestral condition, the newly produced long shoots have inflorescences

borne in the axils of coeval leaves, whereas in the derived condition the entire inflorescence system is borne on short shoots and the inflorescences are borne in the axils of aborted leaves or underdeveloped leaves. *Caragana* has the derived condition, but the situation is not so apparent, because the proximal one to two nodes on the short shoot bear a fully expanded leaf, and the more distal nodes bear only long-pedicellate solitary flowers from the axils of stipular bracts.

27. Inflorescence position: hornotinous stems terminated by inflorescence (often paniculate) = 0; hornotinous stems terminated by vegetative growth (inflorescences usually axillary) = 1. *Hebestigma* and *Olneya*, for example, have the derived condition, even though the inflorescence of these genera appears superficially terminal and paniculate. In *Hebestigma*, the terminus of the shoot bearing racemes becomes senescent but strictly vegetative; the small leaves covering the apical meristem often expand after flowering into fully developed leaves and the apical meristem then continues vegetative growth (this is further described in the discussion of *Hebestigma*). In *Olneya,* the terminal portion of the inflorescence bears vegetative shoots, which commonly abort. Generally, taxa with terminal panicles have axillary ones also; therefore, the states of character #26 are variable with regard to the states of this particular character.

28. Flowers per node of the inflorescence rachis: single, as the "simple" raceme = 0; two or more and borne jointly from a short shoot, brachyblast, or "swollen" (nectariferous) node, the inflorescence thus a pseudoraceme or pseudopanicle (Tucker 1987b) = 1; exactly two flowers borne on a common pedicel (biflorous) at each node along the inflorescence rachis = 2. See discussion of the outgroup genus *Platycyamus* (above) for how this genus is coded. The "biflorous" pseudoraceme, where two flowers are borne on a common pedicel, is characteristic of *Lonchocarpus* and close allies (Sousa & de Sousa 1981), but evidently is found also in *Kunstleria* and *Paraderris* (Mig.) Geesink (tribe Millettieae). The modification of the biflorous pseudoraceme from the common form of the pseudoracemose inflorescence is that the telescoping down of the secondary inflorescence rachis only occurs between the flower-bearing internodes, and not between the base of the rachis and the first flower node. In the common form of the pseudoracemose inflorescence, all internodes, including that from the base to the first flowering node, are telescoped down. This inflorescence character was unordered during analysis because of the uncertain relationship between the two types of pseudoracemes and the simple raceme.

29. Rachis: elongate, racemose or paniculate, with more than two nodes = 0; absent or with one node so that flowers are solitary in axils of leaves, as in *Sphinctospermum* and *Genistidium* = 1. The inflorescence of *Peteria* is very likely homologous with those of *Sphinctospermum* and *Genistidium*, though it was coded as an uncertain state taxon. As such, *Peteria* was grouped mostly in an unresolved position with all other genera of the *Robinia*-group or sometimes in particular with *Genistidium* and *Sphinctospermum*. This latter grouping was the result of interpreting the inflorescence of *Peteria* as comprising solitary flowers. Such an interpretation is consistent with the discussion of the inflorescence of *Peteria* in the taxonomy section. In the genus *Poitea*, the *Poitea galegoides*-alliance has the ancestral state, which is also considered the ancestral state for the genus in Lavin (1993), whereas the *Poitea florida*-alliance has the derived one. This genus was therefore coded as a multistate taxon. Notably, it consistently grouped within a subgroup of genera of

Robinieae, including *Hybosema* and *Gliricidia*, all of which have an elongate inflorescence. Therefore, *Poitea* could have been coded for the ancestral condition.

VI. Vegetative Characters

30. Stipitate glands: absent = 0; present and associated with at least the inflorescence = 1. The genus *Wisteria* is coded as a multistate taxon because only the one species native to the New World, *W. frutescens*, has trichomes (especially on the calyx) that discharge a glandular exudate in long spaghetti-like strands. This condition is not found in other genera, and even the condition in the *Robinia*-group, where glandular exudate is discharged as a droplet at the distal end of the hair, is uncommon in papilionoid legumes. It has been suggested that, among potentially several functions (see Schrire 1989), glandular trichomes serve a function similar to extrafloral nectaries (Polhill 1994; Delgado, unpubl.). Ants and certain other insects feed on the glandular exudate and indirectly reduce the herbivory of other insects on the inflorescence branches.

31. Stipule morphology: absent or consisting of a small nubbin of tissue = 0; herbaceous and with a distinct lamina, caducous or persistent = 1; spinescent or at least lignescent, linear and persistent = 2. Spinescent stipules, which are uncommon in papilionoid legumes, are distinctive of the *Robinia*-group; however, within this group, certain genera such as *Genistidium* and *Sphinctospermum*, as well as many species of *Coursetia*, have persistent, narrowly lanceolate lignescent stipules. These types of lignescent stipules are very similar and considered homologous to the strongly spinescent stipules of *Robinia*, *Olneya*, *Peteria*, and several species of *Coursetia*. These two conditions describe the extremes of a continuum. Though stipule morphology is often fixed within a given higher taxon, the overall pattern of variation of stipule morphology in Leguminosae is highly variable, and the character states of stipules were thus unordered in this analysis.

32. Stipule-pairs subtending each leaf: free from each other = 0; connected by a lamina that runs between the petiole and the main stem = 1. The connected stipule-pairs of *Caragana* form nearly a single laminar structure that serves as a bud scale for developing leaves or flowers on the short shoot. This is similar to the condition in *Poitea* and *Sesbania*, but these two have a distinct stipular lamina running along the edges of the connecting lamina.

33. Leaflet nyctinasty: forward = 0; downward = 1. These types have been described in Lavin (1987, 1988). The forward type is found in Dalbergieae, Aeschynomeneae, and the genus *Sesbania*, and is considered plesiomorphic, in part, because this type predominates in the legume subfamilies Caesalpinioideae and Mimosoideae. Yet, this type is otherwise uncommon in papilionoid legumes, such as in the genera of Millettieae, so this character is only tentatively polarized as such.

34. Patchy leaflet tannins: absent = 0; present and deposited along veins rather than in lamina = 1; present and forming patches visible on the abaxial leaflet surface = 2. All the genera of Robinieae are distinct from relatives in Millettieae and Dalbergieae in having leaflets, especially as seen on herbarium sheets, that are discolored on the adaxial surface by finely scattered tannin deposits. A subgroup of Robinieae (including *Hybosema*, *Poitea*, *Gliricidia*, and *Coursetia*) are notable for having distinctly patchy tannin deposits on the abaxial surface as well. Because

there is no evidence to suggest that character state 1 is a necessary precursor to state 2, this multistate character was unordered during analysis.

35. Leaf rachis: cylindrical (terete in cross section; this is often correlated with an alternate arrangement of leaflets and an absence of stipels) = 0; canaliculate (often including the presence of stipels and an opposite leaflet arrangement) = 1. Though Geesink (1984) used the presence of stipels to infer relationships among the genera of Millettieae, we consider this trait to be a variably present subcharacter of a larger character complex related to the presence of a canaliculate leaf rachis and opposite leaflets.

36. Leaves: imparipinnate = 0; paripinnate = 1. The derived condition occurs uniformly in *Caragana*, *Olneya*, and *Sesbania*, and in some species of *Coursetia* and *Poitea*. Analyses of the latter two genera (Lavin 1988, 1993, respectively) have shown that the paripinnate condition is derived within each genus; however, in this analysis both genera are coded as multistate taxa for this particular character. The interpretation that the imparipinnate leaf condition is ancestral in *Poitea* and *Coursetia* is supported in this analysis.

VII. Seedling Characters

37. Cotyledons: epigeal cotyledons absent or, if present, not functioning primarily as a storage organ = 0; epigeal cotyledons present and serving in part as a storage organ (remaining thick and fleshy) = 1. Though the conditions encompassed in the putative plesiomorphic state are multifaceted and not likely homologous, it is the derived condition found among certain genera of Robinieae that is considered a derived homologue and thus useful for grouping.

Results

A total of 1224 most parsimonious cladograms each with a length of 111, a consistency index of 0.411, a retention index of 0.654, and a rescaled consistency index of 0.269 was discovered using the heuristic search option in PAUP, with the random addition sequence invoked, starting with all initial trees, steepest descent enacted, and global branch swapping on multiparsimonious trees. A strict consensus tree (Fig. 1) reveals the clades that were resolved consistently in all of these most parsimonious cladograms. The Adams consensus (Fig. 2) reveals those groups that were consistently present in the most parsimonious set without consideration of the variably placed taxa. The strict consensus cladogram suggests that the tribe Robinieae may not be monophyletic, since *Schefflerodendron*, *Millettia*, *Wisteria*, and *Caragana* stem from an ancestry common to all of the genera of Robinieae. The Adams consensus tree reveals that a subgroup of genera of Robinieae, *Hebestigma*, *Lennea*, *Robinia*, *Coursetia*, *Olneya*, *Peteria*, *Genistidium*, and *Sphinctospermum*, are consistently grouped together (Fig. 2). The Adams consensus also reveals that *Sesbania* is grouped consistently with only *Caragana* and that *Hybosema*, *Poitea*, and *Gliricidia* consistently form a group. Identical tree statistics were obtained using the mhennig and branch breaking search options (mh*, bb*) in Hennig86 by recoding all multi- and uncertain-state taxa to missing data (?). Permutation tests (Källersjö et al. 1992; Farris 1994) that involved 1000 randomizations of the data matrix yielded an alfa value of 0.0010.

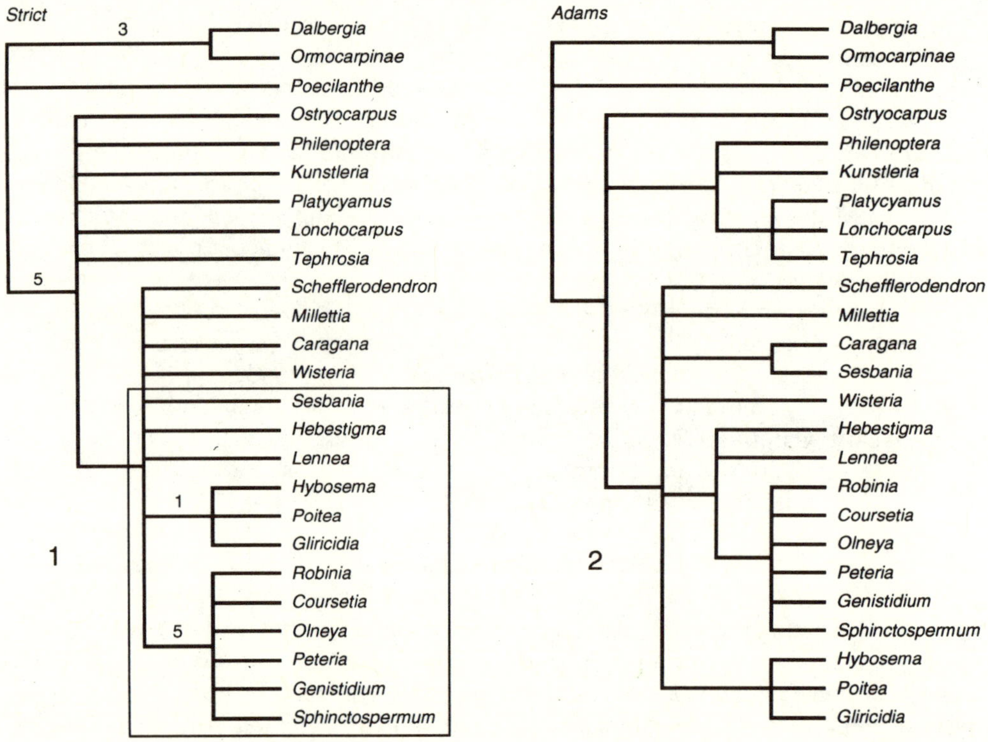

FIGS. 1–2. Consensus cladograms of the genera of Robinieae and putative sister groups. 1. Strict consensus tree of the 1224 most parsimonious cladograms, each with 111 steps, a consistency index of 0.411, and a rescaled consistency index of 0.269. Numbers along nodes represent Bremer support indices. The box encloses the genera of Robinieae. 2. Adams consensus tree of the 1224 most parsimonious cladograms.

The consistent grouping of *Caragana*, *Wisteria*, *Schefflerodendron*, and *Millettia* among the genera of Robinieae weakens the hypothesis that Robinieae is monophyletic. This pattern is more consistent with a new hypothesis that suggests the largely temperate herbaceous Old World legumes (represented by *Caragana* and *Wisteria*) were derived from within the Robinieae. *Caragana* stands out in showing many distinctive similarities with certain genera of Robinieae. In particular, its persistent calyx tube (#9), fused stipule-pairs (#32), and paripinnate leaves (character #36) are found variously in *Sesbania* and *Poitea*, and it is these characters that most strongly suggest that *Caragana* is derived from within the Robinieae lineages (Figs. 3–4). Likewise, the lack of floral bracteoles (character #2, state 2), the well-developed calyx lobes (#7, state 1), and seed accumulating canavanine (#21) are the critical characters of *Wisteria* that suggest its relationship from within the lineages of Robinieae (Figs. 3–4). The pseudoracemose inflorescence (character #28) and bracteolate flowers (#2), though both are reduced in *Millettia* sect. *Podocarpae*, in addition to the absence of canavanine accumulated in seed (#21) and the lack of any modified stipule morphologies (e.g., #31 and 32) most strongly argue against the derivation of this genus from within Robinieae (Fig. 3).

Figures 3 and 4 depict the most different sets of groupings with regard to the

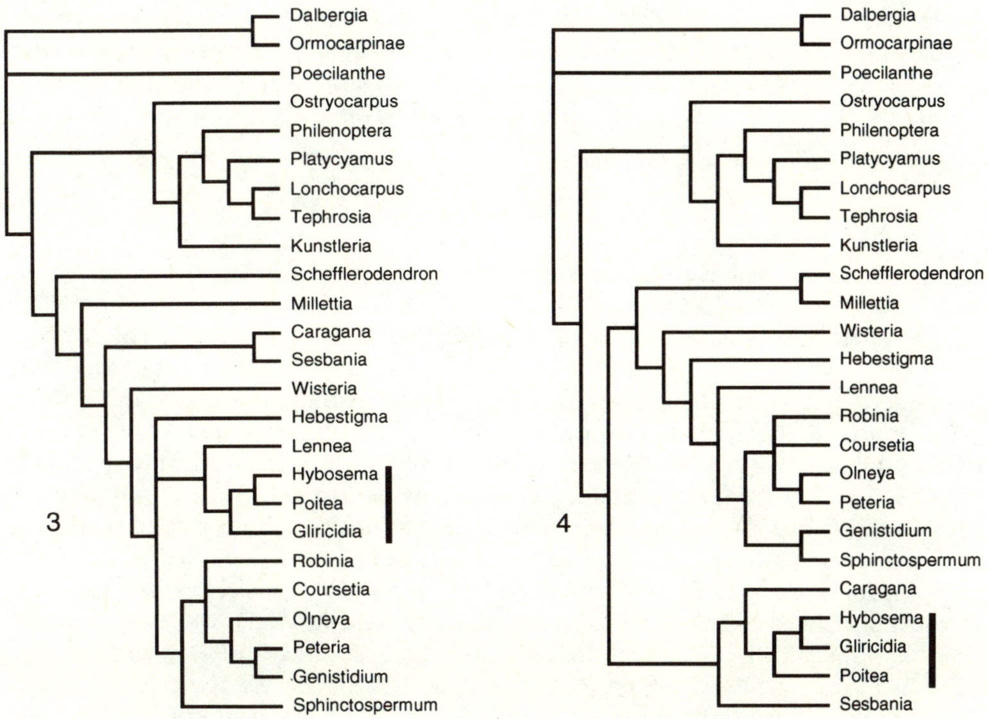

FIGS. 3–4. Two of the most different topologies among the most parsimonious set with respect to the generic relationships of Robinieae (tree to tree distance = 21). 3. One of the 1224 most parsimonious cladograms depicting *Lennea* as the sister group to the clade with *Hybosema*, *Poitea*, and *Gliricidia* (indicated by the vertical line). 4. One of the 1224 most parsimonious cladograms and depicting *Caragana* as the sister group to the clade with *Hybosema*, *Poitea*, and *Gliricidia* (indicated by the vertical line).

genera of Robinieae, as discovered using the tree to tree distances option in PAUP. These trees reveal that the phylogenetic positions of *Wisteria*, *Schefflerodendron*, and *Millettia* are the most variable, followed next by the position of the clade containing *Gliricidia*, *Hybosema*, and *Poitea*. An insightful discussion of the relationships of the genera *Wisteria*, *Schefflerodendron*, and *Millettia* requires a thorough systematic revision of the large and heterogeneous Paleotropical genus *Millettia* and all of its taxonomic segregates. Thus, the discussion is best focused on the relationships of the genera of Robinieae.

Intergeneric relationships of the genera traditionally included in Robinieae are mostly equivocal based on the most parsimonious reconstructions from the morphological and phytochemical data set. The exceptions are the monophyly of the *Robinia*-group (i.e., *Robinia*, *Coursetia*, *Olneya*, *Peteria*, *Genistidium*, and *Sphinctospermum*) and the *Gliricidia*-group (i.e., *Hybosema*, *Poitea*, and *Gliricidia*). These two monophyletic groups are resolved in all most parsimonious reconstructions (Figs. 1–4), and Bremer support indices of d=5 for the *Robinia*-group and d=1 for the *Gliricidia*-group indicate that they are two of the most robust groupings resolved by just the morphological data set. The monophyly of the *Robinia*-group is supported by a style with pollen brush (character #15, state 2), a bulbous style base

(#16), mottled seed testa (#24), stipitate glands (#30), and lignescent to spinescent stipules (#31, state 2). The *Gliricidia*-group is consistently supported by stipitate ovaries (character #18), thin seed cotyledons (#25), and leaflets with patchy leaflet tannins (#34, state 2). It is around these two monophyletic groups that the classification of the genera of Robinieae is here built.

Discussion

Intertribal Relationships: the Monophyly of Robinieae

Evidence has been accumulating from phylogenetic studies of the chloroplast genome (Lavin & Doyle 1991; Liston 1995; Lavin, unpubl.), ITS sequence data from nuclear ribosomal loci (Sanderson pers. comm.), and phytochrome DNA data (Mathews et al. 1995) that calls into question the validity of certain of the above relationships that were consistently resolved among the most parsimonious set (Figs. 1–4). Therefore, a parsimony analysis was performed in which monophyly constraints were imposed in accordance with the findings from these molecular systematic studies (using the constraints tree option, as implemented in PAUP). First, the monophyly of Robinieae is strongly supported by a restriction site analysis of the chloroplast DNA inverted repeat region (Lavin & Doyle 1991; Lavin, unpubl.). Specifically, *Wisteria*, *Millettia*, and *Caragana* lack these marker mutations that are considered unique to the genera of Robinieae. Accordingly, cladogram topology during the parsimony analysis was constrained to enforce ingroup monophyly. The second constraint that was invoked comes from the finding that *Caragana* and *Wisteria* are representatives of a primarily temperate monophyletic clade, which is marked by the loss of the chloroplast DNA inverted repeat and by ITS sequence data (Liston 1995; Sanderson pers. comm.). Certain species of *Millettia*, now all segregated as the genus *Callerya* (Geesink 1984; Schot 1994), are included in this clade that lacks the chloroplast DNA inverted repeat. This is perhaps of no consequence in this analysis, because the species of *Millettia* considered as potential sister groups to Robinieae are all placed within sect. *Podocarpae*, which is not particularly closely related to the *Callerya* group (Dunn 1912; Geesink 1984; the nature of the chloroplast DNA of section *Podocarpae*, however, is yet unknown).

With the above two constraints invoked, a parsimony analysis was performed using the heuristic search option and enacting the random addition sequence and steepest descent, as described above. This analysis yielded 206 most parsimonious trees with a length of 115, a consistency index of 0.398, a retention index of 0.636, and a rescaled consistency index of 0.253. The strict consensus tree (Fig. 6) summarizes the resolution of clades from this most parsimonious set (in this case, the strict consensus tree is identical to the semi-strict and Adams consensus tree). By invoking these constraints, the length was increased by four steps, and the consistency and rescaled consistency indexes were decreased by 0.013 and 0.016, respectively. If the many chloroplast DNA characters (from Lavin & Doyle 1991), which support the monophyly of Robinieae, are combined with the data set presented in this analysis (i.e., in a total evidence analysis), Robinieae is strongly supported as monophyletic (Lavin, unpubl.; this analysis will be presented formally when all chloroplast DNA, ITS, and phytochrome gene family data are obtained for each of

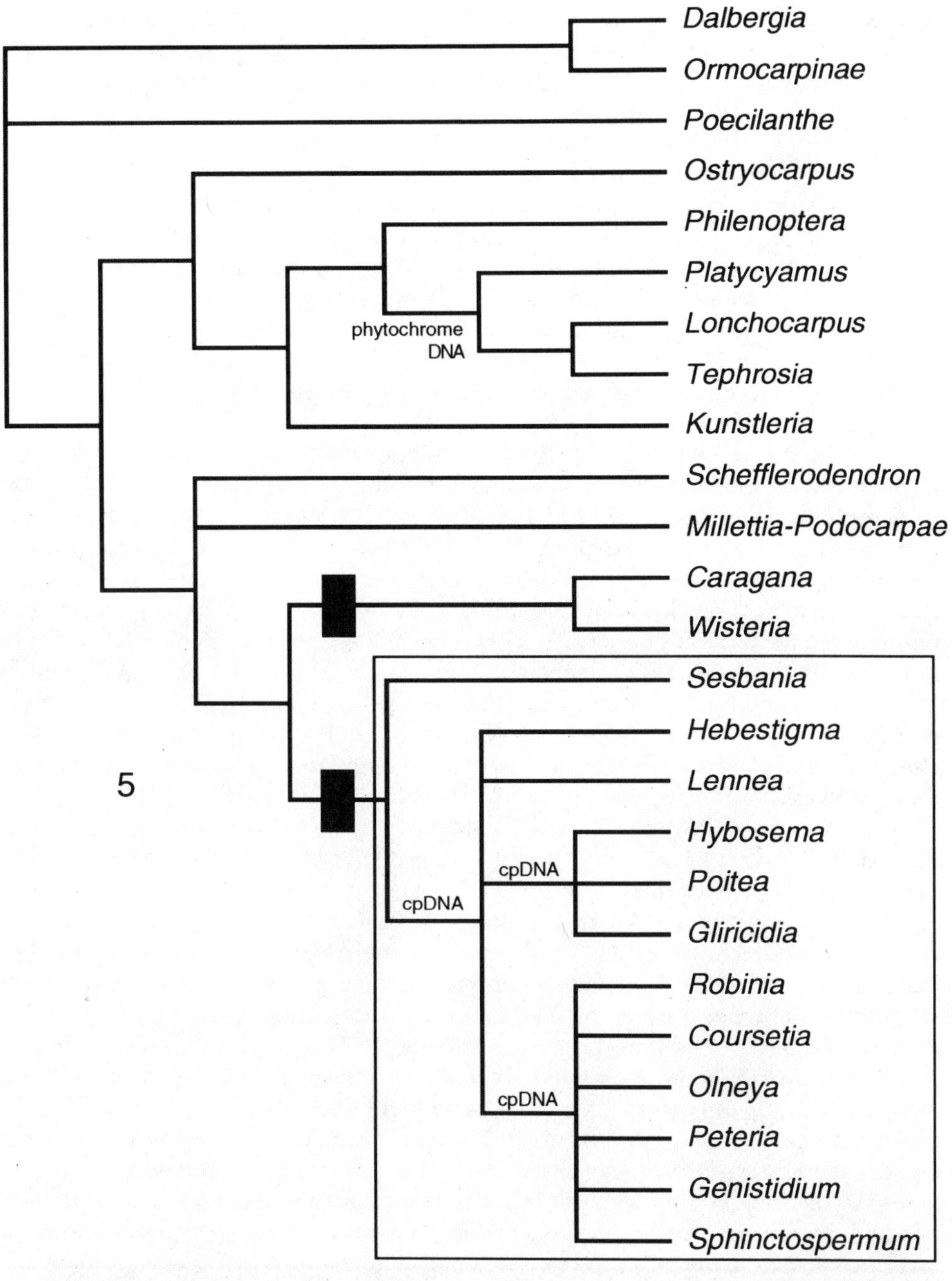

FIG. 5. Strict consensus tree of the 206 cladograms generated with the following monophyly constraints: (genera of Robinieae) and (*Caragana, Wisteria*). The black boxes indicate the nodes that were constrained during the parsimony analysis. In this case, a strict consensus yields the same cladogram as the semi-strict and Adams consensus. Each of the 206 trees had 115 steps, a consistency index of 0.398, and a rescaled consistency index of 0.253. cpDNA refers to chloroplast DNA data (Lavin & Doyle 1991; Lavin 1993; Lavin, unpubl.) that strongly support that clade. Likewise, phytochrome DNA refers to the studies of this gene family (Mathews et al. 1995; Lavin, unpubl.) that strongly support that clade. The box encloses the genera of Robinieae.

the terminal taxa used in this analysis). In this regard, the small decreases in the consistency and rescaled consistency indexes, which result from invoking monophyly constraints, suggest that the morphological data set is not significantly inconsistent with the findings from the DNA data sets.

It should be noted that no non-molecular characters support the monophyly of Robinieae in the topologically constrained analysis. Before the inclusion of *Caragana* and *Wisteria*, certain characters, such as the presence of canavanine (character #21) would have been considered defining autapomorphies for the tribe. The inclusion of these two new outgroup genera has implied that these previous morphological and phytochemical markers of monophyly do not arise uniquely in the ancestry of the genera of Robinieae. Certain morphological characters that occur very infrequently in the Papilionoideae, such as stipitate glands (found in *Wisteria* and the *Robinia*-group) or fused stipule-pairs (found in *Caragana*, *Sesbania*, and *Poitea*), suggest a relationship of the tribe Robinieae to *Wisteria* and *Caragana*.

It is notable that enforcing monophyly constraints of Robinieae and of the *Caragana-Wisteria* clade gives an overall resolution that is consistent with the phylogenetic analysis of chloroplast (Lavin & Doyle 1991; Lavin 1993; Lavin, unpubl.) and phytochrome DNA data (Mathews et al. 1995). For example, *Sesbania* is shown as the sister group to the rest of the tribe (Fig. 5); this relationship is strongly supported by chloroplast DNA data. The monophyly of the *Gliricidia*- and *Robinia*-groups is likewise strongly supported by cpDNA data. The inferred relationships of *Hebestigma* and *Lennea*, as revealed by either morphology or DNA data, are with the *Gliricidia*- and *Robinia*-groups, though in an unresolved position (Fig. 5; see discussion below). Additionally, phytochrome DNA data (Mathews et al. 1995) strongly supports the monophyly of the clade that includes *Platycyamus*, *Lonchocarpus*, and *Tephrosia* (Fig. 5).

The strong support for the monophyly of the tribe Robinieae that comes from restriction site data of the chloroplast DNA inverted repeat (Lavin & Doyle 1991; Lavin, unpubl.) is potentially problematic. That the homologous portion of this genome in *Caragana* and *Wisteria* (and other legumes that have lost the inverted repeat) does not show the "Robinieae" markers (see Lavin & Doyle 1991) could be due to high rates of sequence or structural divergence in this region of the chloroplast genome; the loss of the inverted repeat could be an indirect cause of such high rates of evolution (e.g., see Palmer 1987). To support a hypothesis of monophyly for the tribe Robinieae, it must be shown that the chloroplast DNA markers found in only Robinieae (i.e., those identified in Lavin & Doyle 1991) did not originally occur in *Wisteria* and *Caragana* (and have been modified subsequently to the point that they are no longer identifiable). Given the extreme variability of this region in the legumes that have lost the inverted repeat, this alternative hypothesis can only be addressed with other than restriction site data from the inverted repeat. Currently, DNA sequence data from the nuclear genome are under analysis to address these hypotheses, and the monophyly of Robinieae is finding additional support (Mathews et al. 1995; Lavin, unpubl). Of note here, Sanderson (pers. comm.) reports that ITS sequences from legumes that have lost the inverted repeat (which includes *Wisteria* and *Caragana*) were difficult to align with sequences of *Robinia*. It appears that the ITS sequence data set is strongly supporting the monophyly of the legume lineage marked by the loss of the chloroplast DNA inverted repeat.

Intratribal relationships: the major groups within Robinieae

The differences between the results of the unconstrained analysis presented here (Figs. 1–4) and the results of the chloroplast DNA studies (Lavin & Doyle 1991; Lavin, unpubl.), where Robinieae is determined to be monophyletic, are not necessarily due to fundamental incongruities between the different data sets, which may arise due to processes such as introgressive hybridization of chloroplast genomes, retention of ancestral polymorphisms, or scoring suites of homoplasious correlated characters in the data matrix (where correlation is a result of scoring different descriptors of the same trait erroneously as different characters). The seeming incongruence may be the result of differences in the levels of phylogenetic resolution (e.g., Olmstead 1989), in which morphology happened to capture little, and chloroplast DNA sequence variation much, of the evolutionary event that led to the unique ancestry of Robinieae.

Two subgroups of Robinieae that are consistently well resolved, in both morphology and chloroplast DNA restriction site analysis, are the *Gliricidia-* and *Robinia*-groups. With regard to the morphological data set, a Bremer support analysis reveals that these two generic groupings are the most robust found among all possible groupings that include genera of Robinieae (Fig. 1). Notably, these same generic relationships, which are well resolved with the morphological data set, are also well resolved by a chloroplast DNA restriction site data set. The monophyly of the *Robinia*-group is well corroborated by chloroplast DNA data (Lavin & Doyle 1991; Lavin, unpubl.), as is the monophyly of a group comprising *Hybosema*, *Poitea*, and *Gliricidia* (Lavin, unpubl.; see also Lavin 1993, and Lavin et al. 1991). In addition to these two groupings, the sister group relationship of the genus *Sesbania* to the rest of the tribe Robinieae is supported by the morphological data set when ingroup monophyly constraints are imposed (Fig. 5), and also very well supported by chloroplast DNA data (Lavin & Doyle 1991).

Thus, if the monophyly of Robinieae is assumed based on chloroplast DNA evidence (in addition to the developing phytochrome and ITS DNA sequence evidence), then the following major subgroups and their inter-relationships are most strongly supported by morphological and phytochemical data, and by chloroplast DNA data: 1) the *Sesbania*-group, which comprises just the genus *Sesbania*, is most likely a sister group to the rest of the tribe because of its bracteolate flowers (character #2) and leaflets that lack distinctive tannin deposits (#34); 2) the *Hebestigma*-group, comprising just the genus *Hebestigma*, is very likely an isolated lineage within Robinieae because there are no distinctive similarities of this genus to any small subgroup of other Robinieae genera; 3) the *Gliricidia*-group unequivocally includes the monophyletic assemblage of *Hybosema*, *Poitea*, and *Gliricidia*, and tentatively *Lennea*, because in some phylogenetic reconstructions (Fig. 3), this last genus is the sister group to this subclade due to its bilobed calyx (character #7, state 2) and pseudomonadelphous staminal tube (#12, state 1); and 4) the *Robinia*-group, which includes *Robinia*, *Coursetia*, *Olneya*, *Peteria*, *Genistidium*, and *Sphinctospermum*, also is monophyletic (see characters listed in results section). The details of individual generic relationships are presented further in the discussion of each genus in the taxonomic section.

In summary, the evolution of Robinieae very likely involves the evolution of the predominantly Old World temperate legumes that have lost the chloroplast

DNA inverted repeat region (i.e., tribe Galegeae and close relatives). Definitely, the sister group to the entire Robinieae lineage, whether or not it is eventually shown to include the *Caragana-Wisteria* line, is a *Millettia*-like lineage, one that probably does not include genera of Aeschynomeneae or Dalbergieae. Polhill and Sousa (1981) had suggested that the genera of the tribes Aeschynomeneae, Dalbergieae, and Millettieae were closely related to those of Robinieae; here, we consider just the possibility of Millettieae, in addition to a subgroup of the tribe Galegeae (represented by *Caragana* and *Wisteria*). The relationship of the primarily New World northern hemisphere Robinieae to Old World predominantly northern hemisphere legumes strongly suggests that the ancestor of Robinieae and its sister groups were widespread throughout the northern hemisphere. All putative South American sister groups to the tribe Robinieae are shown to be distantly related and to belong to other legume lineages (e.g., *Poecilanthe* and especially *Platycyamus*). Within the tribe Robinieae, constituent South American members are very likely of recent occurrence on that continent, as in the genera *Coursetia* (Lavin 1988) and *Gliricidia* (see biogeographic section). The only other South American representatives within the tribe are human-introduced or pantropical weedy species of *Sesbania*. Phylogenetic diversity and endemism in *Sesbania* really occur only in Africa and North America (see taxonomy section).

This brings us to the biogeography of Robinieae. The reason that genera traditionally included in the predominantly temperate Old World tribe Galegeae (e.g., *Caragana*) were not previously included as outgroups was that conventional wisdom has held that close relatives of a North American tropical plant group would most likely be found in tropical South America (Lavin & Luckow 1993). Thus, for a predominantly North American tropical group, why should one consider a sister group centered in tropical, let alone temperate, regions of the Old World? Yet, as fossil evidence accumulates (e.g., Herendeen 1992; Herendeen et al. 1992; Wheeler & Baas 1992), and as more phylogenetic analyses are performed (e.g., Luckow 1995), it is becoming evident that studies of North American tropical plant groups need to consider the potential Old World temperate, as well as Paleotropical, relatives in order to evaluate best all potential sister group relationships.

BIOGEOGRAPHY

Historical biogeographic analysis requires that chance biogeographic patterns be distinguished from those that have common causes (Simberloff et al. 1981; Simberloff 1987). Chance dispersal, as an agent of pattern, requires little if any common phylogenetic and biogeographic patterns among a set of taxa. Invoking a common cause (e.g., historical proximity of land masses), however, requires a general pattern. In this regard, the criteria that are used to determine whether two or more similar patterns are indeed the same have to be established. Here we discuss the phylogenetic and biogeographical pattern of the tribe Robinieae. We evaluate this pattern in terms of how common it is and consider one method for determining what constitutes sameness of pattern among a set of cladograms. This method involves the determination of ancestral areas, though not in the realm of vicariance biogeography (Bremer 1992; Ronquist 1994). We begin by presenting causal hypotheses that are derived from geological and paleontological data; specifically, these are the Geoflora (Chaney 1940, 1947; Chaney & Axelrod 1959) and the Boreotropical flora (Wolfe 1975) hypotheses. We then evaluate the biogeographic pattern of Robinieae against these hypotheses.

The Tertiary Geoflora and the Boreotropical flora hypotheses identify cosmopolitan biotas (or large scale areas of endemism) from the Tertiary period. Tertiary land bridges and climates, which generally differ from those of today, are the postulated causes of these past regionally differentiated biotas. The biogeographic patterns displayed by the tribe Robinieae are, in part, the result of Tertiary land bridges and climates. Chance dispersal may be also explanatory. As we will argue later, we can predict a biogeographic pattern of a taxonomic group that was influenced mostly by land bridges and climates assumed for the Boreotropical flora hypothesis versus one for the Geoflora hypothesis. Thus, chance dispersal will be considered only if predictions are not borne out.

The Tertiary Geoflora Hypothesis

Chaney (1947) circumscribed the Arcto-Tertiary Geoflora, the centerpiece of his Geoflora concept (Chaney & Axelrod 1959), as a floristic fossil assemblage resembling very closely the present eastern temperate deciduous forest of the United States. It was thought to be dominated by such genera as *Acer* (Aceraceae), *Alnus*, *Betula*, and *Carpinus* (Betulaceae), *Cercidiphyllum* (Cercidiphyllaceae), *Cornus* (Cornaceae), *Castanea*, *Fagus*, and *Quercus* (Fagaceae), *Fraxinus* (Oleaceae), *Gingko* (Gingkoaceae), *Juglans* (Juglandaceae), *Pinus* (Pinaceae), *Platanus* (Hamamelidaceae), *Crataegus* (Rosaceae), *Populus* and *Salix* (Salicaceae), *Sequoia* and *Taxodium* (Taxodiaceae), *Tilia* (Tiliaceae), and *Ulmus* and *Zelkova* (Ulmaceae). The Arcto-Tertiary Geoflora was considered Holarctic; thus, it is used commonly today to explain the floristic similarities between temperate regions in southeastern United States and those in eastern Asia. In contrast, the Neotropical-Tertiary Geoflora (generally occurring to the south of the fossil beds containing Arcto-Tertiary elements) was circumscribed to include fossil assemblages dominated by *Cordia* and *Ehretia* (Boraginaceae), *Canarium* (Burseraceae), *Terminalia* (Combretaceae), *Ceratozamia* and *Dioon* (Cycadaceae), *Actinidia*, *Davilla*, *Dillenia*, and *Tetracera* (Dilleniaceae), *Acalypha*, *Alchornea*, *Aporusa*, *Drypetes*, *Mallotus*, and *Sapium* (Euphorbiaceae), *Acrodiclidium*, *Cinnamomum*, *Cryptocarya*, *Laurus*, *Lindera*, *Machilus*, *Nectandra*, *Neolitsea*, *Ocotea*, and *Persea* (Lauraceae), many woody and viney genera of Leguminosae, *Carapa* and *Cedrela* (Meliaceae), *Siparuna* (Monimiaceae), and *Meliosma* (Sabiaceae).

One assumption of the Geoflora hypothesis is that climatic tolerances of plant taxa do not generally change markedly through time (Wolfe 1972, 1977). Chaney (1947) assumed that the taxonomic composition of a fossil flora, based on modern analogs, indicated the vegetation and climate of the period and locality. He suggested that Tertiary Geofloras existed largely intact throughout the Cenozoic, and were represented by climax communities that would shift in latitude depending on climate change, analogous to the change in modern vegetation along a latitudinal gradient but with a large time dimension. According to Chaney, during the very warm periods of the early Tertiary (i.e., Eocene) the Neotropical Tertiary Geoflora extended at least as far north as present day northwestern United States, while during the cooler climates of the mid- to late-Tertiary, the Arcto-Tertiary Geoflora extended south into present-day southern United States and even the mountains of northeastern Mexico.

A second assumption of the Geoflora hypothesis is that South America was the ultimate source area for the North American portion of the Neotropical Tertiary Geoflora. Chaney (1947) reported that Early Tertiary fossil floras from South America were similar to those from North America. A third assumption of the Geoflora hypothesis is that the land connection between North America and Eurasia, the Beringian land bridge, facilitated an interchange that involved primarily temperate floras.

Chaney's concept was developed without regard to plate tectonic theory, and regardless of this, or of how unsubstantiated one may consider the above assumptions, the Geoflora concept still greatly influences our modern view of the New World. For example, the Arcto-Tertiary Geoflora concept is used by many botanists to explain the floristic similarities between temperate deciduous forests in southeastern United States and those in eastern Asia. Sclater's (1858) biogeographic regions of the world, which are still reproduced in many modern books (e.g., Simpson 1965; MacFadden 1992; Pianka 1994), treat the tropics in North and South America as if they belonged to a single cosmopolitan Neotropical biogeographic region; yet the Paleotropics in Sclater's depiction are divided into the Ethiopian, Oriental, and Australian. Cronquist (1988), Forero and Gentry (1988), Gentry (1982), Raven and Axelrod (1974), Raven and Polhill (1981), and many others have strongly emphasized, implicitly or explicitly, that indigenous tropical North American plants stem entirely, or nearly so, from Tertiary and Quaternary immigrations from South America.

The Boreotropical Flora Hypothesis

Wolfe (1972, 1977) strongly questioned the Geoflora hypothesis. His main criticism stemmed from Chaney's inference that the taxonomic composition of a flora, fossil or extant, was directly related to the vegetation and climate. Wolfe's approach to reconstructing the Early Tertiary vegetation and climate of North America was not to give any weight to the taxonomic composition of a fossil flora, but rather to consider only factors better correlated with climate, such as leaf physiognomy (e.g., Wolfe & Wehr 1991; Chaney also used leaf physiognomy data but apparently gave primacy to taxonomic composition for climatic inferences). By using factors independent of taxonomy to reconstruct paleo-vegetation, Wolfe rejected the Arcto-Tertiary Geoflora concept as put forth by Chaney, because his interpretation of strictly leaf-physiognomic data from early Tertiary fossil floras suggested that tropical vegetation during the Eocene extended to very northern latitudes, even into present day Alaska (Wolfe 1972, 1977). Thus, according to Wolfe, the temperate Arcto-Tertiary Geoflora in North America was much more limited in distribution than reported by Chaney. For example, tropical rain forests and tropical semi-deciduous forests predominated throughout much of North America during the Eocene (Upchurch & Wolfe 1987).

Wolfe (1972) strengthened this idea by determining that some of the putative indicators of the Arcto-Tertiary Geoflora, for example, *Alnus*, *Betula*, *Carpinus*, *Carya*, *Platycarya*, *Pterocarya*, *Cercidiphyllum*, *Tetracentron*, and the hamamelids *Liquidambar*, *Fothergilla*, and *Fortunearia*, attained their widest geographic distribution in the Eocene tropical and subtropical forests. Furthermore, the few North American Eocene fossil assemblages that were deemed warm-temperate lacked many taxa supposedly characteristic of the Arcto-Tertiary Geoflora, especially Salicaceae, Fagaceae, Hamamelidaceae, and Aceraceae. In the supposed warm-temperate Eocene Republic flora (Wolfe & Wehr 1991) of Washington state, for example, members of these families are rare to absent; presumably, they had yet to radiate into temperate climates (Wolfe 1972). As another example, *Platycarya*, a supposed indicator of the Arcto-Tertiary Geoflora, co-occurred in abundance in the early Eocene deposits with the palm genus *Nipa*, as in the London Clay fossil flora (Reid & Chandler 1933). This, plus leaf physiognomic evidence, suggested that *Platycarya* inhabited tropical vegetation during the Eocene (Wolfe 1977). Chaney, in contrast, explained these co-occurrences of "tropical" and "temperate" taxa in Early Tertiary fossil floras as being due to secondary mixing (e.g., elements of an upland temperate flora being transported along a water course to an adjacent lowland tropical flora), or to the former existence of an intermediate climatic regime that would allow lineages like *Nipa* and *Platycarya* to co-exist.

Wolfe (1975) challenged the Neotropical Tertiary Geoflora hypothesis by postulating a primary early Tertiary relationship of North American tropical floras with those in Europe and Asia, instead of the putative early Tertiary primary relationship between the tropical floras of North and South America (Chaney 1940, 1947). Although he considered Beringia to be the main corridor and the North Atlantic a secondary one, Tiffney (1985a, 1985b) subsequently argued that the early Tertiary North Atlantic connection could have been a significant corridor between North America and Eurasia during the Eocene. In support of Wolfe and Tiffney's ideas, Lavin (1995), Lavin and Luckow (1993), Madison (1977), Rzedowski (1993), Sousa and Delgado (1993), VanderHammen & Cleef (1983), Wendt (1988, 1989, 1993), Zona (1990), and others have posited that the effects of this early Tertiary corridor are still manifest in the modern floristic composition of tropical North America. Direct support for a North Atlantic tropical interchange continues to come from recent fossil discoveries; for example, leaf impressions from Early Tertiary fossil beds in Greenland have characters seen only in the Heliconiaceae, Musaceae, and Strelitziaceae (Boyd 1992).

The Boreotropical flora hypothesis is supported by the view of mammalogists, who long have considered a Tertiary circum-northern hemisphere fauna that reached the southern hemisphere secondarily (e.g., Matthew 1915). Simpson's (1980) argument that the mammalian faunas of North and South America were very distinct throughout the Paleogene underscores Wolfe's (1975) distinction of the early Tertiary Boreo- and Austral-tropics (see Vrba, 1993, and MacFadden, 1990, for recent support of Simpson's view). That the North Atlantic was a significant Eocene corridor for faunistic interchange between North America and Europe (e.g., Simpson 1953; McKenna 1972, 1975, 1983a, 1983b) is also consistent with Wolfe's hypothesis, at least as modified by Tiffney (1985b). Though mammalogists have described this Tertiary trans-Atlantic fauna as Holarctic, others have been very specific about its tropical nature (e.g., West & Dawson 1978; Prothero 1994).

One difference between Chaney's Geoflora and Wolfe's Boreotropical flora hypotheses is that the former posits the relative isolation of the Neotropics during the Tertiary, whereas the latter suggests the isolation of the North American tropics from South America, and its contiguity via the North Atlantic or Beringia with the

Paleotropics. In Wolfe's hypothesis, tropical climates extended to very northern paleolatitudes during the Eocene, whereas in Chaney's hypothesis, temperate climates predominated in northern latitudes so as to preclude significant migrations of tropical taxa across available land bridges, particularly those of the North Atlantic or Beringia. It is important to note that there are many components to each of these hypotheses, and that some components of the Geoflora hypothesis are not necessarily mutually exclusive to certain parts of the Boreotropical flora hypothesis. For example, the North American tropics could have been influenced by Tertiary migrations from South America, as well as possibly by remnants of a once more widespread northern tropical flora. The important question here pertains to the relative influence of each on the modern tropical flora in North America. Also, no doubt at least some early Tertiary temperate and tropical taxa had similar ecological tolerances to their modern descendants. The real question here is what influence did the North Atlantic or Beringian land bridges have on the evolution of modern temperate and tropical North America biotas.

One prediction of Chaney's hypothesis is that examples of northern hemisphere vicariant patterns should be found in only temperate taxa, such as the disjunct distributions of groups occurring in deciduous forests of southeastern U.S.A. and eastern Asia. Such vicariant biogeographic patterns for tropical North America and the Paleotropics would be precluded by Chaney's Geoflora hypothesis. Another prediction of Chaney's hypothesis would be that any taxon concentrated in tropical North America should have a sister group relationship with a Neotropical group, very probably from South America. In contrast, a northern hemisphere tropical vicariant pattern would be predicted by the Boreotropics hypothesis, and a group concentrated in tropical North America could very likely have a Paleotropical sister group. It is reasoned that the tribe Robinieae and its close allies (i.e., genera of the tribe Millettieae) provide a model group for evaluating these particular predictions (Lavin & Luckow 1993). Robinieae and certain other legume groups are centered in tropical North America, but also occur in tropical South America and in the Paleotropics. Furthermore, the tribe Robinieae has an early Tertiary fossil record from North America (Wheeler & Baas 1992). Therefore, the modern distribution of this group had to have been affected by the Tertiary land bridge and climatic constraints, which are proposed for each hypothesis.

Biogeographic Patterns in the Tribe Robinieae

A phylogenetic analysis of the tribe Robinieae has been estimated with morphological (Lavin 1987, 1988, 1990, 1993, this analysis; Lavin & Luckow 1993), phytochemical (Lavin 1986; Evans et al. 1985), chloroplast DNA (Lavin et al. 1991; Lavin 1993; Lavin & Doyle 1993; Lavin, unpubl.), and phytochrome DNA (Mathews et al. 1995; Lavin et al., unpubl.) data. Together these studies suggest that Robinieae is monophyletic, and that all the potential sister groups of the tribe are strictly from the Paleotropics (Fig. 5). The lack of endemic South American genera of Robinieae and the occurrence there of only 19 endemic species restricted to just two genera, *Coursetia* and *Gliricidia*, stands in contrast to the nine genera and 55 species endemic to North America. Within the genus *Sesbania* alone, the four endemic North American species each belong to a different section of the genus. In contrast, all species of the genus occurring in South America are pantropical weeds and represent just two sections of the genus (see taxonomy section). The Paleotropics also contains little of the taxonomic diversity of the tribe with only the genus *Sesbania* occurring there naturally. Clearly, the main diversification of the tribe Robinieae occurred in tropical to warm-temperate North America, where all 12 genera of the tribe are native.

Ancestral area analysis. When pantropical taxa are subjected to vicariance biogeographical analysis, it is assumed that they were also cosmopolitan during the late Cretaceous to Tertiary (i.e., primitive cosmopolitanism of Nelson & Platnick 1981), and that late Tertiary to Quaternary tectonic and climatic changes resulted in vicariant events and thus regional differentiation of tropical floras (e.g., Grimaldi 1988; Liebherr 1986). However, there exists much evidence to suggest that neither the biotas of North and South America (e.g., Simpson 1980; Wolfe 1975) nor of Africa and South America (e.g., Thorne 1973; Cracraft 1994) were cosmopolitan during the Tertiary. Indeed, both the Boreotropical flora and Geoflora hypotheses posit that Tertiary tropical biotas were regionally differentiated across continents. If the well-known Eocene fossil records of mammals are used to identify Tertiary areas of endemism (cf. fossil areas of endemism of Grande, 1985), then North America during this period was very likely contiguous with the Eurasia via at least a North Atlantic land bridge (e.g., McKenna 1972, 1975, 1983a), and South America and Australia were most likely isolated islands (Simpson 1980; Marshall & Sempere 1993; Vrba 1993). An assumption we are going to make here is that the pattern of large-scale Tertiary areas of endemism for mammals may be generalized to other taxonomic groups, such as angiosperm families like Leguminosae (see Parrish 1987, 1993a, 1993b, and Pitman et al. 1993, for corroborating geological evidence). If it could be considered a modern cosmopolitan biota, the Neotropics has existed as such for only 2.5 million years, or since the formation of the Panamanian Isthmus (e.g., see Marshall & Sempere 1993; and Vrba 1993, for recent reviews). These large-scale Tertiary areas of endemism are significant, because it was precisely during this period, according to the fossil record, that many modern groups, particularly Leguminosae, were diversifying. Thus, the existence of large-scale Tertiary areas of endemism, such as North America with Eurasia and Africa, and South America and Australia as isolated continents, should have constrained the distribution of modern plant and animal groups that were undergoing diversification during the early Tertiary (see discussion for mammals by Vrba, 1993). This pattern of Tertiary cosmopolitan biotas, if it involved tropical elements, is much different from the modern Neo- and Paleotropical pattern, and it explains the findings of Sousa and Delgado (1993), who have posited that many Mexican legume lineages have Paleotropical relationships independent of South America.

Testing the above hypothesis was proposed in a preliminary way by Lavin and Luckow (1993) whereby extant pantropical groups with early Tertiary fossil records would be subjected to phylogenetic analysis. The approach does not entail vicariant biogeographic analysis, because it assumes from the outset that tropical biotas of the world were not cosmopolitan during the Tertiary. The reason for selecting extant tropical groups with early Tertiary fossil records in North America is that, during this period, they would have more likely occurred in the Old World than they would have in South America, if the Boreotropics hypothesis is true. The evolutionary outcome of this would be that the vicariant event beginning with the spreading of the North Atlantic sea floor, and ending with global cooling during the Oligocene, would result in sister group relationships between tropical North American and Paleotropical taxa. Of course, some groups evolved tolerance to temperate climates subsequently and persist today in the vicariant temperate biotas of southeastern U.S.A. and eastern Asia (Tiffney 1985a, b). This northern tropical vicariant pattern, if it did exist, was obscured beginning in the late Pliocene, unlike for temperate taxa, because the formation of the Panamanian Isthmus allowed directed dispersal of northern tropical lineages southward into South America. Thus, a phylogenetic analysis would have to provide evidence, in some way, of an ancestral area in North America and a recent occurrence in South America, or vice versa in support of the Neotropical Tertiary Geoflora hypothesis. Within the vicariant biogeographic realm, determining ancestral areas has been criticized (e.g., Page 1990; Cracraft 1988) because of inappropriate analogies between taxa on an area cladogram and characters on a taxon cladogram; in particular, taxa can have only one history whereas areas can have multiple histories. Also, ancestral area analysis might be confused with centers of origins concepts. Questions regarding ultimate origins, however, are irrelevant to understanding modern day distributions (Croizat et al. 1974); whether Robinieae initially arose in North America or Eurasia has no bearing on understanding the geological and climatic constraints imposed on Tertiary biotas. In this study, we will estimate only one ancestral area, that at the base of the Robinieae clade, and outside the contexts of vicariance biogeography and centers of origin concepts.

Criterion for pattern similarity: ancestral area indices. The tribe Robinieae meets the above requirements of an early Tertiary North American fossil record (Wheeler & Baas 1992) and a pantropical distribution. The current best estimate of the phylogeny (Fig. 5) can be converted into an area cladogram in a trivial fashion (Fig. 6). The areas of endemism are the Paleotropics (designated by P), the North American tropics (N), and the South American tropics (S). The standard assumptions of vicariance biogeography for converting a taxon cladogram into an area cladogram (e.g., Nelson & Platnick 1981; Page 1990; Nelson & Ladiges 1991) need not be applied. First, this is not a vicariance analysis, and, second, the ancestral area to the New World clade could be polymorphic. Widespread taxa should be maintained as such without making any assumptions (see Ronquist 1994; cf. Bremer 1992). The only exception is the area for *Sesbania*, which is designated as PN because there are no endemic South American species, and because it is very likely that all or nearly all species of this genus occur there because of human introduction. Also, redundant distributions are so common (Fig. 6) that area relationships would be difficult to infer from this area cladogram. Ultimately, the determination of an ancestral area is the goal, not area relationships.

With the area cladogram (Fig. 6), we can apply a step matrix of areas (Fig.7; cf. Ronquist 1994) in order to determine what is most likely the ancestral area for Robinieae. Ronquist (1994) applied reversible parsimony, which is really Wagner parsimony, but called as such to contrast it to Bremer's (1992) method of determining ancestral areas with the use of Camin-Sokal parsimony. A reversible parsimony (RP) index is calculated for each putative ancestral area as follows. Posit that South America (S) is the ancestral area of the tribe Robinieae. Utilizing the step matrix (Fig. 7), three steps are required to account for the PN distribution of *Sesbania*, and

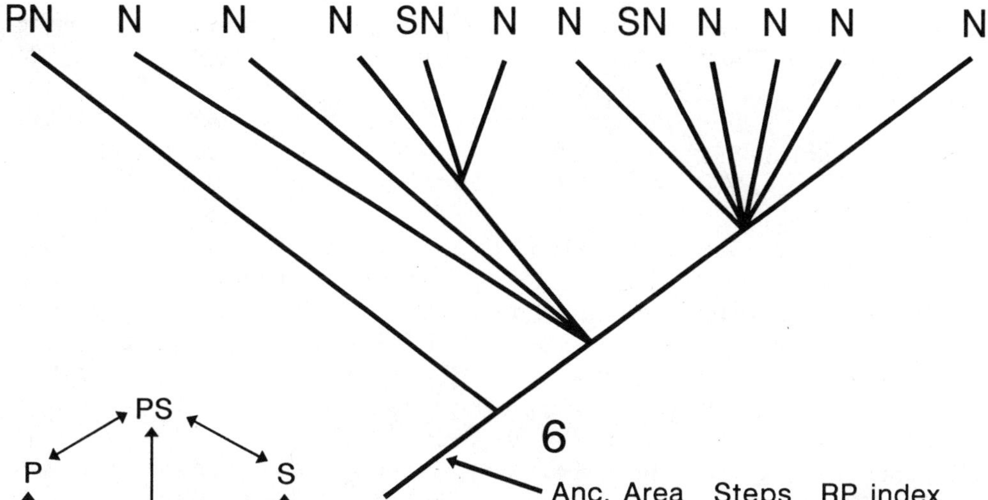

6

Anc. Area	Steps	RP index
S	5	0.60
P	4	0.75
N	3	1.00
PN	3	1.00
PS	5	0.60
SN	4	0.75
PSN	4	0.75

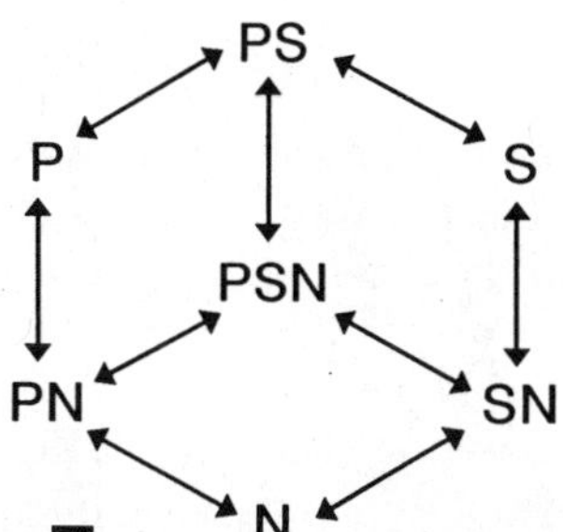

7

FIGS. 6–7. Ancestral area analysis. 6. Area cladogram derived from Fig. 5, but only including that part of the cladogram showing the relationships of the genera of Robinieae (i.e., the portion enclosed in the box in Fig. 5) and with the geographical distribution replacing the generic name. The table indicates the number of steps that would be required to account for the distribution of terminal taxa for each putative ancestral area. Anc. area = putative ancestral area, Steps = the number of steps required on the illustrated cladogram to account for the distribution of the terminal taxa, RP index = reversible parsimony index (see Ronquist 1994; and described in text). P = Paleotropical, S = South American, N = North American, PN = Paleotropical and North American, SN = South and North American, PS = Paleotropical and South American, and PSN = pantropical. 7. Step matrix utilized in the determination of the number of steps required to account for the distribution of terminal taxa on the cladogram (Fig. 6) given a putative ancestral area. Abbreviations are the same as those given in Fig. 6.

an additional two steps are required for two independent reversals back to an SN distribution. A total of five steps is required to account for the area distributions of the terminal taxa if South America is posited as the ancestral area. Next, posit that North America is the ancestral area. Using the step matrix once again (Fig. 7), one step is required at the base of the clade to account for the PN, and two addition steps are required for two independent gains of the SN distributions. A total of three steps is required in this case. Continuing as such, the number of steps is calculated for each putative ancestral area and for all putative polymorphic ancestral areas (the table in Fig. 6). Then, the RP index (Ronquist 1994) is calculated for each by taking the reciprocal of the number of steps required and multiplying it by the smallest number of steps required by any one of these putative ancestral areas. In this case, that number is three (see Steps column in the Fig. 6 table). Thus, the RP index for South America as the ancestral area is 0.6 (3×⅕), whereas that for North America is 1.0 (3×⅓). All values are provided for each possible ancestral area (the table in Fig. 6). Thus, North America (N) or the Paleotropics and North America (PN) are the most likely ancestral area of Robinieae, according to this type of analysis. Of note, the calculation of RP indices of Robinieae plus the sister group clade *Caragana-Wisteria* (Fig. 5; Paleotropical-North American in distribution) results in the Paleotropics-North America being the most likely ancestral area (RP = 1.00), followed by North America (0.80) and then South America (0.67).

A potential criticism of ancestral area analysis is that it assumes that early divergent, or plesiomorphic, lineages do not disperse relative to those shown as more derived in the cladogram. The areas represented by the earliest divergent lineages on a cladogram are more likely to have the highest RP index; however, Ronquist (1994) has shown empirically that this is not necessarily the case, as in the mtDNA phylogeny of humans. Also, the RP index approximates a clade dispersion index, which is directly correlated with a minimum number of monophyletic groups (as shown on the cladogram) that occur in an area. The assumption here is that the longer a lineage is in residence on a particular continent, the more diverse (and thus early branching) will be the sublineages on that continent, providing extinction has been minimal. (Unfortunately, the determination of ancestral areas from extant distributions does not accommodate high extinction rates.) Yet, we still think that this type of analysis has merit, particularly for legumes, because legume lineages are not phylogenetically diverse on land masses that are relatively recently emergent, thus implying long periods of time required to evolve phylogenetic diversity (there are few, if any, cases of adaptive radiation of legumes on oceanic islands). As an example from a continental island, the Greater Antilles are remarkably diverse in legumes with the exception of Jamaica. So poor is the phylogenetic diversity (and endemism) of legumes on Jamaica that Sousa and Delgado (1993) do not include it in their Mexican phytogeographical legume area, a region which comprises Cuba, Hispaniola, Puerto Rico, and the adjacent mainland centered in and around Mexico. The most likely reason for this is that Jamaica, unlike the other continental islands of the Greater Antilles, was inundated during the Oligocene, and is thus relatively recently emergent (e.g., see Buskirk 1985). Regarding North and South America, both are extremely rich in legume groups. So the question here is which groups show high lineage diversity in North America and correspondingly poor lineage diversity in South America. Likewise, which groups show high lineage diversity in South America and accordingly poor lineage diversity in North America. We think there is such a pattern in legumes, and predict that few, if any, legume groups will show equal diversification of lineages in North and South America.

Identifying a common pattern: predictions from the fossil record. From the ancestral area analysis, it is inferred that the tribe Robinieae has an ancient occurrence on the North American continent and a more recent one in South America. As with any pattern, it is meaningless if not found to be general. Any strength that the ancestral area approach has in a biogeographical analysis lies in the ability to predict such an area for a given group. For the family Leguminosae, it is possible to establish some criteria by which predictions of ancestral areas can be made. Herendeen (1992), Herendeen et al. (1992), Buzek (1992), and Wheeler and Baas (1992) report the following genera of the subfamily Papilionoideae have unequivocal fossil records from the early Tertiary of North America: *Swartzia* Schreb.,

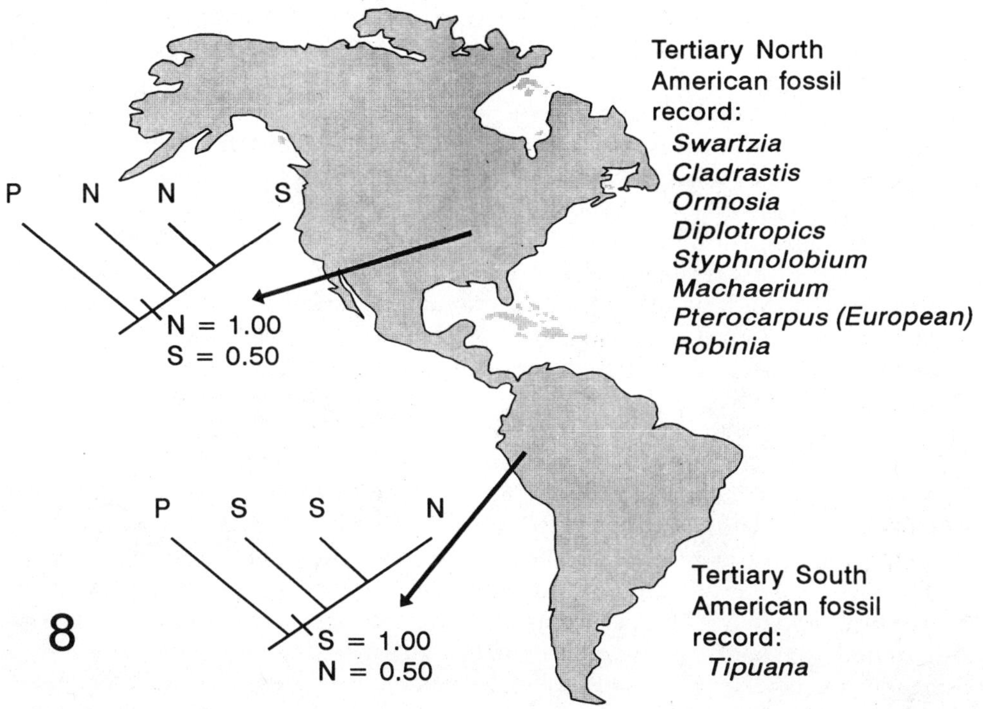

FIG. 8. Predicting ancestral areas for taxa with fossil records in North or South America. The genera of Papilionoideae represented by Tertiary North American fossil records (European in the case of *Pterocarpus*) were identified by Herendeen (1992), Herendeen et al. (1992), Buzek (1992), and Wheeler and Baas (1992). The one genus of Papilionoideae represented by Tertiary South American fossils, *Tipuana*, was identified by Burnham (1994). If North and South American tropical biotas were strongly differentiated during the Tertiary, then extant taxa that have North American fossil records should have an ancestral area in North America relative to South America, as illustrated on the cladogram adjacent to North America. Likewise, extant taxa that include *Tipuana* should have an ancestral area from South America relative to North America, as illustrated on the cladogram adjacent to South America. These genera are used here for illustrative purposes only; the predictions extend to any extant taxon with a Tertiary fossil record in North or South America. Failure to find such a pattern would weaken support for the Boreotropical flora hypothesis.

Cladrastis Raf., *Ormosia* G. Jacks., *Diplotropis* Benth., *Styphnolobium* Schott ex Endl., *Machaerium* Pers., cf. *Pterocarpus* Jacq. (from the Tertiary of Europe), and *Robinia* (Fig. 8; there are many other reports of North American early Tertiary fossils of Mimosoideae and Caesalpinioideae in Herendeen et al. 1992). In contrast, Burnham (1994) reports the genus *Tipuana* (Benth.) Benth. from the Miocene of Ecuador (Fig. 8). Unlike other reports of fossil legumes, these identifications were made with great attention to detail and have a very high probability of being correct. Since fossils of *Robinia* were present in North America during the late Eocene, the Boreotropics hypothesis predicts that the lineage including *Robinia* should have an ancestral area in North America. Future species-level phylogenetic work on the other seven genera (or monophyletic groups that include each of these genera) represented by early Tertiary North American fossil records should very likely reveal a North American ancestral area (at least within the New World), as established by some criterion such as RP values (Fig. 8). Likewise, a phylogenetic analysis of *Tipuana* and its close generic relatives of the tribe Dalbergieae should reveal a South American ancestral area using the RP index (Fig. 8). Notably, *Tipuana* and its closest relatives are nearly confined to South America (Polhill 1981a), and *Styphnolobium* is found in tropical to warm-temperate North America and southeast Asia (Sousa 1990); such superficial patterns are consistent with their respective fossil records. Also, fossil fruits of something very similar to *Pterocarpus* are known from lower Miocene in central Europe (Buzek 1992); the fossils include stipitate fruits of a kind that are found in *Pterocarpus* from Africa and Asia, and from one species in Mexico. Other Mexican and all South American species have sessile fruits (Sousa & Delgado 1993). Polarization of this fruit character will have to be determined before it can be suggested whether the South American species of *Pterocarpus* are all from a derived lineage that is marked by sessile fruits. Already, however, it appears the species-level phylogenetic and ancestral area analysis on these genera with New World (or northern latitude) fossil records will be revealing.

Because ancestral areas can be predicted for a Neotropical group depending on whether a fossil history is from the Tertiary of North or South America, chance matching of ancestral areas (or relative clade dispersion values) among a set of cladograms can be evaluated by the use of a priori probabilities (cf. Rosen 1978). The chance probability of a particular ancestral area for a single cladogram can be determined by generating random trees from the single cladogram of interest (e.g., using the area labels in Fig. 2 and using the random trees option as implemented in COMPONENT, Page 1993). For each random tree, the most likely ancestral area could be calculated. Then, the percentage of random trees that had a particular most likely ancestral area could be determined. Given the set of labels for the area cladogram shown in Fig. 6, however, the chance probability that North America would be more likely an ancestral area than South America is 0.95 (using the random model, equiprobable trees option in COMPONENT and ntrees=100). Thus, it would be highly likely even by chance that North America should be involved in the ancestral area. Two factors suggest that this approach still has merit for ultimately evaluating the relationship between ancestral areas and timing of fossil records. First, the above estimation of chance probability should ultimately factor in the chance probability that most labels on a set of cladograms should be North American. Second, even if it is ultimately determined that North America has a high chance probability of being the ancestral area for the cladogram of Robinieae, the chance probability of other phylogenies of taxa with Tertiary North American fossil records could also be calculated. Then, all probabilities could be multiplied to derive a single probability of chance matching of ancestral areas among a set of cladograms. Rosen (1978) proposed this sort of chance probability estimation for matching sets of cladograms, and Simberloff et al. (1981) criticized it, because if one searched long enough, a set of matching cladograms (matching in topology, or in this case, estimated ancestral area) could be found. This criticism does not apply here since it is identified beforehand which groups should have a phylogeny that results in an estimated ancestral area of North (or South) America.

In summary, the phylogenetic and biogeographic patterns of the tribe Robinieae are seemingly as well explained by chance as they are by the Boreotropics or Geoflora hypotheses. In other words, a single migration event from anywhere in the world to North America followed by adaptive radiation could account for the pattern seen in North American Robinieae. However, given the antiquity of Robinieae in North America and correspondence in the family Leguminosae of age and phylogenetic diversity, the pattern observed for this tribe may be best explained by the Boreotropical flora hypothesis. Regardless, single phylogenies cannot provide the ultimate test of biogeographic hypotheses, and a method is provided above to test components of the Boreotropics and Geoflora hypotheses with additional phylogenetic studies of specific legume taxa. The isolation of the Neotropics throughout the Cenozoic, as one component of Chaney's Neotropical Geoflora hypothesis, predicts that ancestral area analysis for Neotropical taxa should yield equivalent probabilities for North and South America, or with a bias of higher values for South America. As exemplified by the phylogeny of Robinieae, we doubt this will be the case for Neotropical Leguminosae. Robinieae are typical of Neotropical legumes in showing a great asymmetry in the distribution of the major clades between North and South America. Thus, Wolfe's Boreotropical flora hypothesis is just as explanatory. One component of Chaney's Geoflora hypothesis, that disjunct distributions in North America and Asia should be restricted to temperate biotas, does not find support. This biogeographic pattern found in Robinieae is also found commonly in tropical floristic studies (e.g., Wendt 1993), though much more phylogenetic work is needed to ensure that putative disjunct sister group relationships are well supported. Did the existence of a northern latitude tropical biota during the Tertiary have a great influence on modern North American tropical floras? As will be determined by future studies, the extent of the phylogenetic patterns discussed for Robinieae among other Neotropical taxa should be directly correlated with the relative degree of this influence. How many North American tropical taxa have sister groups from the Paleotropics versus from the Neotropics? If eventually it is shown that the magnitude of the former is large, then components of the Neotropical Tertiary and Arcto-Tertiary Geoflora concepts will need to be greatly modified.

TAXONOMY

Robinieae (Bentham) Hutchinson, Gen. fl. plants 1: 366. 1964. Galegeae subtribe Robiniinae Bentham in Bentham and Hooker, Gen. pl. 1(2): 445. 1865.—TYPE: *Robinia* L.

Sesbanieae (Rydberg) Hutchinson, Gen. fl. plants 1: 401. 1964. Galegeae subtribe Sesbaniinae Rydberg, Amer. J. Bot. 10: 495. 1923.—TYPE: *Sesbania* Adanson.

Galegeae subtribe Corynellinae Rydberg, Amer. J. Bot. 11: 471. 1924.—TYPE: *Corynella* DC. [=*Poitea* Ventenat].

Trees and large shrubs, secondarily shrublets and herbs. Leaves pulvinate, mostly imparipinnate but also paripinnate (simple in *Sphinctospermum*, unifoliolate in two species of *Coursetia* and on distal branch ends in *Genistidium);* leaflets mostly opposite, occasionally alternate; tannins usually accumulated in distinct lines or patches, often on the abaxial surface; stipules reduced to small nubbins, herbaceous, or spinescent; stipels usually present, sometimes reduced to small nubbins or absent. Inflorescences of racemes in axils of coeval leaves on long shoots or of aborted leaves on short shoots, flowers always single at a node;

pedicels confluent or articulated with the calyx; bracteoles essentially absent, but present in *Sesbania*; calyx with a well-developed hypanthium, the lobes very short or well developed in the *Robinia*-group. Standard petals glabrous on the abaxial surface or tomentose in *Olneya*; standard callosities usually well developed in the region of the nectar-guide; claw of standard usually abruptly contracted into a claw; inflexed auricles subtending the juncture of the standard claw and blade; wing petals free from the keel except in a subgroup of *Poitea*; keel petals falcate and distally fused usually to near the tip. Staminal column diadelphous, secondarily pseudomonadelphous in *Lennea* and *Hybosema*, and in some species of *Robinia* and *Coursetia*; basal fenestrae well developed with callus tissue; intrastaminal nectariferous disk absent except in one species of *Gliricidia*. Ovary sessile to subsessile, secondarily stipitate in some species of *Sesbania*, *Hybosema*, *Poitea*, *Gliricidia*, and *Peteria*, hairy or secondarily glabrous; ovules nearly always at least six per ovary; style glabrous, ciliate in *Lennea*, with a pollen brush in the *Robinia*-group; stigmas short-ciliate, long-ciliate in *Hebestigma*. Mature legumes elastically (explosively) dehiscent, or very tardily dehiscent in *Sesbania*; seeds mostly lenticular or nearly so, orbicular in *Olneya*; hilum apical to subapical and attached to a short funiculus, or hilum lateral and attached to a long funiculus in *Robinia*; canavanine accumulated in seed.

The tribe Robinieae comprises 12 genera concentrated in tropical North America including the Greater Antilles, with secondary radiations into South America (species of *Sesbania*, *Gliricidia*, and *Coursetia*) and into warm- and cool-temperate North America (genera of the *Robinia*-group). *Sesbania* is the only genus with native species in both the Neo- and Paleotropics. All of the South American species of *Sesbania* are widespread weeds or introduced by humans.

Three genera, *Gliricidia* (Withington et al. 1987; Hughes & Styles 1989), *Robinia* (Hanover 1992), and *Sesbania* (Macklin & Evans 1990), have or had worldwide economic value especially for revegetation projects, the production of wood for construction, livestock fodder, and shade trees for agricultural purposes. Each of the rest of the genera of Robinieae has at least one species that is cultivated locally as an ornamental for its showy blossoms. The Sonoran Desert Ironwood, *Olneya*, has been used by the Seri Indians of northwestern Mexico in recent times to produce wood carvings and sculptures that depict their environment (Felger & Moser 1985).

The tribe Robinieae traditionally has been distinguished from similar tribes of Papilionoideae by its ebracteolate flower (except *Sesbania*) with a well-developed hypanthium, free wing petals, diadelphous to pseudomonadelphous staminal tube with well-developed basal fenestrae, and ovary with generally six to many ovules. Also, it is distinguished by its inflorescence that bears flowers singly at each node, and its woody legumes that are elastically dehiscent (except in *Sesbania*) and with seeds that accumulate canavanine. These traditional distinguishing features remain useful for separating Robinieae from similar tribes, such as Aeschynomeneae, Amorpheae, Dalbergieae, Indigofereae, and Millettieae. Aeschynomeneae have bracteolate flowers with staminal tubes that lack well-developed basal fenestrae and legumes that are lomented and indehiscent. Amorpheae and Dalbergieae have mostly bracteolate flowers with staminal tubes that lack well-developed basal fenestrae and legumes that are indehiscent and 1-seeded to few-seeded. Indigofereae have flowers with staminal tubes that lack the well-developed basal fenestrae and standard petals that lack calluses and inflexed auricles at the junction of the claw and blade. Millettieae have flowers that are bracteolate or arranged in pseudoracemes and pseudopanicles; species of this tribe lacking both of these traits are very few (e.g., in *Wisteria*) but have been maintained in that tribe because of the terminal position of the inflorescence. According to the results in the phylogenetic analysis section, as well as other studies, such as Geesink (1984), Lavin et al. (1990), and Liston (1995), Millettieae is paraphyletic, and is thus in need of systematic study in order to reconcile the taxonomy of the tribe with the new findings that suggest some genera of the tribe are more closely related to genera of other tribes.

Herbaceous species of Robinieae, as in *Coursetia* or *Peteria*, are superficially very similar to herbaceous genera of the tribe Galegeae (e.g., *Astragalus* L.); however, the pulvinate leaves with usually stipellate rachises, leaflets with tannin deposits, wing petals free from the keel petals, and an indumentum consisting of, in part, stipitate glands, collectively distinguish these herbaceous Robinieae from herbaceous Galegeae. Robinieae may be a New World lineage that was (or is) on its way to evolving temperate herbaceous lineages, but may have been usurped by galegeoid lineages that apparently established themselves more rapidly in the Old World and subsequently migrated to the New World. Both groups have evolved from something very similar to the genus *Millettia*, and the herbaceous species of the two tribes are extremely similar. In contrast to the Galegeae, all Robinieae have a chloroplast genome containing the inverted repeat region (Lavin et al. 1990; Liston 1995).

As outlined in the phylogenetic section, four subgroups of Robinieae are recognized, and they are referred to the informal taxon "group" in this treatment. A general morphological description of these groups, aside from the characters discussed in the phylogenetic section, follows:

1. *Sesbania*-group. Leaves alternate; leaflets without distinctive patchy tannin deposits on the abaxial surface, nyctinasty such that the leaflets fold forward clasping the rachis; stipules herbaceous; inflorescence lacking stipitate glands; flowers bracteolate; style glabrous to puberulous; stigma ciliate to glabrous.

2. *Hebestigma*-group. Leaves opposite to subopposite; leaflets without distinctive patchy tannin deposits on the abaxial surface, nyctinasty such that the leaflets fold downward from the rachis; stipules herbaceous (if perceptible); inflorescence lacking stipitate glands; flowers ebracteolate; style glabrous; stigma penicillate with trichomes distinctly surpassing the length of the stigma.

3. *Gliricidia*-group. Leaves alternate; leaflets with distinctive patchy tannin deposits on the abaxial surface, nyctinasty such that the leaflets fold downward from the rachis; stipules herbaceous; inflorescence lacking stipitate glands; flowers ebracteolate; style glabrous or ciliate in *Lennea*; stigma ciliate to glabrous.

4. *Robinia*-group. Leaves alternate; leaflets with distinctive patchy tannin deposits in *Coursetia* only, nyctinasty such that the leaflets fold downward from the rachis, or in some species of *Coursetia* backward along the rachis; stipules spinescent to lignescent, less commonly herbaceous; inflorescence often with stipitate glands; flowers ebracteolate; style with a pollen brush; stigma penicillate with trichomes as short as the stigma.

The following key distinguishes these major groups and the genera of the tribe Robinieae.

Key to the Genera of Robinieae

1. Style glabrous or ciliate, without a pollen brush; calyx lobes minute, 2 or 4–5 but then much shorter than the calyx tube; inflorescence rachis, pedicels, calyx, or ovary without stipitate glands; stipules always herbaceous, though sometimes inconspicuous; ovaries mostly stipitate, the stipe as long as or longer than the calyx tube.
   2. Flowers bracteolate; leaves paripinnate; leaflet nyctinasty such that leaflets fold forward; legumes inertly dehiscent and lignescent, modified into long linear pods (with 20–80+ ovules per ovary), 4-winged pods (with 4–8 or 25–30 ovules per ovary), torulose pods (with 8–17 ovules per ovary), or bladdery inflated pods (with 2 ovules per ovary); diverse endemics from Africa and mainland North America, widespread weeds in the Antilles, South America, and Asia. 1. *Sesbania*.
   2. Flowers ebracteolate; leaves imparipinnate (paripinnate in six species of *Poitea*); leaflet nyctinasty such that leaflets fold downward; legumes elastically dehiscent and the valves woody (with 4–30 ovules per ovary); native distribution only in the New World.
      3. Leaves and branches opposite to subopposite; leaflets (6–) 9–18 cm long at maturity, (3–) 5–9 (–11) per leaf, without distinctive patchy tannin deposits on the abaxial surface of dried leaflets; calyx with 5 equally spaced short lobes; stigma penicillate; legumes very thick and woody, 2.5–3.7 cm wide at maturity; ovary with 4–6 ovules; endemic to Cuba. 2. *Hebestigma*.
      3. Leaves and branches alternate (rarely subopposite in *Gliricidia*); leaflets less than 7 cm long at maturity, usually many more than 11 per leaf, with distinctive patchy tannin deposits on the abaxial surface of dried leaflets; calyx lobes absent or displaced to the ventral and dorsal side so as to give the calyx a bilobed appearance; stigma glabrous or inconspicuously ciliate; legumes woody but less than 2.5 cm wide at maturity; ovary with 4–30 ovules. *Gliricidia*-group.
         4. Stigma subterminal on a distally introrse (incurled) subclavate to narrowly tapered style; staminal tube pseudomonadelphous; ovary with 4–9 ovules; seed lenticular to suborbicular, at least 2 mm thick; pedicels confluent with the calyx; tropical dry and wet forest of Mexico and Central America. 3. *Lennea*.
         4. Stigma terminal on a distally erect narrowly tapered style (subterminal in two species of *Poitea*, and then the style strongly clavate); staminal tube 9+1 diadelphous except pseudomonadelphous in *Hybosema*; ovary with (5–) 7–30 ovules; seeds lenticular but very thin and flat, less than 2 mm thick; pedicels mostly articulated with the calyx.
            5. Standard and wing petals shorter than the keel petals; stipule-pairs adnate, each broadly attached to the stem and base of petiole, stipules distichously arranged on short shoots; ovary with 15–30 ovules; legumes with thin woody valves, rapidly maturing and readily explosively dehiscent, mature seeds soon dispersed and thus only shortly retained by parent plant; inner epidermis of legume valves smooth and never spongy; Cuba, Hispaniola, Puerto Rico, Dominica. 5. *Poitea*.
            5. Standard and wing petals as long or longer than the keel petals; stipule-pairs free, each stipule narrowly attached to stem in axil of leaf; short shoots not evident; ovary with 5–16 ovules; legumes with thick woody valves, slowly maturing, explosively dehiscent but not readily so, mature seeds retained for some time and thus often found on the parent plant; inner epidermis of legume valves commonly spongy.
               6. Calyx bilobed, the tube narrowly cylindrical; standard petals reflexed at most 90°, with a linear callus along the midrib in the region of the nectar-guide; staminal tube pseudomonadelphous; ovary with 5–7 or 13–16 ovules; montane and seasonally dry tropical forests of interior Mexico and Central America, absent from the Yucatan Peninsula and South America. 4. *Hybosema*.
               6. Calyx truncate, the lobes very reduced, the tube campanulate; standard petals reflexed nearly 180°, with a callus on either side of the midrib adjacent to the region of the nectar-guide; staminal tube diadelphous; ovary with 7–15 ovules; seasonally dry tropical forests, especially of coastal Mexico and Central America, present in the Yucatan Peninsula and South America. 6. *Gliricidia*.
1. Style with a pollen brush; calyx lobes 4–5 and usually as long or longer than the tube; inflorescence rachis, pedicels, calyx, or ovary with stipitate glands; stipules spinescent to

lignescent, or less commonly herbaceous; ovaries nearly sessile, the ovarian stipe shorter than the calyx tube. *Robinia*-group.

7. Legumes with valves laterally and regularly impressed between seeds, forming uniform seed compartments; ovary with (6–) 12–30 ovules; inflorescences of axillary racemes with mostly 1–2 flowers reaching anthesis simultaneously on a rachis; style with a latrorse to extrorse pollen brush; stipules mostly not spinescent or lignescent; keel petals acute, often rostrate; from northern Mexico and adjacent U.S.A. (Arizona and Texas) to South America. 8. *Coursetia*.
7. Legumes not regularly impressed between seeds and not forming uniform seed compartments (except in *Sphinctospermum*); ovary with 4–15 (–17) ovules; inflorescences of solitary flowers, or of terminal or axillary racemes, but if the latter, then with many flowers reaching anthesis simultaneously on a single rachis; style with a pollen brush uniformly surrounding the distal end; stipules spinescent or lignescent; keel petals blunt; mostly northern Mexico and the U.S.A.
    8. Tip of the leaf rachis terminated by a spinose seta; inflorescences of axillary racemes, these often clustered on short shoots with abortive leaves and appearing as panicles terminating a branch; calyx lobes broad and overlapping even in bud; dorsal surface of standard tomentose; legumes terete; seeds spheroidal; deserts of Sonora, Baja California, and Baja California del Sur, Mexico, and southern Arizona and adjacent California, U.S.A. 9. *Olneya*.
    8. Tip of the leaf rachis terminated by a well-developed leaflet (except in some specimens of *Peteria scoparia*); inflorescences of elongate racemes or flowers solitary in leaf axils and subtended by fully expanded leaves; calyx lobes not broad nor overlapping; dorsal surface of standard glabrate; legumes laterally flattened; seeds lenticular.
        9. Inflorescences of terminal racemes; stems completely herbaceous or lignescent only at base, arising from large tuberous rootstocks; desert grasslands and scrub in the western U.S.A. and northern Mexico (principally of the Great Basin and Chihuahuan Desert and vicinity). 10. *Peteria*.
        9. Inflorescences axillary; stems woody, or if herbaceous, annual and not arising from large tuberous rootstocks.
            10. Inflorescences of pendent to lax axillary racemes; legumes not explosively dehiscent; seeds obliquely oriented, hilum distinctly lateral and attached to a relatively long curved funiculus; trees or large shrubs; montane regions of southern U.S.A., and Chihuahua and Sonora, Mexico. 7. *Robinia*.
            10. Inflorescences of solitary flowers; legumes explosively dehiscent; seeds transversely oriented, hilum terminal or subterminal and attached to a very short straight funiculus; small shrubs and herbs; Mexico, and adjacent Texas and Arizona, U.S.A.
                11. Small bushes to straggling shrubs with trifoliolate, sometimes unifoliolate, leaves; flowers yellowish; ovary with 4–6 ovules; legumes without uniform seed compartments; seeds lenticular; Chihuahuan Desert of Coahuila, Mexico and adjacent Texas, U.S.A. 11. *Genistidium*.
                11. Annual herbs with simple leaves; flowers pinkish to mauve; ovary with 10–12 ovules; legumes with uniform seed compartments formed by the expansion of the papery inner epidermis; seeds squarish, with a central constriction; Sonoran Desert of Arizona and California, U.S.A., south along the Pacific Coastal dry forest to Oaxaca, Mexico. 12. *Sphinctospermum*.

1. **Sesbania** Adanson, Fam. pl. 2: 327. 1763, nomen conserv. (as *Sesban*, but corrected to *Sesbania* by Scopoli, Intr. hist. nat. 308. 1777; see Gillett 1963).—Type: *Sesbania sesban* (L.) Merrill.

Single-trunked trees, large shrubs, and perennial and annual herbs. Stems erect, branches many from above, alternate; bark none to smooth or superficially furrowed. Roots deep and not producing leafy sprouts. Leaves alternate, pulvinate, paripinnate; leaflets opposite, elliptic to narrowly elliptic, rounded to emarginate, mostly persistent on the disarticulating rachis during dry periods or with age, adaxial surface glabrous to sericeous, without finely scattered patchy tannin deposits, abaxial surface glabrate to sericeous, without finely scattered patchy tannin deposits; leaf rachis canaliculate, the groove without a dense indumentum, stipels inconspicuous; stipules adnate, narrowly to broadly attached to the stem adjacent to the petiole, herbaceous, generally persistent. Inflorescences of pedunculate to lax axillary racemes borne in axils of coeval leaves, with a generally synchronous maturation of flowers on a single rachis, each with mostly 5–50 nodes, stipitate glands absent. Flowers bracteolate, with a well-developed hypanthium; floral bracts conspicuous, caducous to persistent; pedicels confluent with the calyx, straight and ascending during anthesis and seed set, falling with the aborted flower. Calyx campanulate, generally attenuate at base, glabrate to sericeous or tomentose, stipitate glands absent, with five evenly spaced inconspicuous lobes that are at most shorter than the tube, the lobes, if developed, triangular and not overlapping at anthesis. Standard orbicular to oblong, emarginate, reflexed about 90–120° at anthesis, dorsal surface glabrous, subequal in length to the keel petals, with two prominent callosities about the midrib and in the region of the nectar-guide, claw abruptly to gradually contracted from the blade, inflexed auricles present, sometimes inconspicuous; wing petals sculptured, commonly projecting beyond the keel at anthesis, and loosely attached to it by a fold in the basal auricle, the distal ends glabrous, the blades parallel to perpendicular to the plane of the keel; keel valvately connate distally to near the tip, apex blunt to rarely rostrate (in *Sesbania grandiflora* and relatives). Staminal tube diadelphous, with basal fenestrae; anthers included in the keel tip, basally attached to the filaments, the broad connective reddish brown; intrastaminal nectariferous disk absent. Pollen spheroidal, tricolporate, mesocolpium rugulate-perforate, polar regions smooth. Ovary sessile to stipitate, the stipe as long as the calyx tube, generally glabrous, rarely sericeous, eglandular, ovules 2 (sect. *Glottidium*), 4–8 or 25–30 (sect. *Daubentonia*), 8–17 (sect. *Daubentoniopsis*), or 20+ (sect. *Sesbania*); styles glabrous to puberulous (but then not considered a pollen brush), the base thin and imperceptibly confluent with the distal portion of the ovary, not becoming displaced to the carinal side during fruit maturation; stigma ciliate to glabrous, terminal on a distally tapered style. Legumes mostly stipitate, glabrate, eglandular, modified into long linear pods (with 20–80+ ovules per ovary), 4-winged pods (with 4–8 or 25–30 ovules per ovary), torulose pods (with 8–17 ovules per ovary), or bladdery inflated pods (with 2 ovules per ovary), inner epidermis of the endocarp of adjacent valves spongy and abutting between seeds, forming uniform seed compartments (except in sect. *Glottidium*), the valves lignescent (i.e., very thinly woody) at maturity, inertly dehiscent, the dorsal and ventral margins generally parallel to subsinuous in fruit. Seeds transversely oriented, thickly lenticular to subcylindrical, not constricted in the middle, accumulating canavanine, uniformly light to dark brown, many with darkish purple mottling, hilum apical and attached to a very short funiculus, testa firmly adherent. Seedlings with foliar epigeal cotyledons, eophyll simple and opposite the metaphyll, metaphylls paripinnate and alternate (Gillett 1963; Lavin 1987). Base chromosome number x = 6 (Turner 1955; summary in Gillett 1963; Sachet 1987). Figs. 9–12.

The distribution of *Sesbania* is distinctive among the genera of Robinieae in

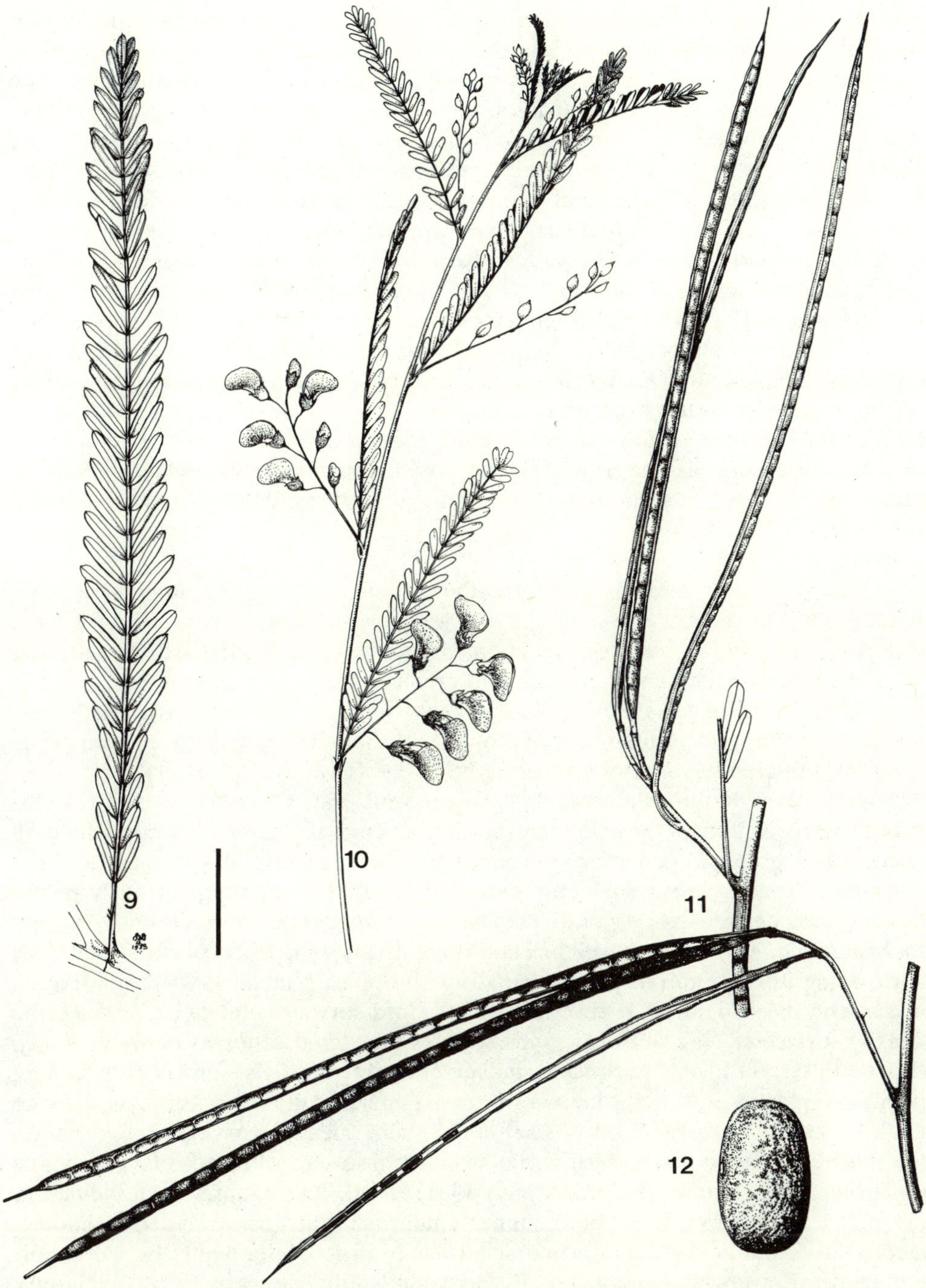

FIGS. 9–12. *Sesbania* sect. *Sesbania* (*Sesbania herbacea*). 9. Leaf. 10. Flowering branch. 11. Fruiting branches. 12. Seed. Reproduced from vol. 5 of *Flora Novo-Galiciana* by Rogers McVaugh. (9, 11, bar = 2 cm; 10, bar = 1 cm; 12, bar = 4 mm).

that it is the only one to have a native occurrence of species in the Old World, particularly in tropical Africa and southeastern Asia into northeastern Australia; in the Old World, members of the sections *Daubentonia*, *Daubentoniopsis*, and *Sesbania* are native. In the Neotropics, the genus has greatest phyletic diversity in North America where the sections *Glottidium*, *Daubentonia*, *Daubentoniopsis*, and *Sesbania* are native; South America contains just human introductions or the widespread New World or pantropical weedy species, which belong to sections *Sesbania* or *Daubentonia*. That *Sesbania* has a Paleotropical-North American center of present day distribution and is the putatively earliest offshoot in Robinieae (i.e., the sister group to the rest of the tribe; see Fig. 5) is consistent with the hypothesis that the group is of Boreotropical origin (see biogeography section). This implies that the evolution of the sectional diversity in *Sesbania* predated the vicariant event that separated the New and Old World species. This phylogenetic position as the earliest offshoot is based on the bracteolate flowers of *Sesbania*, in addition to chloroplast DNA data (Lavin & Doyle 1991). If just genera of Dalbergieae and Aeschynomeneae were used as outgroups in the phylogenetic analysis, this sister group position would also be supported by the forward nyctinastic leaflet movements that are found in *Sesbania* and these outgroup tribes, and in no other genus of Robinieae.

Gillett (1963) summarized the broad economic uses of *Sesbania* in Africa and Asia. Such uses range from shade plants, wind breaks, cover crops, ornamentals, fish poisons, fiber sources, construction materials, to food for both human and domesticated livestock consumption, and green manure. Its use as a fish poison is notable in that it suggests the presence of isoflavones (e.g., rotenones), which have not been reported from members of Robinieae, though they have been commonly cited from related legumes of the tribe Millettieae (Gomes et al. 1981).

Sesbania as herein circumscribed corresponds exactly to Rydberg's (1923b) tribe Galegeae subtribe Sesbaniinae, but it is circumscribed broadly to include all previously segregated genera; the synapomorphies marking the monophyly of the constituent species are many compared to those that mark the previously recognized segregate genera, *Agati*, *Daubentonia*, *Daubentoniopsis*, *Glottidium*, and *Pterosesbania*. The genus is well marked vegetatively by the pair of stipules at each node being fused, and paripinnate leaves with opposite leaflets, which undergo a nyctinastic movement such that the leaflets fold forward and twist, leaving the abaxial surfaces facing outward. Such stipules are found otherwise only in *Poitea* (Lavin 1993), and such nyctinastic movements are otherwise uncommon among papilionoid legumes, but are known from the genera of the tribe Aeschynomeneae and the genera *Dalbergia* and *Machaerium* (Dalbergieae). Karyologically, the genus is unique among tropical papilionoids in having a base chromosome number of 6 (Turner 1955; Gillett 1963; Goldblatt 1981). Also, the legumes of *Sesbania* are distinctive in having a very spongy inner epidermis that forms the seed compartments; this condition is approached superficially in Robinieae only by *Coursetia*. Finally, the flowers of *Sesbania* are distinctive in Robinieae by being conspicuously and uniformly bracteolate, although, in contrast to the traits listed above, this trait is presumably plesiomorphic and is found in nearly all outgroups. In fact, the monophyly of the rest of the tribe Robinieae is marked, in part, by ebracteolate flowers (see phylogeny).

No formal treatment of *Sesbania* is provided here, because the genus is under systematic study by Dr. Joanne Chamberlain at Oxford University. A brief synopsis is presented, however, to emphasize the four major subgroups within *Sesbania*. These groups are well marked morphologically and also by chloroplast DNA characters (Lavin & Doyle 1991; species of sect. *Daubentoniopsis* have not been analyzed with regard to DNA). A tentative key to these sections follows.

KEY TO THE SECTIONS OF SESBANIA

1. Legumes with parallel or slightly sinuous dorsiventral margins, neither bladdery inflated nor with winged margins; seeds cylindrical or nearly so; testa light brownish or finely mottled with purplish striations; each legume with 20–80+ seeds, borne on a stipe 0–0.5 cm long; distribution worldwide. 1.1. *Sesbania* sect. *Sesbania*.
1. Legumes torulose, with sinuous dorsiventral margins, or the valves bladdery inflated, or with winged margins; seeds orbicular to lenticular; testa uniformly light to dark brown; each legume with 2–17 or 25–30 seeds, borne on a stipe mostly 1–4 cm long.
   2. Legumes 4-winged or 4-angled, with 4–8 seeds (25–30 seeds in African species), inner epidermis not becoming separated from the mesocarp; legume tardily dehiscent; style occasionally ciliate; standard petals usually reflexed greater than 90° at anthesis; Africa, southeastern Asia to Australia, U.S.A. and adjacent Mexico (Tamaulipas, San Luis Potosí, and Veracruz), and sporadically in South America. 1.3. *Sesbania* sect. *Daubentonia*.
   2. Legumes not 4-winged or 4-angled, with 2 seeds (the legume with an inner epidermis that becomes separated from the mesocarp), or 8–17 seeds (the legume torulose); legume dehiscent by one or two sutures; style not ciliate; standard petals reflexed 90° at anthesis.
      3. Legumes torulose, distinctly constricted dorsiventrally and laterally between the 8–17 seeds, inner epidermis not separating from the meso- and exocarp at maturity; legume dehiscent along both margins; Africa, central Mexico (Guanajuato, Jalisco, Michoacán). 1.2. *Sesbania* sect. *Daubentoniopsis*.
      3. Legumes bladdery inflated, terete in cross section, with 2 seeds, inner epidermis separating from the meso- and exocarp at maturity; legume dehiscent along the carinal margin; southeastern U.S.A. 1.4. *Sesbania* sect. *Glottidium*.

1.1. Sesbania sect. **Sesbania.**

Agati Adanson, Fam. pl. 2: 326. 1763. *Sesbania* subg. *Agati* (Adanson) Baker in J. D. Hooker, Fl. Brit. Ind. 2: 115. 1876.—TYPE: *Agati grandiflora* (L.) Desvaux [=*Sesbania grandiflora* (L.) Poiret].

Resupinaria Rafinesque, Sylva tell. 115. 1838.—TYPE: *Resupinaria grandiflora* (L.) Rafinesque [=*Sesbania grandiflora* (L.) Poiret].

Darwinia Rafinesque, Fl. ludov. 106. 1817, non *Darwinia* Rudge, 1815 [Myrtaceae]. *Monoplectra* Rafinesque, Fl. ludov. 106. 1817, pro. syn.—TYPE: *Darwinia exaltata* Rafinesque [=*Sesbania emerus* (Aublet) Urban].

Separation of the genus *Agati* was based only on large flower size, but this renders *Sesbania* s.s. without autapomorphies, or as "*Sesbania* not *Agati*," and thus paraphyletic. *Agati*, therefore, should be included within sect. *Sesbania*.

1.2. Sesbania sect. **Daubentoniopsis** Lavin, comb. nov. *Daubentoniopsis* Rydberg, Amer. J. Bot. 10: 497. 1923.—TYPE: *Sesbania longifolia* (Cavanilles) DC.

Though traditionally characterized only by the torulose legume, sect. *Daubentoniopsis* is here recognized because it is morphologically distinct from sect. *Sesbania*. Whereas sect. *Sesbania* has been modified to include all species with legumes having numerous seeds, sect. *Daubentoniopsis* is characterized to include species with torulose legumes that are long-stipitate and have 8–17 ovules per ovary. The geographic range suggests an ancient Boreotropical distribution; one arborescent species is endemic to central Mexico (*Sesbania longifolia*), and the rest occur in continental Africa.

1.3. Sesbania sect. **Daubentonia** (DC.) Bentham in Bentham & Hooker, Gen. pl. 1: 502. 1865. *Daubentonia* DC., Mém. Légum. 285. 1823. *Sesbania* subg. *Daubentonia* (DC.) Baker in Oliver, Fl. trop. Afr. 2: 133. 1871.—TYPE: *Sesbania punicea* (Cavanilles) Bentham.

Sesbania subgenus *Pterosesbania* Gillett, Kew Bull. 17: 149. 1963.—TYPE: *Sesbania tetraptera* Hochstetter ex Baker.

Gillett (1963) contends that the two species with 4-winged legumes from Africa (*S. tetraptera* and *S. rogersii* Phill. & Hutch. of his subg. *Pterosesbania*) are evolutionarily distinct from American members of *Daubentonia*, which have nearly identical 4-winged legumes. Gillett fails to account for the taxonomic relationships of the Asian and Australian species of *Daubentonia* (e.g., *S. tripetii*) and further suggests that subg. *Pterosesbania* is most closely related to African *S. macrantha* Phill. & Hutch. and *S. pachycarpa* DC. (of sect. *Sesbania*), because the legumes of these species are quadrate in cross section. The chloroplast DNA evidence suggests that North American and Asian-Australian "Daubentonias" are very closely related (these were used in a sample of genera presented in Lavin & Doyle 1991), and morphological analysis (Lavin, unpubl.) reveals that there is no difference between the winged pods of the African, Asian, and American species. Lewis (1988) remarked that *Daubentonia* is as distinct a group as is *Glottidium*. The characteristics he provides of *Daubentonia*, such as the reniform seeds and standard petals with only minute appendages (callosities) near the juncture of the blade and claw, are in need of investigation and may serve as further evidence for the recognition of this subgroup of *Sesbania*.

Sesbania virgata Poir., though also in scattered localities in South America, including northern Argentina, occurs as a native in swampy localities in Veracruz, Mexico (Sousa, pers. observ.). The only other species of sect. *Daubentonia* to occur in South America, other than those that are cultivated, such as *Sesbania tetraptera*, is *Sesbania punicea*, a pantropical species that occurs from coastal southeastern U.S.A. to South America along the Argentine coast, and also in coastal regions of Africa.

1.4. Sesbania sect. **Glottidium** (Desvaux) Lavin, comb. nov. *Glottidium* Desvaux, J. Bot. Agric. 1: 119. 1813.—TYPE: *Sesbania vesicaria* (Jacquin) Elliott.

The circumscription of *Sesbania* presented here deviates in a significant way from the important contributions of Rydberg (1924b), Gillett (1963), and Polhill and Sousa (1981): *Glottidium* is not recognized as distinct from the rest of *Sesbania*. Gillett listed the following ten characters of *Glottidium* as evidence for its distinction from *Sesbania*: 1) vexillary filament straight at the base, 2) filaments greatly tapered distally, 3) tip of style incurved, 4) stipe of pod up to 2 cm long, 5) pod not septate, 6) pod twice as wide as seeds, 7) endocarp separating from meso-and exocarp, 8) ovary with just 2 ovules, 9) seeds up to 12 mm in diameter, 10) hilum linear-oblong. Except for character #5, each of these traits listed by Gillett is a derived character (i.e., autapomorphy) of *Glottidium*, and thus the opposing conditions are plesiomorphic and not useful for grouping the species of *Sesbania* excluding *Glottidium*. Consistent with this is a chloroplast DNA phylogeny of Robinieae presented by Lavin and Doyle (1991) that shows *Glottidium* as derived out of the *Sesbania* clade. This chloroplast DNA phylogeny suggests that the inner epidermis of the legume of *Glottidium*, which is expanded and papery like in the rest of *Sesbania*, is homologous to the expanded inner epidermis of the rest of *Sesbania*. That *Glottidium* does not form uniform seed compartments (character #5, above) could be a secondary modification to the evolution of the legume of *Glottidium*, which bears just two large seeds.

2. **Hebestigma** Urban, Symb. antill. 2: 289. 1900.—TYPE: *Hebestigma cubense* (H. B. K.) Urban.

Single-trunked trees. Stems erect, branches opposite to subopposite; bark smooth to superficially furrowed, more deeply cracked in the oldest of specimens. Roots apparently deep and not producing leafy sprouts. Leaves opposite to subopposite, pulvinate, imparipinnate; leaflets opposite, broadly elliptic, acuminate, caducous from the more persistent rachis during dry periods or with age, adaxial surface glabrous, often with finely scattered patchy tannin deposits, abaxial surface glabrate, without finely scattered patchy tannin deposits; leaf rachis subcanaliculate, the groove without a dense indumentum, stipels absent; stipules inconspicuous to absent. Inflorescences of pedunculate lax axillary racemes, mostly at nodes with slightly or undeveloped leaves and clustered at branch ends, rarely in axils of coeval leaves, with a synchronous maturation of flowers on a single rachis, each with mostly 20–50 nodes, stipitate glands absent. Flowers ebracteolate, with a well-developed hypanthium; floral bracts conspicuous, persistent; pedicels confluent with the calyx, straight and ascending during anthesis and seed set, falling with the aborted flower. Calyx campanulate, generally attenuate at base, sericeous to tomentose, stipitate glands absent, with five evenly spaced lobes that are mostly shorter than the tube, the lobes triangular and not overlapping at anthesis. Standard orbicular, emarginate, reflexed about 90° at anthesis, dorsal surface glabrous, subequal in length to the keel petals, with two prominent callosities about the midrib and in the region of the nectar-guide, claw abruptly contracted from the blade, inflexed auricles present; wing petals sculptured, commonly projecting beyond the keel at anthesis, and loosely attached to it by a fold in the basal auricle, the distal ends glabrous, the blades parallel, or nearly so, to the plane of the keel; keel valvately connate distally to nearly the tip, apex blunt. Staminal tube diadelphous, with basal fenestrae; anthers included in the keel tip, basally attached to the filaments, the broad connective reddish brown; intrastaminal nectariferous disk absent. Pollen spheroidal, tricolporate, mesocolpium rugulate-perforate, polar regions smooth. Ovary subsessile, stipe much shorter than calyx tube, tomentose, eglandular, with 4–6 ovules; styles glabrous, the base thin and imperceptibly confluent with the distal portion of the ovary, not becoming displaced to the carinal side during fruit maturation; stigma penicillate, terminal on a distally tapered style. Legumes subsessile, glabrate, eglandular, inner epidermis of the endocarp of adjacent valves spongy and abutting between seeds but not forming uniform seed compartments, the valves thickly woody at maturity, elastically and explosively dehiscent, the dorsal and ventral margins generally parallel to subsinuous in fruit. Seeds transversely oriented, thickly lenticular, not constricted in the middle, accumulating canavanine, uniformly light to dark brown, without dark mottling, hilum apical and attached to a very short funiculus, testa loosely to firmly adherent. Seedlings with thick and fleshy hypogeal cotyledons, eophylls of two pairs of decussately arranged simple and cataphyll-like leaves, metaphylls unifoliolate and alternate. Base chromosome number x = 10 (Goldblatt 1981).

Hebestigma is easily distinguished in fruit from all other genera of Robinieae by its thick-walled legumes that measure (9–) 11–22 cm long by 2.0–3.7 cm wide. Such legumes are, in overall dimensions, the most robust in the tribe, as are likewise its seeds, which measure 14–18 mm in diameter. Notably, legumes of similar dimensions and that have a spongy inner epidermis are found in the genus *Millettia* (tribe Millettieae) from the Old World tropics. Other characters that distinguish *Hebestigma* from the rest of Robinieae include its large leaves, 18–45 cm long, with mostly 5–9 large glabrous concave leaflets distinctively borne on long pulvinules 5–10 mm long, an opposite to subopposite branching pattern, and the distinctly penicillate stigma. Also, the highly reduced, often imperceptible, stipules of *Hebestigma* are distinctive.

The pattern of growth and dormancy of branch systems in *Hebestigma* is also diagnostic and not found, except for occasional individuals of *Gliricidia sepium*, in any other woody papilionoid legume. During vegetative growth, the nodes are subopposite but otherwise well spaced along the elongating branch. During dry seasons, the apical meristem becomes dormant, newly developing leaves are aborted, and branch growth thus stops. During a rainy period, the apical meristem becomes active again but does not necessarily resume long-shoot vegetative growth. It often produces short-shoot growth with underdeveloped leaves in whose axils are borne inflorescence rachises (or branch systems bearing inflorescences in the axils of underdeveloped leaves). During this period, the inflorescence superficially appears to be a terminal panicle, but it is actually not terminal because the vegetative apical meristem subtended by unexpanded leaves is continually present. Following the flowering season, this apical meristem begins to produce long-shoot growth with fully developed leaves at each node and without fertile axillary branches. As observed in field and greenhouse plants, *Hebestigma* is very drought-tolerant. The arrested development of this very hardened apical tissue very likely facilitates drought resistance.

The phylogenetic and biogeographic position of the monotypic *Hebestigma* is remarkable. It is tentatively regarded as an early offshoot of the strictly New World native members of Robinieae (i.e., a possible sister group of the rest of the tribe that is restricted to the New World) and is endemic to Cuba. The phylogenetic position suggests that *Hebestigma* is an ancient lineage of North American Robinieae, and its restriction to Cuba, if not due to over-water dispersal, could be explained by Caribbean vicariance models (Rosen 1976, 1985) that depict Cuba as having drifted eastward as two separate blocks during the Eocene from its place of origin near the Mayan Block (northwestern Central America). The same model explains, in part, the occurrence of *Poitea* in the Greater Antilles (Lavin 1993).

2. Hebestigma cubense (H. B. K.) Urban, Symb. antill. 2: 289. 1900. *Robinia cubensis* H. B. K., Nov. gen. sp. 6: 392. 1823. *Gliricidia cubensis* (H. B. K.) C. Wright, Anal. Acad. Ci. Habana, Rev. 5: 332. 1869.—TYPE: CUBA. La Habana: prope Batabanó, *Humboldt & Bonpland 1348* (holotype: P-HBK, microfiche: IDC 6209.161.II.6!).

Lonchocarpus latifolius A. Richard ex R. de la Sagra, Hist. phys. Cuba, Bot. Pl. vasc. part 2, 10: 179. 1841. *Gliricidia latifolia* (A. Richard ex R. de la Sagra) Grisebach, Cat. pl. Cub. 70. 1866. *Hebestigma cubense* var. *latifolium* (A. Richard ex R. de la Sagra) Urban, Symb. antill. 2: 290. 1900.—TYPE: CUBA. Guantánamo: Oriente, in littoribus prope Baracoa, *Wright s.n.* (holotype: not located).

Gliricidia platycarpa Grisebach, Cat. pl. Cub. 70. 1866.—TYPE: CUBA. Pinar del Río: Loma de Rangel, "occidental," 18 May 1863, *Wright 2349* (lectotype, designated by Howard, 1988: GOET!; isotypes: GH! GOET! HAC! MO!).

Trees 2–13 m tall. Stems erect and single, without major branches from near base; bark light to dark grayish brown, occasionally mottled, minutely furrowed toward the base on older stems; branches glabrate, only sericeous to tomentose on very young stems or toward the very distal tip with silvery or brownish trichomes. The larger leaves 18–45 cm long; leaflets (3–) 5–9 (–11) per leaf, occasionally distally accrescent, broadly elliptic, the larger (6–) 9–18 cm long, (4–) 5–9 mm wide, both surfaces glabrous and occasionally obscurely darkened with brownish tannin deposits (younger developing leaves strigose mainly along the margins, and lower midrib and secondary veins). Inflorescence rachises 6–15 (–21) cm long, slightly pedunculate (sterile basal portion less than one-tenth the rachis length), strigose to sericeous with brownish trichomes, with 20–50 nodes, internodes mostly 0.1–4.0 mm long; floral bracts 0.4–1.0 mm long, triangular, sericeous, persistent; pedicels 5–6 mm long at anthesis, tomentose. Calyx campanulate, attenuate at base, generally tomentose with brownish and whitish trichomes, cup 3.5–5.0 mm long, the lobes 1–2 mm long, the vexillary lobe bifid. Standard pale pinkish, aging whitish with a tinge of purple, nectar-guide greenish, blade reflexed 90°, (13–) 14–19 mm long, 14–19 mm wide, orbicular, emarginate; wing petals 13–18 mm long, parallel to the plane of the keel, pale pinkish; keel petals 13–18 mm long, adnate for much of the distal end, straight, with a blunt tip, pale pinkish. Staminal tube straight, pseudomonadelphous, filaments subequal, free for about one-fourth the length. Style 5–6 mm long at anthesis; ovary tomentose with fine whitish to brownish trichomes, with 4–6 ovules. Legume (9–) 11–22 cm long, 2.5–3.7 mm wide, strongly laterally compressed, light to dark brown, mostly glabrate after dehiscence, but commonly villous before; seeds 3–6, lenticular, 14–18 mm diameter, uniformly light brown, aging dark brown. Figs. 13–15.

Phenology. Flowering specimens come from February through July; mature fruiting specimens mostly from February through August, rarely to October, November, or December. Flowers are fully open by mid-morning and are heavily visited by *Xylocopa*-sized bees.

Distribution (Fig. 16). Cuba; lowland seasonally deciduous forests; 0–120 m. Particularly abundant on calcic substrates and co-occurring most commonly with other legumes, such as *Caesalpinia* and *Acacia*.

FIGS. 13–15. *Hebestigma cubense*. 13. Flowering branch mostly in bud. 14. Young leafy branch tip. 15. Mature legume valve with seed attached. (13, bar = 20 mm; 14, bar = 10 cm; 15, bar = 3 cm).

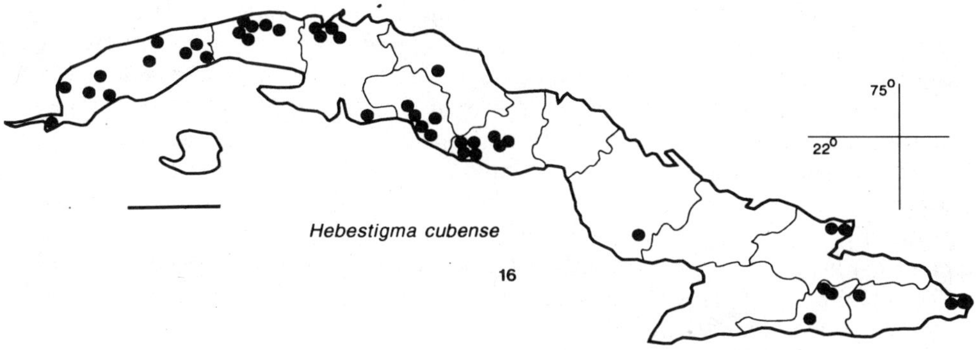

FIG. 16. Distribution of *Hebestigma cubense* (bar = 200 km).

ADDITIONAL SPECIMENS EXAMINED. **Cuba.** CAMAGÜEY: mogote pequeño en la Sierra de Najasa, dentro de la Finca Belén, *Areces et al. 31705* (HAJB).—CIENFUEGOS: Manigua seca al este del faro de la bahía, *Bisse et al. 36381* (HAJB); Punta Colorada, Cienfuegos Bay, *Britton & Wilson 5633* (NY); Río San Juan, *Britton et al. 5845* (NY); beyond Faro Villa Nueva, *Combs 600* (MO); Trinidad Mountains, San Blas to Buenos Aires, *Gonzales 95* (GH, NY); limestone outcrop of Potrero Seboruco between Tierra Roja and mangrove at Gavilán, *Howard et al. 11* (GH, UC); Gavilán, *Howard 4916* (GH, NY, US); Cienfuegos, near shore of bay, *Jack 4957* (GH, NY); Harvard Tropical Garden, Soledad, *Jack 5569* (NY); Limones, Soledad, *Jack 6039* (GH); Limones, Soledad, *Jack 6562* (GH, US); Vega, Soledad, *Jack 7797* (GH); between Soledad and Guajimico, *Moore 6077* (GH); Limones, *Pringle 99* (GH, MO, NY); Central Soledad, *Smith 782* (HAC); coral rock formation near the coast, Gavilán, *Smith & Hodgdon 3187* (GH, NY); seashore, Guajimico Bay, 25 km SE of Cienfuegos, *Webster et al. 81* (GH).—GUANTÁNAMO: Mesa del Chivo, Maisí, *León & Victorin 17180* (HAC); orillas del Río Maya, Maisí, *León 18124* (HAC); Bayate, Monte Oscuro, *Ekman 4622* (HAC).—HOLGUÍN: Mpio. Banes, 2–3 km NW of Cañadón toward Guardalavaca, *Lavin et al. 7144* (MONT); Jagüeyes de Mulas, Banes, *Roig 6526* (HAC).—LA HABANA: Guatao, *Alain 1948* (HAC); on Mt. Santa Barbara, near Bajnéal, *Van Hermann 748 & 749* (NY); Santa Barbara, *Van Hermann 5039* (NY); National Botanic Gardens, *Lavin 5611* (TEX); Orillas del Sta. Ana Corrales, Cangrejeras, *León 4501* (HAC); Loma de Guerrero, cerca del Ariquauabo, *León 10698* (HAC); woody banks of Ariguanabo River, San Antonio de los Baños, *León & Alain 5026* (HAC, NY); Jardín Botánico de la Habana, *Santiago & Ramona* 27590 (HAC); Bosque Cienega, P. Grande, *Suárez 14-I–51* (HAC); Loma de Esperón, NW outskirts of Caimito, *Webster 3682* (MICH); Sierra de Anafe, *Wilson 11346* (NY).—MANTANZAS: vicinity of Matanzas, *Britton et al. 33* (NY); entre Matanzas y Varadero, junto a Finca Carbonera, *Casa & Valverde 10673* (HAC); Managua Costera, Guasasas a Playa Girón, *Oviedo 27992–27995* (HAC); Loma de Coliseo, *Medina 821* (HAC); in sylvis prope Matanzas, *Rugel 265* (GH, GOET, MO, NY).—PINAR DEL RÍO: Top of El Mogote, Soroa, *Alain 2773* (GH); barranca del Río Taco-taco, Santa Cruz de los Pinos, *Alain & Clemente 1426* (GH, HAC); San Diego de los Baños, *Britton et al. 6699* (NY); Santa María a La Vigia, Gibara, *Díaz F. s.n.* (HAC); Barrio de Taeronas, *Díaz P. s.n.* (HAC); Guama, borders of Río Portales, at La Ullóa, in limestone hills, *Ekman 18712* (HAC); Valle del Taco-taco, *León 12749* (HAC); Bosque, Finca Imposible, Las Pozas, *León 17090* (HAC); Costa de las Peñas, 20 km NE de Jaimanitas Cabo Corrientes, *Lepper et al. 34358* (HAC); between Río Cuyaguateje and Sierra Guane, *Shafer 10454* (NY); La Guira, N of Sumidero, *Shafer 13744* (HAC, NY); Galalón de Caiguanabo, *M. Yero s.n.* (HAC).—SANCTI SPÍRITUS: Trinidad, loma del Burro, *Areces 25031* (HAJB); Arroyo Grande to Trinidad, *Britton & Wilson 5475* (NY); La Vigia Hill, Trinidad, *Britton & Wilson 5517* (GH, NY, US); Río Tuyaba, Trinidad, *Britton & Wilson 5561* (NY); Lomas de Banao, *Luna 576* (HAC); Loma del Puerto, Trinidad, *Roig 5728* (HAC); vicinity of Pitajones, along Río Unimazo, *Shafer 12287* (GH, MO, NY); La Ceiba to Rincón, *Shafer 12302* (NY).—SANTIAGO DE CUBA: Santa María, La Vigía, Gibara, *Díaz P. 453* (HAC); Bayate, Monte Oscuro, prope Sierra de Nipe, *Ekman 4622* (HAC); Alto Cedro to Paso Estancia, *Shafer 1616* (NY).—VILLA CLARA: La Sierra, *Jack 8100* (GH).—PROVINCE UNKNOWN: vicinity of Piedra Gorda, Oriente, *Shafer 3271* (NY); *Jack 4542* (GH); Las Ninfas, *Hioram 1969* (NY).

Hebestigma cubense is superficially very similar to *Gliricidia sepium*, which has been widely introduced to Cuba. Notably, *G. sepium* occasionally displays subopposite leaves similar to those of *H. cubense*, and, furthermore, the inflorescence of *G. sepium* approximates that of *H. cubense* in that it occasionally comprises axillary racemes clustered along a short shoot near the terminus of an otherwise long shoot. The inflorescence rachis, persistent floral bracts, pedicels, and calyx are densely sericeous to tomentose in *H. cubense* and glabrate to sparsely strigose in *G. sepium*. Furthermore, the calyx lobes are small but distinct in *H. cubense* and inconspicuous in *G. sepium*. If leaves are available, those of *H. cubense*, with (3–) 5–9 (–11) large leaflets, are very different from those of *G. sepium*, which usually have many more than 9 (and up to 25) leaflets.

Hebestigma cubense is a taxonomically isolated species with presumably no close relative from among the genera of Robinieae. Very likely, if *Hebestigma* had been collected from Africa or southeast Asia, it would have been described as a species of *Millettia* and without subsequent controversy about such a generic placement (much in the same way *Gliricidia sepium*, introduced to the Philippines, was described by Asa Gray as *Millettia luzonensis*; see synonymy of *G. sepium*). The flowers of *Hebestigma,* which lack bracteoles and are arranged in simple racemes, would be somewhat unusual in *Millettia* but easily accommodated in *Millettia* sect. *Podocarpae* (Dunn 1912). The distinction between *Millettia* and the genera of Robinieae is discussed in the phylogeny section. Thus, *Hebestigma cubense* must be the least morphologically modified member of the tribe Robinieae, i.e., relatively unmodified compared to the putative common ancestor of the tribe.

The binomial *Gliricidia sagraei* Bentham, though often cited and used to annotate herbarium specimens (e.g., Ramon de la Sagra's specimens from "La Havanah" at GH and MO), apparently was never validly published. Urban (1900) included it in the synonymy of *Hebestigma cubense* but noted that the name came only from specimens annotated by Bentham in the de Candolle herbarium. Also, Micheli (1894) had alluded to the name and the characters of the specimen when distinguishing it from the newly described *Gliricidia guatemalensis* (=*Hybosema ehrenbergii*).

3. Lennea Klotzsch in Link, Klotzsch & Otto, Icon. pl. rar. 2: 65. pl. 26. 1842.— TYPE: *Lennea melanocarpa* (Schlechtendal) Vatke ex Harms.

Single- and multistemmed trees and shrubs. Stems erect to ascending or splayed laterally, branching alternate; bark smooth to superficially furrowed. Roots commonly superficial and often producing leafy sprouts. Leaves alternate, pulvinate, imparipinnate; leaflets opposite, narrowly to broadly elliptic, retuse to cuspidate, caducous from the more persistent rachis during dry periods or with age, adaxial surface glabrous, often with finely scattered patchy tannin deposits, abaxial surface glabrous to densely sericeous, usually without finely scattered patchy tannin deposits; leaf rachis canaliculate, the groove without a dense indumentum, stipels present; stipules free, narrowly attached to the stem adjacent to the petiole, herbaceous, at least the basal portion persistent. Inflorescences of subsessile to pedunculate pendulous to lax axillary racemes, in the axils of coeval leaves or occasionally from leafless nodes of older stems, with a synchronous maturation of flowers on a single rachis, each with mostly 10–85 nodes, stipitate glands absent. Flowers ebracteolate, with a well-developed hypanthium; floral bracts conspicuous, at least the basal portion persistent; pedicels confluent with the calyx, straight and ascending during anthesis and seed set, falling with the aborted flower. Calyx campanulate, generally attenuate at base, glabrate to sericeous, stipitate glands absent, with two prominent lobes mostly shorter than the tube, the vexillary lobe bifid and the carinal one trifid, the ultimate lobes linear to triangular, not overlapping at anthesis. Standard orbicular, emarginate, reflexed about 90° at anthesis, dorsal surface glabrous, subequal in length to the keel petals, with two prominent callosities about the midrib and in the region of the nectar-guide, claw abruptly contracted from the blade, inflexed auricles present; wing petals sculptured, commonly projecting beyond the keel at anthesis, and loosely attached to it by a fold in the basal auricle, the distal ends glabrous, the blades parallel, or nearly so, to the plane of the keel; keel valvately connate distally to nearly the tip, apex blunt. Staminal tube pseudomonadelphous, with basal fenestrae; anthers included in the keel tip, basally attached to the filaments, the broad connective reddish brown; intrastaminal nectariferous disk absent. Pollen spheroidal, tricolporate, mesocolpium rugulate-perforate, polar regions smooth. Ovary subsessile, the stipe much shorter than calyx tube, glabrous to densely sericeous especially along the placental margin, eglandular, with 4–9 ovules; style with crisped uniseriate trichomes along the ventral (inner) surface from the placental margin of the ovary to near the stigma, the base thin and imperceptibly confluent with the distal portion of the ovary, not becoming displaced to the carinal side during fruit maturation; stigma glabrous, subterminal on a distally introrse (incurled) subclavate to narrowly tapered style. Legumes subsessile, glabrous, villous, or sericeous to tomentose, eglandular, becoming glabrate with age, inner epidermis of the endocarp of adjacent valves thin and not abutting, or spongy and abutting between seeds, but not forming uniform seed compartments, the valves woody at maturity, elastically and explosively dehiscent, the dorsal and ventral margins generally parallel in fruit. Seeds transversely oriented, thickly lenticular to suborbicular or cylindrical at maturity, not constricted in the middle, accumulating canavanine, uniformly light to dark brown, without dark mottling, hilum apical and attached to a very short funiculus, testa loosely to firmly adherent. Seedlings with thick and fleshy epigeal cotyledons, eophylls opposite, 5-foliolate, metaphylls and mature leaves 5–many-foliolate, alternate (*Magallanes 1662*, MEXU; Sousa & de Sousa 1981). Base chromosome number unknown.

Lennea is distinguished from the other continental North American genera of Robinieae by its flowers in which the style is incurled at the distal end so that the stigma is introrse. The incurled style is not so prominently clavate as in certain species of *Poitea*, a related genus native to the Antilles (Lavin 1993). The ciliate style of *Lennea* is distinctive in that uniseriate trichomes line the ventral side of the style in a manner analogous to a pollen brush, but the trichomes are loose, crisped, and not appressed. Though these stylar trichomes of *Lennea* are not the dense stiff ascending type characteristic of the pollen brush (Lavin & Delgado 1990), they could be homologous (Fig. 4 in phylogeny section). Another distinctive trait of *Lennea* are the diminutive flowers, which have petals measuring less than 12 mm long and are borne on slender raceme rachises. This slender flower and inflorescence architecture stand in great contrast to a robust woody growth form of *Lennea* (generally trees). The small flowers with a pseudomonadelphous staminal tube, persistent conspicuous floral bracts, and woody legumes with fewer than 10 ovules form a distinctive combination found only in *Lennea*, but it is approached most closely in Robinieae by the genus *Hybosema*. This latter genus has flowers of similar size with a pseudomonadelphous staminal tube and subtended by persistent generally conspicuous floral bracts, but the ovules number up to 16 per ovary.

The geographical distribution *Lennea*, in southern and eastern Mexico and Central America, lies in the heart of the distribution for all of the New World species of the tribe Robinieae. *Lennea* is apparently closely related to the genera *Hybosema* (southern Mexico and adjacent Central America), *Gliricidia* (native to much of Mexico and Central America, but with one species from southern Ecuador and adjacent Peru), and *Poitea* (from the Greater Antilles). This relationship is based on very few morphological characters, such as the occurrence of bilobed calyces and pseudomonadelphous staminal tubes (as in Fig. 3 of the phylogeny section), and is, therefore, only provisionally recognized. There is evidence of an alternative relationship of *Lennea* to the *Robinia*-group, based on the possibility that the ciliate style of *Lennea* is homologous to the pollen brush found among the species of the *Robinia*-group (Fig. 4). The relationship of *Lennea* with the *Gliricidia*-group is favored tentatively over the one with the *Robinia*-group because of the larger number of discrete derived similarities. A close relationship of *Lennea* with *Poitea*, based on the presence of incurled styles with introrse stigmas, is not supported here, because this character occurs only in two highly modified species of the genus *Poitea* (Lavin 1993).

Although species of Robinieae are primarily centered in tropical forests with a pronounced dry season, species of *Lennea* are notable in the tribe Robinieae for being largely restricted to lowland evergreen or riparian forests, and are locally common only there. Such localized distributions may explain why *Lennea*, like other genera of Robinieae, is seldom utilized by humans, even though it has several potentially valuable uses. We have observed *Lennea viridiflora* as a robust constituent of a living fence system in Veracruz, Mexico, along with *Diphysa americana* (Miller) M. Sousa and *Gliricidia sepium*. All species of *Lennea* have showy clusters of deep red-rose or pale yellow flowers, and are occasionally cultivated as ornamentals (according to label data). In addition, all species of the genus root-sprout and easily can inhabit unstable soils on steep slopes. Such an attribute has considerable value in revegetation or reforestation projects.

KEY TO THE SPECIES OF LENNEA

1. Petals predominantly pale yellow, in relatively short lax axillary racemes, the rachis 1–5 (–7.5) cm long; ovaries with 7–9 ovules; mature legume strongly laterally compressed, villous, or sericeous to tomentose; the inner epidermis of the mature legume spongy, forming seed compartments; seeds generally lenticular, testa firmly adherent; lower leaflet surface strigose to sparsely or densely sericeous, new vegetative branch growth sericeous to tomentose, with mostly brownish trichomes.
   2. Leaflets (9–) 11–19 (–23) per leaf, not distally accrescent, strigose to sparsely sericeous on abaxial surface, apex rounded, rarely emarginate; stipels 0.9–4.1 mm long; legume villous prior to dehiscence; Gulf Coastal and southern Mexico, Central America south to Panama. 3.1a. *L. viridiflora* var. *viridiflora*.
   2. Leaflets 5–7 (–9) per leaf, distally accrescent, strigose to densely sericeous on abaxial surface, apex acuminate or rarely emarginate; stipels 0.3–0.9 mm long; legume sericeous to tomentose prior to dehiscence; Mexico (Jalisco, Colima, and Michoacán). 3.1b. *L. viridiflora* var. *novogaliciensis*.

1. Petals predominantly deep red-rose, very rarely greenish yellow, in elongate pendent axillary racemes, the rachis 4–20 cm long; ovaries with 4–6 ovules; mature legume terete, glabrate; the inner epidermis of the mature legume smooth, not forming seed compartments; seeds suborbicular to cylindrical, testa loosely adherent; lower leaflet surface glabrous, new vegetative branch growth glabrate to strigose with mostly whitish trichomes.
   3. Leaflets 7–17 per leaf, not or slightly distally accrescent, the larger 25–70 (–93) mm long, 12–35 (–43) mm wide, apex rounded to emarginate; pulvinules 1.0–2.5 (–3.0) mm long; stipels longer than diameter of adjacent pulvinule; raceme rachis pedunculate, the peduncle 1–4 cm long, fertile portion with 3–6 nodes/cm; generally Atlantic coastal Mexico (Campeche, Hidalgo, Puebla, Veracruz, Oaxaca, Quintana Roo, and San Luis Potosí) and adjacent Belize and northern Guatemala. 3.2. *L. melanocarpa*.
   3. Leaflets 3–5 per leaf, distinctly distally accrescent, the larger 75–200 mm long, 39–85 mm wide, apex cuspidate, with a drip tip, rarely rounded; pulvinules 4–6 mm long; stipels indistinct; raceme rachis subsessile, with 8–12 nodes/cm; mostly interior southern Mexico (Chiapas and adjacent areas in Veracruz, Tabasco, and Oaxaca), central Guatemala, and Honduras. 3.3. *L. modesta*.

3.1. Lennea viridiflora Seemann, Bot. voy. Herald 3: 107. 1853.—TYPE: PANAMA. Veraguas: Santiago de Veraguas, *Seemann 1189* (holotype: K!, photos: F! MONT!).

Trees and large shrubs 1–12 m tall. Stems erect to ascending, 2–18 cm diameter at base; bark gray to grayish brown, not mottled, minutely furrowed toward the base on older stems; branches sericeous to tomentose with mostly brownish trichomes. The larger leaves 5–19 cm long; leaflets 5–23 per leaf, distinctly distally accrescent only in Jalisco, Michoacán, and Colima, narrowly to broadly elliptic, the larger 25–51 mm long, 10–18 mm wide, (43–90 mm long, 20–45 mm wide in Jalisco, Michoacán, and Colima), adaxial surface of lamina glabrous to puberulous and slightly to not darkened with diffuse tannin deposits, abaxial surface strigose to sparsely or densely sericeous, apex rounded or rarely emarginate (acuminate or rarely emarginate in Jalisco, Michoacán, and Colima); pulvinules 1.7–4.0 mm long; stipules 3–4 (–7) mm long, linear-lanceolate, 3–4 (–7) mm long, the basal 1–2 mm persistent; stipels well developed and longer than diameter of subtending pulvinule, except in Jalisco, Michoacán, and Colima. Inflorescences of lax racemes in axils of coeval leaves or at leafless nodes, rachises 1.0–5.0 (–7.5) cm long, sessile (essentially no sterile basal portion), strigose to sericeous with brownish trichomes, with 10–45 nodes, 8–12 nodes/cm; floral bracts 1–2 mm long, linear-lanceolate, strigose, persistent; pedicels 2–3 (–4) mm long at anthesis, strigose. Calyx campanulate, attenuate at base, sericeous with brownish and whitish trichomes, cup 2.0–3.5 (–4.0) mm long, the lobes 1.5–2.0 mm long, the vexillary lobe entire to bidentate, the carinal one tridentate. Standard pale yellow (also in age), nectar-guides greenish, blade reflexed 90°, 8–9 mm long, 11–12 mm wide, orbicular, emarginate; wings 10–11 mm long, parallel to the plane of the keel, pale yellow; keel petals 9–10 mm long, adnate for much of the distal end, straight, with a blunt tip, pale yellow. Staminal tube straight, pseudomonadelphous, filaments subequal, free for about one-sixth the length. Style 5–6 mm long at anthesis, arcuate, incurled and gradually tapered to the subapical introrse stigma, included in the keel tip; ovary densely whitish sericeous, especially on the placental margin (densely sericeous all over the ovary in Jalisco, Michoacán, and Colima), with 7–9 ovules. Legume 5–11 cm long, 1.2–2.0 cm wide, strongly laterally compressed, pale to dark brown, mostly glabrate after dehiscence but commonly villous (or sericeous to tomentose in Jalisco, Michoacán, and Colima) before, inner epidermis of mature legume spongy, forming seed compartments; seeds 5–8, lenticular, 9–12 mm diameter, uniformly light brown, testa firmly adherent. Chromosome number unknown.

Lennea viridiflora stands apart from the other two species in the genus owing to its mature legumes and seeds, which are strongly laterally compressed, flowers with predominantly pale yellow petals, sessile raceme rachises most commonly much shorter than 5 cm (rarely to 7.5 cm), ovaries with 7–9 ovules, abaxial leaflet surfaces strigose to sparsely or densely sericeous, and a much denser indumentum of predominantly brownish trichomes on the young stems and inflorescences. In contrast, *L. melanocarpa* and *L. modesta* have suborbicular to cylindrical seeds and terete legumes at maturity, flowers with predominantly deep rose-red petals, inflorescence rachises commonly more than 7.5 cm long, ovaries with 4–6 ovules, glabrous abaxial leaflet surfaces, and a sparse indumentum of a few whitish trichomes on the young stems and inflorescences. A comparative analysis of the species of *Lennea*, and using *Hebestigma* as an outgroup, suggests that *L. viridiflora* represents the unspecialized species in the genus relative to those traits discussed above, especially with regard to fruit and seed characters (Lavin, unpubl.).

Two varieties of *L. viridiflora* are here recognized.

3.1a. Lennea viridiflora var. **viridiflora.**

Lennea brunnescens Standley, Contr. U.S. Natl. Herb. 23: 480. 1922.—Type: Mexico. Veracruz: Carrizal, 12–14 May 1901, *Goldman 701* (holotype US!, photos: MEXU! NY!).

Lennea salvadorensis Standley, Publ. Field Mus. Nat. Hist., Bot. Ser. 4(8): 213. 1929.—Type: El Salvador. La Libertad: Hacienda Corinto, Zaragosa, 19 Jul 1928, *Calderón 2394* (holotype: F!, photo: MEXU!).

Gliricidia malacocarpa A. Molina, Ceiba 3: 93. 1952.—Type: Honduras. Cortés: bosque de la Presa Tulián, 16 Jun 1952, *Molina R. 5201* (holotype: US!, photos: F! GH! MEXU! NY! TEX! UC!; isotype: GH!).

Leaves (9–) 11–19 (–23) cm long; leaflets 9–23 per leaf, not distally accrescent, narrowly to broadly elliptic, 25–51 mm long, 10–18 mm wide, apex rounded, rarely emarginate, adaxial surface glabrous to puberulous and slightly to not at all darkened with diffuse tannin deposits, abaxial surface strigose to sparsely sericeous; pulvinules 2.0–4.0 mm long; stipels 0.9–4.1 mm long. Ovary glabrous to densely sericeous along the placental margin and distal end. Legume pale to dark brown, commonly villous, becoming glabrate after dehiscence.

Phenology. Flowering and fruiting throughout the year, apparently without a predominant flowering season.

Distribution (Fig. 25). Locally common in Panama, Costa Rica, Nicaragua, Honduras, El Salvador, and Mexico (Oaxaca, Guerrero, and Veracruz); riparian habitats in lowland seasonally dry or evergreen rain forest, and co-occurring commonly with *Diphysa* and *Gliricidia*; 0–640 m.

Additional Specimens Examined. **Mexico.** Guerrero: Laguna Colorada, Mpio. Petatlán, *Diego 5183* (MEXU).—Oaxaca: presa Temazcal, al NW sobre la cortina de la presa, *Cortés & Torres C. 18* (MEXU); parte norte de la cortina de La Presa Temazcal, Mpio. Soyaltepec, *Cortés & Torres 660*

(MEXU); 3 km S de Temazcal, camino al Vertedero la presa, *García et al. 2754* (MEXU); 2 km E de Temazcal, camino a Los Vertederos, Dtto. Tuxtepec, *Martínez S. & Ramos 24032* (MEXU); 7 km al NW de Río Grande, Dtto. Juquila, *Sousa et al. 7076* (MEXU); Vertederos, Presa Temazcal, *Sousa et al. 13103, 13095* (MEXU); lado N cortina Presa Miguel Alemán, Temazcal, *Sousa et al. 13096* (MEXU).—VERACRUZ: 3.5 km al S de la Rancheria "Caballo Blanco" SE del Pueblo Atoyac, Mpio. Cuitlahuac, *Acevedo & Vázquez 415* (MEXU); El Salto de Eyipantla, a 8 km del Pueblo de Sihuapan, San Andrés Tuxtla, *Calzada 1619* (MEXU); 2 km antes de Jalcomulco, sobre la Carretera Tuzamapan–Jalcomulco, Coatepec, *Castilla C. & Tápia 796* (MO); El Salto de Eyipantla, entre carr. Catemaco–San Andrés Tuxtla, *Cedillo Trigos 3047* (MEXU); cerca de Puente Nacional, Rinconada, *Hernández M. & Dorantes 1813* (F, MEXU, MO, NY); Puente Nacional, *Hernández A. & Figueroa N. 9* (F); 2 km NW of Jalcomulco toward Tuzamapan, *Lavin & Delgado 8228* (MEXU, MONT); near Puente Nacional off Hwy 140, S bank of Río Santa María, *Maxwell 151* (MEXU, MO); Baños del Carrizal, *Purpus 6083* (F, MO, NY, US); Plan del Río, *Rosas R. 612* (F, GH); Tamarindo, Mpio. de Puente Nacional, *Ventura A. 8178* (CAS, MEXU, MO); La Bandera, Mpio. Actopan, *Ventura 10961* (MEXU); Paso de La Milpa, Mpio. de Actopan, *Ventura 12674* (MEXU); Los Baños Termales, Mpio. Apazapan, *Ventura A. 15184* (MEXU, MO).

El Salvador. LA LIBERTAD: Zaragosa, *Calderón 2518* (F, US); Zaragosa, Hacienda Corinto, *Calderón 2519* (F, GH, NY). **Nicaragua.** ZELAYA: Caño Hormiguero, SE of Cerro La Pimienta, *Pipoly 5062* (MEXU). **Costa Rica.** ALAJUELA: Upala, Colonia Puntarenas, Del Salón Comunal al Río Chimuria, *Herrera 1223* (MEXU). **Panama.** CANAL ZONE: shoreline, Farfan Beach area, *D'Arcy & D'Arcy 6060* (MO); Fort Clayton, *Dwyer 9575* (MEXU, MO); Madden Dam, *Ebinger 874* (F, MO, US); Farfan Beach Rd, *Gentry 4900* (MEXU, MO, WIS); area around Farfan Beach, *Huft 1899* (MEXU, MO, TEX); Gaillard hwy just N of Paraíso, *Lewis et al. 5367* (MO, NY, UC); Boy Scout Rd, Madden Dam area, *Porter et al. 4025* (F, MEXU, MO, NY, UC); Farfan Beach area, *Tyson et al. 3164* (MO); Miraflores locks area near hwy and Fort Clayton, *Tyson 5340* (MO); just N of Miraflores rd at railroad, *Tyson & Gentry 6409* (F, GH, MO, NY); Pedro Miguel on rd to garbage dump, *Tyson 6625* (MO).—COCLÉ: between Paso del Arado and Olá, banks of Río Chico, *Pittier 5101* (US).—DARIÉN: near Río Chucunaque, ca. 4 mi below Yaviza, *Duke 4873* (GH, MEXU, MO, NY); 15–21 mi N of Santa Fé, *Duke 10311* (MO); Río Sabana above Santa Fé, *Duke 14086* (F, GH, MO, NY, TEX); Patiño, southern Darién, *Pittier 6609* (GH, US); Río Cingo, *Pittier 6976* (US).—PANAMÁ: sea level to 10 m, along stream side, *Kennedy et al. 2017* (MO).

Lennea viridiflora includes a pale-yellow-flowered variant, named *L. brunnescens,* that is commonly known from Mexico. Traditionally, *L. brunnescens* was distinguished by supposedly producing inflorescences on leafy branches (Rydberg 1924b); however, this character does not stand up to scrutiny. Most specimens from Mexico to Panama produce inflorescences in the axils of coeval leaves; occasional specimens of *L. viridiflora* from throughout the range, but more commonly in Central America, have some, but not necessarily all, inflorescences borne at leafless nodes. In fact, Central American populations of *L. viridiflora* display a restricted degree of morphological variation relative to that found among the populations from Mexico. For example, the mature legumes from Mexican populations vary in color from light (most of Mexico) to dark reddish brown (southern Oaxaca and possibly Guerrero) and in indumentum from villous (most of Mexico) or sericeous to tomentose (the Pacific coastal variety *novogaliciensis*). The mature legumes from Central American populations are mostly dark reddish brown with villous indumentum. Furthermore, the leaf morphology of Mexican populations varies from forms with 5–7 distally accrescent leaflets (the Pacific coastal variety *novogaliciensis*) to forms with mostly 9–17 leaflets that are not distally accrescent (most of Mexico) to forms that have 15–23 leaflets that are also not distally accrescent (northern Oaxaca in the District of Tuxtepec). The leaflets from populations in Central America consist of just the forms with an intermediate number of leaflets that are not distally accrescent. It appears that the Central American populations have been derived from very few Mexican populations of *L. viridiflora*; notably, the collection from southern Oaxaca (*Sousa et al. 7076*, MEXU) is most similar to the predominant Central American form in having legume valves that are reddish and inflorescences borne at leafless nodes.

A similar scenario has been suggested for Central American populations of *Gliricidia sepium* (Lavin et al. 1991). Southern Mexican and Guatemalan populations show the greatest amount of chloroplast DNA (and morphological) variation, and populations more to the southeast in Central America show a uniform and much more restricted chloroplast DNA variation. Such patterns are consistent with the presumptive Mexican origin of the tribe Robinieae, or at least its ancient occurrence there (Lavin & Luckow 1993). Also, the inflorescences of the Panamanian populations of *L. viridiflora* var. *viridiflora* display a slightly longer raceme rachis, up to 7.5 cm long, relative to populations to the northwest in Central America and into Mexico, where the inflorescence rachis measures 1–5 cm long. It is unknown whether this is an intrinsic attribute of certain Panamanian populations or an environmentally induced extreme yet to be detected elsewhere.

Populations of *L. viridiflora* are common in Mexico, but few in Chiapas and adjacent Central America, and then again common in southeastern Central America (in this case Panama). Such a geographical pattern is not uncommon in Robinieae and is known from the species of *Coursetia* sect. *Madrenses* (Lavin 1988).

3.1b. Lennea viridiflora var. **novogaliciensis** Lavin & Sousa, var. nov.—TYPE: MEXICO. Michoacán: El Ojo de Agua de San Telmo, Mpio. Coahuayana, 2 Sep 1980, *Soto N. & Cortés A. 2599* (holotype: MEXU!; isotypes: MEXU! MO! TEX!).

Var. viridiflorae similis sed foliis plerumque 5–7-foliolatis, foliolis distaliter accrescentibus, apicibus acuminatis, raro rotundatis, foliolis majoribus 43–90 mm longis, 20–45 mm latis, stipellis 0.3–0.9 mm longis, ovario ubique dense sericeo et legumine uniformiter sericeo vel tomentoso.

Leaves 7–17 cm long; leaflets 5–7 (–9) per leaf, distally accrescent, broadly elliptic, the larger 43–90 mm long, 20–45 mm wide, apex acuminate or rarely emarginate, adaxial surface glabrate to puberulous, slightly darkened with diffuse tannin deposits, abaxial surface strigose to densely sericeous; pulvinules 2.5–3.0 mm long; stipels 0.3–0.9 mm long. Ovary densely sericeous over its entire surface. Legume pale brown, sericeous to tomentose while maturing and well past dehiscence. Figs. 17–24.

Phenology. Flowering and fruiting throughout the year, perhaps without a predominant flowering season.

Distribution (Fig. 25). Pacific coastal Mexico in Colima, Jalisco, and adjacent Michoacán; along arroyos and banks of rivers near the coast, riparian evergreen forests, commonly co-occurring with *Gliricidia* and *Lonchocarpus*; 0–100 m.

ADDITIONAL SPECIMENS EXAMINED. **Mexico.** COLIMA: Km 30 brecha Manzanillo–Cuyutlán, salino próximo a la Laguna, *Díaz L. 3340* (MICH); Los Reyes (Rancho), S de Manzanillo, carr. Manzanillo–Armería, Mpio. Manzanillo, *Magallanes & Basurto 1830* (MEXU); a la salida de Santiago camino a Chandiablo, Mpio. Manzanillo, *Magallanes & Basurto 1835* (MEXU); Mpio. Manzanillo, 4 km al N de

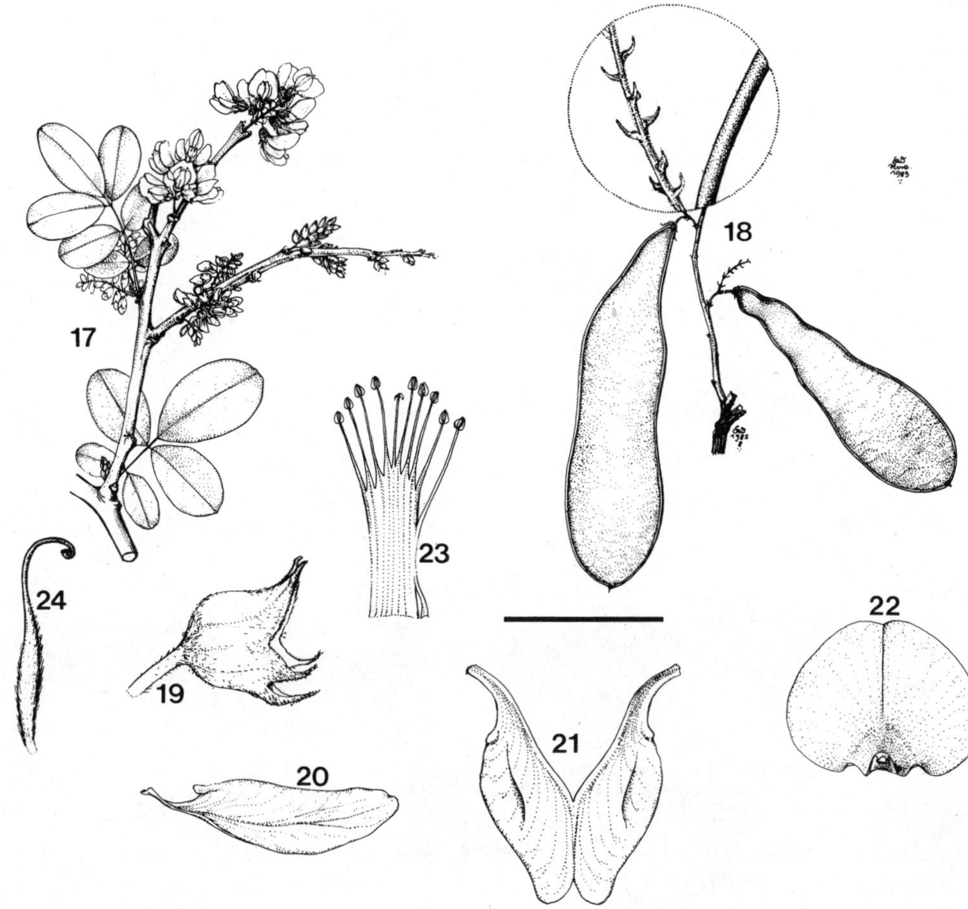

FIGS. 17–24. *Lennea viridiflora* var. *novogaliciensis*. 17. Flowering branch. 18. Fruiting branch. 19. Calyx. 20. Wing petal. 21. Keel petals. 22. Standard petal. 23. Androecium. 24. Gynoecium. Reproduced from vol. 5 of *Flora Novo-Galiciana* by Rogers McVaugh. (17, 18, bar = 10 cm; 19, 21, 23, 24, bar = 7 mm; 20, 22, bar = 10 mm).

Santiago, brecha Chandiablo–Huiscolote, *Santana & Cervantes* 321 (MEXU).—JALISCO: El Aguacate, Mpio. Cihuatlán, 5 km desv. a Barra de Navidad, carr. Barra de Navidad–Cihuatlán, *Magallanes 1144* (MEXU); 2 km al W de El Aguacate, ca. 10 km al SW de Barra de Navidad, carr. a Cihuatlán, *Magallanes 969* (MEXU); El Aguacate, 5 km al S de Barra de Navidad, *Magallanes 1593* (MEXU); El Aguacate en la desv. del puente, Mpio. Cihuatlán, *Magallanes 1662* (MEXU, MICH); Arroyo Seco, Mpio. La Huerta, *Magallanes 2044* (MEXU); 2 km al E de la carr. B. de Navidad–Pto. Vallarta, brecha a El Arroyo Seco, cerca de Cuitzmala, Mpio. La Huerta, *Magallanes & Basurto 1885* (MEXU); E of Barra de Navidad, *McVaugh 11855* (CAS, MEXU, MICH); Arroyo Seco, carr. a Manzanillo, camino a Zihuatlán, *Morales 39* (MEXU); a 1 km de El Aguacate, Cihautlán, *Pérez 1726* (MEXU).

Lennea viridiflora var. *novogaliciensis* differs from var. *viridiflora* primarily by leaf morphology and secondarily by the type of indumentum covering the ovary and mature legume. The leaves are often distinct in having distally accrescent leaflets mostly with acuminate apices. Because leaves are more consistently available on herbarium specimens than are either inflorescences or infructescences, a phenetic

FIG. 25. Distribution of *Lennea viridiflora* in Mexico and Central America.

analysis was undertaken on nine leaf measurements (Fig. 26) to quantify the foliar divergence of these northwestern-most populations from the more inland Mexican and Central American populations. Only one phenogram was generated during this analysis, and it depicted the Michoacán, Colima, and Jalisco populations as forming a cluster distinct from the other populations of *L. viridiflora*. Though three collections from Panama are equally divergent (see PANA3164, PANA4900, and PANA4086), this divergence is the result of longer pulvinules (3–4 mm in these specimens compared to mostly less than 3 mm in others) and longer stipels (2.8–4.1 in these specimens compared to mostly less than 2 mm in others). This cluster is referred to var. *viridiflora*, since there is no additional corroborating evidence to suggest that it forms a distinct species. In contrast, for example, var. *novogaliciensis* shows legume and flower distinctions that corroborate the results of this cluster analysis based just on leaf characters.

The mature legume of var. *novogaliciensis* is like that of the northern race of var. *viridiflora* in being pale brown as opposed to the dark to reddish brown forms that predominate in southern Mexico south to Panama; however, the sericeous to tomentose mature legume of var. *novogaliciensis* stands in contrast to the villous one of var. *viridiflora*. Also, the calyx cup of var. *novogaliciensis* generally measures toward the larger end of the variation scale (3–4 mm long) found in *L. viridiflora*. Notably, var. *novogaliciensis* is confined to the Pacific coast of central Mexico at elevations at or less than 100 m. It is thus disjunct from var. *viridiflora*, which occurs more commonly in Gulf-coastal Mexico at elevations up to 640 m and

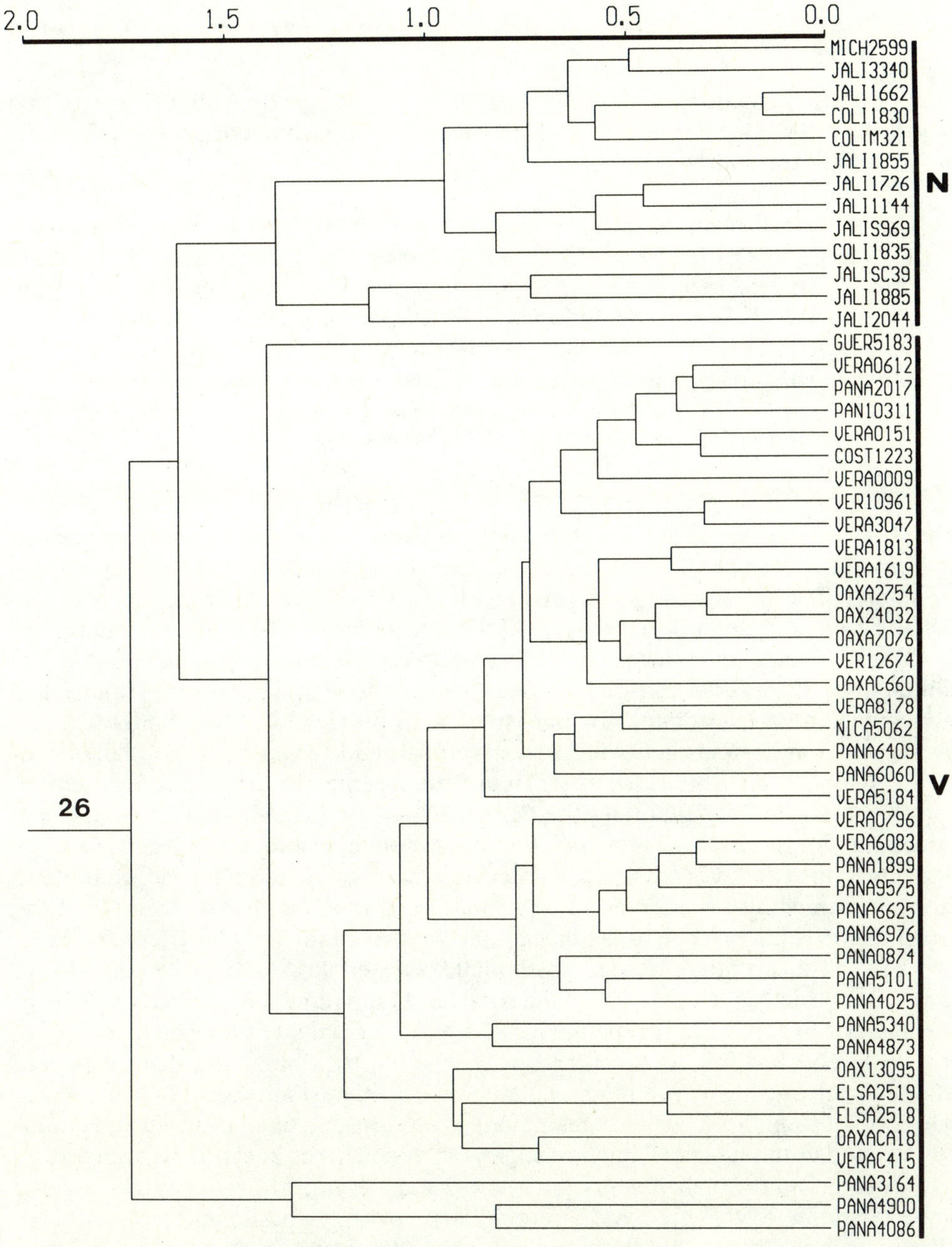

FIG. 26. UPGMA cluster analysis of average taxonomic distances derived from leaf measurements taken from herbarium specimens of *Lennea viridiflora* and with a cophenetic correlation coefficient of 0.7693. This analysis was performed with NTSYS-pc (version 1.80, Rohlf 1993). A single phenogram was generated from nine leaf measurements: 1. leaflet number of largest leaf; 2. leaf rachis length of largest leaf; 3. pulvinule length of largest leaf; 4. abaxial leaflet indumentum: trichomes per 0.25 mm^2; 5. terminal leaflet length of largest terminal leaflet; 6. leaflet width of largest leaflet; 7. stipel length of largest stipel; 8. ratio of the terminal leaflet length of largest leaflet over the length of the adjacent distal-most lateral leaflet; 9. apex: emarginate=1, rounded=2, acute=3. All but the last measurement were standardized before the calculation of pair-wise average taxonomic distances. The OTU labels include an abbreviation for the country or, in the case of Mexico, the state, followed by the collector's number. The abbreviations are: MICH = Michoacán, JALI = Jalisco, COLI = Colima, VER(A) = Veracruz, OAX(A) = Oaxaca, GUER = Guerrero, ELSA = El Salvador, PAN(A) = Panama, COST = Costa Rica, NICA = Nicaragua. See representative collections for corresponding collection number. N = *L. viridiflora* var. *novogaliciensis*; V = *L. viridiflora* var. *viridiflora*.

is not known from along the Pacific coast of central Mexico in Colima, Jalisco, and Michoacán. The two known localities of var. *viridiflora* from along the Pacific coast occur well to the southeast in Guerrero and Oaxaca.

3.2. Lennea melanocarpa (Schlechtendal) Vatke ex Harms, Repert. Spec. Nov. Regni Veg. 19: 68. 1923. *Robinia melanocarpa* Schlechtendal, Linnaea 12(3): 305. 1838.—TYPE: MEXICO. Veracruz: in sylvis Papantlae, Jan 1829, *Schiede & Deppe 284* (holotype: HAL!, photo: MONT!; possible isotypes: BM! HAL! OXF! W!, photos of BM and OXF sheets: MEXU! MONT!).

Lennea robinioides Klotzsch in Link, Klotzsch & Otto, Icon. pl. rar. 2: 65. pl. 26. 1842.—TYPE: MEXICO. Veracruz: in sylvis Papantlae, Jan 1829, *Schiede & Deppe s.n.* (holotype: HAL!).

Trees and large shrubs (0.5–)1–7 (–19) m tall. Stems erect to gradually ascending, 1–15 cm diameter at base; bark light brown, not mottled, with fine vertical fissures at the base on older stems; branches glabrate to strigose with mostly whitish trichomes. The larger leaves 8–20 cm long; leaflets 7–17 per leaf, not or only slightly distally accrescent, elliptic to broadly elliptic, the larger 25–70 (–93) mm long, 12–35 (–43) mm wide, adaxial surface of lamina glabrous and darkened or not with diffuse tannin deposits, abaxial surface glabrous, apex rounded to emarginate; stipules 1–3 mm long, linear-lanceolate, not persistent with older leaves; pulvinules 1.0–2.5 (–3.0) mm long; stipels generally well developed and longer than the diameter of adjacent pulvinule. Inflorescences of pendent racemes in axils of coeval leaves, rachises 4–16 cm long, glabrate, the peduncle 1–4 cm long, with 10–25 nodes, 3–6 nodes/cm; floral bracts 1–2 mm long, linear-lanceolate, glabrate, persistent; pedicels 4–5 mm long at anthesis, glabrate. Calyx campanulate, attenuate at base, glabrate to strigose with whitish trichomes, cup 2–3 mm long, the lobes 0.5–2.0 mm long, the vexillary lobe entire to bidentate, the carinal one tridentate. Standard deep rose-red, aging purplish, nectar-guides greenish, blade reflexed 90°, 8–10.5 mm long, 10–11 mm wide, orbicular, emarginate; wing petals 8–11 mm long, parallel to the plane of the keel, deep rose-red; keel petals 7.5–9.5 mm long, adnate for a small portion of the distal end but free at the very tip, straight, with a blunt tip, deep rose-red. Staminal tube straight, pseudomonadelphous, filaments subequal, free for about one-sixth the length. Style 3.5–4.5 mm long at anthesis, arcuate, incurled and gradually tapered to the subapical introrse stigma, included in the keel tip; ovary glabrate, but sericeous on the placental margin at the distal margin, with 4–6 ovules. Legume 4.0–9.5 cm long, 1.0–1.5 cm wide, terete, dark brown, glabrate, inner epidermis of mature legume smooth, not forming seed compartments; seeds 2–4, suborbicular to somewhat cylindrical, (5–) 6–9 mm in diameter, uniformly dark brown, testa loosely adherent. Chromosome number unknown. Figs. 27–38.

Phenology. Flowering and fruiting throughout the year, perhaps without a predominant flowering season; however, in Hidalgo and San Luis Potosí, the predominant flowering season appears to be May through August.

Distribution (Fig. 39). Mexico (Campeche, Chiapas, Guerrero, Hidalgo, Puebla, Veracruz, Oaxaca, Quintana Roo, and San Luis Potosí) and adjacent Belize and Guatemala; riparian areas of lowland forest with a pronounced seasonally dry period; commonly co-occurring with *Diphysa*, *Gliricidia*, *Calliandra*, and *Andira*; 0–1000 m.

FIGS. 27–38. *Lennea melanocarpa*. 27. Flowering branch. 28. Infructescence and one valve of a mature legume. 29. Leaf. 30. Pattern of tannin deposits on adaxial leaflet surface. 31. Flower. 32. Stamens. 33. Calyx. 34. Standard petal. 35. Wing petals. 36. Keel petals. 37. Androecium. 38. Gynoecium. (bars = 1 cm, except in 32, where bar = 1 mm).

FIG. 39. Distribution of *Lennea melanocarpa* and *L. modesta* in Mexico and Central America.

ADDITIONAL SPECIMENS EXAMINED. **Mexico.** CAMPECHE: 20 km S de Conhuas sobre el camino al Centro Regional de Calakmul, *Cabrera et al. 4484* (MEXU); 55 km al S del Km 98 de la carretera Escársega a Chetumal, sobre el camino a las ruinas de Calakmul, Mpio. Champotón, *Martínez & Téllez 2948* (MEXU).—CHIAPAS: from Mapastepec to Escuipulas along southern slopes of Sierra Soconusco, *Xolocotzi & Sharp X–268* (MEXU).—HIDALGO: en San Nicolas, Mpio. Chapulhancán, *Calzada et al. 2581* (MEXU); límites de Hidalgo–San Luis Potosí, hacia Tamazunchale, *Hernández M. 3917* (MEXU); between Huejutla and Macuxtepetla, on rd to Ixcatlán, *Moore 2929* (GH); Km 317, between Zimapán and Tamazunchale, *Schnooberger 8054* (MICH).—OAXACA: Chiltepec, Tuxtepec, *Calderón 113* (CAS, TEX); Dtto. Tuxtepec, Chiltepec and vicinity, *Calderón 678* (MICH); Dtto. Tuxtepec, Chiltepec and vicinity, *Calderón 841* (GH, MICH); Km 24, carr. Tuxtepec–Jalapa de Díaz, *Chavelas & Pérez 389* (MEXU); S of San José de Chiltepec, *Schultes & Reko 573* (GH); Dtto. Tuxtepec, Tuxtepec, *Sousa 721* (MEXU); (camino Chiltepec a Oaxaca) Cerro Brujo, *Sousa 780* (MEXU); Chiltepec, Ditto. Tuxtepec, *Sousa et al. 7954* (CAS, MEXU); 4 km al ESE de Temazcal, en camino al Vertedero, *Sousa et al. 13046* (MEXU); Mpio. Santa María Jacatepec, subida predio del Aguila, 25 km W de La Reforma, carr. a Ayozintepec, *Torres & Cortés 11519* (MEXU); subida al predio El Aguila, al W de San Agustín, entrando por la Reforma, 28 km al SW de Tuxtepec, carr. Tuxtepec–Matías Romero, *Torres & Martínez 10990* (MEXU); Choapán, *Roble s.n.* (MEXU); Ejido de la Mina, a 10 km del entronque de carr. Tuxtepec–Matías Romero hacia Palomares, Mpio. Tuxtepec, *Vovides 548* (MEXU, MO).—PUEBLA: 3 km al NNE de Mecapalapa, Mpio. Pantepec, *Basurto & Durán 10* (MEXU); 2 km al W de Villa Lázaro Cárdenas, Mpio. Venustiano Carranza, *Basurto & Durán 48* (CAS, MEXU); 2 km al N de Mecapalapa, Mpio. Pantepec, *Basurto & Durán 174* (CAS, MEXU); Yancuitlalpan, Cuetzalan, *Basurto & Patrón 204* (MEXU); 2 km E de Mecapalapa, Mpio. Pantepec, *Basurto & Durán 225* (MEXU); 2 km N de Mecapalapa, Mpio. Pantepec, *Basurto & Durán 249* (MEXU); 5 km N de Mecapalapa, Mpio. Pantepec, *Basurto & Durán 308* (MEXU); 3 km al N de Mecapalapa, Mpio. Pantepec, *Basurto & Durán 366* (CAS, MEXU, MO); 1 km al N de Mecapalapa, Mpio. Pantepec, *Basurto & Durán 465* (CAS, MEXU); Sierra Norte, *Basurto & Durán 529* (MEXU); 41 km al N de Mecapalapa, Mpio. Pantepec, *Basurto &*

Durán 558 (CAS, MEXU); 3 km N de Ameluca, Mpio. Pantepec, *Basurto & Durán 595* (MEXU); 4 km al N de la Pagua, *Basurto & Durán 630* (MEXU); 4 km al E de Apantilla, *Basurto & Durán 710* (MEXU); Lomas de Arena, carr. Tlapacoyan–San José Acateno, Mpio. Hueytamalco, *Márquez et al. 619* (F, MEXU, NY); Km 247.3 de carr. México–Tuxpan, *Sarukhán et al. 2256* (MEXU); Km 271 de la carr. México–Tuxpan, antes del rancho Sta. Maúra, *Sarukhán et al. 2930* (MEXU); Paxta, Mpio. Hueytamalco, *Ventura A. 14950* (MEXU).—QUINTANA ROO: 4 km al NW de Estero Franco, sobre el camino en contrucción a Tomás Garrido, lugar llamado El Danto, *Cabrera C. 8381* (MEXU).—SAN LUIS POTOSÍ: Tamazunchale area, *Clint M–373* (US); Tamazunchale, *Fisher 45171* (US); Tamazunchale, *Kenoyer A53* (F); Tamazunchale, *Kenoyer A638* (F); near Tamazunchale, *Leavenworth & Leavenworth 894* (F); Tamazunchale, in forest at riverside, *Lundell & Lundell 7266* (MICH, NY); Tamazunchale, orilla N río, *Miranda 9013* (MEXU); S of Tamazunchale, *Wheelock M–47A* (US).—VERACRUZ: orillas de la Laguna Catemaco, *Calzada 283* (MEXU); al SE del poblado de Juchique de Ferrer, carretera para Plán de Las Hayas, Mpio. Juchique de Ferrer, *Calzada 3249* (F, MEXU, NY); Rancho Nuevo, Mpio. Cazones, *Cortés 456* (MEXU); Cerro del Carbón, Mpio. Papantla, *Cortés 477* (MEXU); Plán de Hidalgo, Mpio. Papantla, *Cortés 555* (MEXU); region of San Andrés Tuxtla, along river draining Laguna de Catemaco, *Dressler & Jones 54* (GH, MEXU, MICH, NY, US); 500 m S del campo de aviación de Coxquihui, *Evangelista & Mendoza 54* (MEXU); Mpio. Coxquihui, "El Zapote" 2 km SSW de Coxquihui, *Grimes et al. 2723* (MEXU, TEX); 2 km al E de San Pablo, Mpio. Papantla, *Gutiérrez R. 95* (MEXU); Km 46 on route 180 between Papantla and Gutiérrez Zamora at the village of Barriles, *Harriman 15666* (US); Isla de Tenaspi, Mpio. Catemaco, *Hernández 93* (MEXU); Rancho Macarena, Mpio. Gutiérrez Zamora, *Jarquín & Lot 53* (MEXU); 8 mi S of Martínez de la Torre, N of Misantla, *Lavin & Luckow 5789* (TEX); just N of Sarabia, *Lavin & Delgado 8217* (MEXU, MONT); Papantla, *Liebmann 4653* (GOET, UC, US); 500 m S del Campo de aviación de Coxquihui, *Mendoza & Evangelista 41* (MEXU); Ojo de Agua Grande, cerca Potrero (Córdoba), *Miranda 8856* (MEXU); 6.5 km N of Huejutla on rd to Platón Sánchez, Mpio. Chalma, *Nee & Hansen 18455* (MEXU); near Lake Catemaco, *Nelson 423* (F, US); Cebollal, San Andrés Tuxtla, *Ponce C. 131* (CAS, MEXU, TEX); Misantla, *Purpus 5891* (CM, F, GH, MO, NY, UC, US); Cameron, *Purpus 8721* (CAS, GH, MO, NY, UC, US); Salto de Agua arribita de la máquina, *Shapiro 216* (CAS, MEXU); Cerro Espaldilla, about 5 km N of Misantla, *Shatz et al. 1193* (NY); Misantla, *Sousa 768* (MEXU); Playa Azul, orilla de la Laguna de Catemaco, *Sousa 2622* (MEXU); 7 km NE de Catemaco, camino a Dos Amates, Mpio. Catemaco, *Trigos 3852* (MEXU); Isla Tenaspi en la Laguna de Catemaco, Mpio. Catemaco, *Vázquez et al. V–661* (MEXU); El Palmar, Mpio. Dos Ríos, *Ventura 8229* (MEXU); Buenos Aires, Mpio. Misantla, *Ventura 11455* (MEXU); Cuaputojapan, Mpio. Tlapacoyan, *Ventura 12596* (MEXU); La Palmilla, Mpio. Tlapacoyan, *Ventura 14966* (MEXU); Novara, Mpio. Atzalan, matorral en cañada, *Ventura 19730* (MEXU); Coapa, Mpio. Martínez de la Torre, *Ventura 20337* (MEXU); Wartenberg, near Tantoyuca, prov. Huasteca, *Ervendberg 28* (GH).

Belize. BELIZE: Western hwy, Gracie Rocks, *Whitefoord 3160* (MEXU, MO).—EL CAYO: El Cayo, *Chanek 97* (F, MICH); Benque Viejo, 2.5 km on Arenal Rd, *Contreras 7051* (F, MEXU, MO, TEX); Benque Viejo, 2.6 km on Arenal Rd, *Contreras 7053* (CAS, F, MEXU, NY, TEX); Vaca, *Gentle 2583* (F, GH, MEXU, MICH, MO, NY, TEX); Roaring Creek, *Lundell 390* (CAS, F); forest near Mantee River, *Peck 422* (GH, NY); 2 km W de Teakettle, *Sousa et al. 12077* (MEXU); W of Hummingbird hwy at a point 1 mi S of its junction with Western hwy, *Spellman & Newey 2025* (MO). **Guatemala.** PETÉN: Uaxactún, *Bartlett 12224* (F, MICH); Yal'och, *Bartlett 12855* (CAS, F, MICH, NY); Santa Elena, 1 km S on La Cueva de Jobitzinaj rd, *Contreras 3580* (F, MEXU, TEX); Lake Petén Itza, on San Andres rd, *Contreras 9639* (MEXU, NY); en el camino para Uaxactún, ±13 km de la aldea de Tikal, en Parque Nacional de Tikal, *Ortíz 1720* (F, MICH, US).

Populations of *L. melanocarpa* from Hidalgo and San Luis Potosí are somewhat distinct in that they commonly have leaves with 7–9 distally accrescent leaflets, each of which is covered with patchy tannin deposits (at least on dried herbarium leaves); also, the leaflets reach their largest dimensions in this region, up to 93 mm long and 43 mm wide. Distally accrescent leaflets and patchy tannin deposits are not exclusive to these populations, nor are they consistently present in this northernmost region. These attributes are simply more frequent in populations from Hidalgo and San Luis Potosí, compared to those from the south, whereas the sizes of the leaflets are just slightly larger than the largest sizes found to the south. Notably, populations of *L. melanocarpa* rarely grow above 300 m, except in Hidalgo and San Luis Potosí, where they occur between 200–1000 m. This may explain why these northern populations have a more restricted flowering season, May through August, compared to populations from elsewhere to the south, which do not have a predominant flowering season.

The pollen of *L. melanocarpa* (reported as *L. robinioides*) is suggested to be specialized relative to other Robinieae (Ferguson & Skvarla 1981). Whereas the rest of the tribe shows tricolporate pollen, a thin endexine, and an entire foot layer, *L. melanocarpa* is reported to produce commonly 4-colpate pollen with a thickened endexine and a thin fractionated foot layer. Additional surveys of pollen (Lavin, unpubl.), however, reveal that all species of *Lennea*, including *L. melanocarpa*, have tricolporate pollen, at least in part, and the report by Ferguson and Skvarla could mean that pollen dimorphism is occurring in at least one of the species.

3.3. Lennea modesta (Standley & Steyermark) Standley & Steyermark, Fieldiana, Bot. 24: 275. 1946. *Lonchocarpus modestus* Standley & Steyermark, Publ. Field Mus. Nat. Hist., Bot. Ser. 23: 56. 1944.—TYPE: GUATEMALA. Izabal: Bay of Santo Tomás, between Escobas and Santo Tomás, at sea level, 13 Apr 1940, *Steyermark 39238* (holotype: F!).

Trees and large shrubs 3–20 m tall. Stems erect to ascending, 2–30 cm diameter at base; bark grayish white/yellow, mottled with light spots, with fine vertical fissures at the base on older stems; branches glabrate to strigose with mostly whitish trichomes. The larger leaves 14–27 cm long; leaflets 3–5 per leaf, distinctly distally accrescent, broadly elliptic, the larger 75–200 mm long, 39–85 mm wide, adaxial surface of lamina glabrous and darkened with diffuse tannin deposits, abaxial surface glabrous, apex cuspidate, with a drip tip, rarely rounded; pulvinules 4–6 mm long; stipules 2–3 mm long, linear, not persistent with older leaves; stipels indistinct or at best shorter than diameter of subtending pulvinule. Inflorescences of pendent racemes in axils of coeval leaves, rarely on leafless short shoots, rachises (4–) 5–20 cm long, glabrate to strigose, rarely tomentose, subsessile, with (20–) 40–85 nodes, 8–12 nodes/cm; floral bracts 1–2 mm long, linear-lanceolate, sericeous, persistent; pedicels 2.5–4.0 mm long at anthesis, glabrate to strigose. Calyx campanulate, attenuate at base, sparsely strigose to sericeous with whitish trichomes, cup 2.5–3.0 mm long, the lobes 1.0–2.5 mm long, the vexillary lobe entire to bidentate, the carinal one tridentate. Standard deep rose-red (very rarely greenish yellow), aging purplish, nectar-guides greenish, blade reflexed 90°, 9–11 mm long, 10–12 mm wide, orbicular, emarginate; wing petals 10–11 mm long, parallel to the plane of the keel, deep rose-red (very rarely greenish yellow); keel petals 8.5–10.0 mm long, adnate for a small portion of the distal end but free at the very tip, straight, with a blunt tip, deep rose-red (very rarely greenish yellow). Staminal tube straight, pseudomonadelphous, filaments subequal, free for about one-sixth the length. Style 5–6 mm long at anthesis, arcuate, incurled and gradually tapered to the subapical introrse stigma, included in the keel tip; ovary glabrous, sparsely strigose along placental margin at distal end, with 4–6 ovules. Legume 4.0–9.5 cm long, 0.9–1.9 cm wide, terete, dark brown, glabrate, inner epidermis of mature legume smooth, not forming seed compartments; seeds 1–6, suborbicular to cylindrical, 9–10 mm in diameter, uniformly dark brown, testa loosely adherent. Chromosome number unknown.

Phenology. Flowering and fruiting throughout the year, perhaps without a predominant flowering season; collected in flower mostly from March through August.

Distribution (Fig. 39). Mexico (Chiapas and adjacent Tabasco, Veracruz, and Oaxaca), Guatemala, and Honduras; riparian areas of lowland evergreen rain forest, commonly co-occurring with *Acacia*, *Dialium*, *Heliocarpus*, *Erythrina*, *Pinus*, *Quercus*, *Liquidambar*, *Inga*, *Clethra*, *Eugenia*, *Lonchocarpus*, and *Chrysophyllum*; 0–1100 m.

ADDITIONAL SPECIMENS EXAMINED. **Mexico.** CHIAPAS: slopes near Habenal, paraje of Mahben Chauk, Mpio. Tenejapa, *Breedlove 7664* (F, US); wooded slope along the Tanate River at Habenal, paraje of Mahben Chauk, Mpio. Tenejapa, *Breedlove 12877* (F, US, WIS); slope along stream, Paraje of Mahosik, Mpio. Tenejapa, *Breedlove 29312* (MICH, MO); slopes along stream, montane rain forest 3 km north Bachajón–Ocosingo rd on rd to Palenque, Mpio. Bachajón, *Breedlove 50902* (MEXU, NY); Yajalón, *Enríquez 7612* (MEXU); Cascadas de Agua Azul, *Góngora 591* (MEXU); 7 km past Ocosingo toward Palenque, *Lavin & Delgado 8210* (MEXU, MONT); 2 km N of Yajalón turnoff toward Palenque, *Lavin & Delgado 8212* (MONT); slope in the barrio of Ti Ha', paraje of Mahbenchaok, Mpio. Tenejapa, *Shilom Ton 2262* (US); Río Cruz Pilal, Tenejapa, *Shilom Ton 4243* (MEXU); Arroyo Jeshab, Mpio. Tenejapa, *Shilom Ton 5863* (MEXU, TEX); Pie del Cerro Ak'bana, Mpio. Tila, *Shilom Ton 5976* (MEXU); El Azufre, Mpio. Yajalón, *Shilom Ton 6221* (MEXU); Río Pupitillo, Mpio. Yajalón, *Shilom Ton 6246* (MEXU, TEX); Ch'ijtaltik, Mpio. Yajalón, *Shilom Ton 6289* (MEXU); Cerrito T'izton, Mpio. Yajalón, *Shilom Ton 6734* (MEXU); El Río del Paraíso, Mpio. Yajalón, *Shilom Ton 6776* (MEXU, NY); Rancho en la carr. Chibal a Ocosingo, Km 14, Mpio. Ocosingo, *Shilom Ton 8985* (MEXU); Col. Maravilla, Tenejapa, Mpio. Margaritas, *Shilom Ton 9069* (MEXU, NY); Las Lajas, a 6 kms al NW de Ocosingo, Mpio. Ocosingo, *Sousa et al. 11382* (MEXU, MO).—OAXACA: Mpio. Sta. María Chimalapa, Paso Lagarto del Río Corte, ca. 5 km al N de Sta. María, *Hernández G. 81* (CAS, MEXU, MO); Mpio. Sta. María Chimalapa, Paso Lagarto del Río Corte, ca. 5 km al N de Sta. María, *Hernández G. 82* (CAS, MEXU, MO); Mpio. Sta. María Chimilapa, ca. 5 km al NE de Sta. María, *Hernández G. 1348* (MEXU).—TABASCO: orilla Río Puyacatengo, Mpio. Teapa, *Magaña & Guadarrama 1225* (MEXU); orilla del Río Puyacatengo, en el Cruce Uach, Mpio. Teapa, *Ortiz 737* (MEXU); Río Puyacatengo, camino al Rancho San Antonio, Mpio. Teapa, *Rico & Martínez S. 741* (CAS, MEXU, WIS); Puyacatengo, Mpio. Teapa, *Ventura 20646* (MEXU); orilla del río cerca de Puyacatengo, Mpio. Teapa, *Ventura 21226* (MEXU).—VERACRUZ. Mpio. Hidalgotitlán, Campamento Hermanos Cedillo, *Calzada 1725* (CAS, MEXU, TEX); Km 12 del camino Cedillo La Escuadra por el Río Soloxuchil, *Dorantes 3555* (MEXU, MO); 1 km SE of Agustín Melgar, *Nee 29778* (MEXU); Río Soloxuchil, al SW del Camp. Hnos. Cedillo, *Ponce C. 256* (MEXU, TEX, UC); Río Soloxuchil, entre Hnos. Cedillo y La Escuadra, *Vázquez et al. V–369* (MEXU, MICH, TEX, UC); Brecha Hnos. Cedillo–Agustín Melgar, *Vázquez et al. V–493* (MEXU, MICH, TEX, UC); Brecha Hnos. Cedillo–La Escuadra, por el Río Soloxuchil, *Vázquez et al. V–567* (CAS, MEXU, MICH, TEX, UC); La Escuadra (por el río Soloxuchil y desembocadura del río Las Cuevas, *Vázquez et al. V–1762* (MEXU).

Guatemala. ALTA VERAPAZ: Chahal, on Sepur Ranch, bordering Río Chahal, 3 km S in clearing, *Contreras 7869* (CAS, F, MEXU, MO, NY, TEX); Chahal, on Sepur Ranch, bordering Río Chahal, 3 km S in clearing, *Contreras 7873* (F, MEXU, MO, TEX); Chahal, bordering Chiyu River, in high forest, 3 km W of airport, *Contreras 7979* (F, MEXU, MO, TEX); vicinity of Cubilgüitz, 1.5–2 mi S of Cubilgüitz, *Steyermark 44432* (F); along Río Sehachachoj, halfway between Cobán and Finca Cubilgüitz, *Steyermark 45719* (F)—IZABAL: Cárdenas, on Saliela Rd bordering the Cienega River, 22 km in clearing, *Contreras 10037* (F, MEXU, MO, TEX); Puerto Méndez, 26 km on Río Dulce rd, *Contreras 10329* (CAS, F, MEXU, MO, TEX); Puerto Méndez, 26 km on Río Dulce rd, *Contreras 10331* (F, MEXU, TEX); Bay of Santo Tomás, between Escobas and Santo Tomás, *Steyermark 39238* (F). **Honduras.** COPÁN: 5 km S de Sta. Rita, *Molina R. 6649* (F)—CORTÉS: edge of river of Río Lindo Peña Blanca near Lake Yojoa, *Williams & Molina R. 14578* (F, TEX); Río Lindo, near Carrizal, *Williams & Molina R. 17813* (F, GH).

Although most collectors cite the flower color as deep rose-red (or some near equivalent), one collection (*Vásquez et al. V–567*, MEXU), including both the actual specimen and label description, clearly suggests that the flowers have greenish yellow petals. Because this color-morph is otherwise absent among the collections cited, and because other specimens collected at this locality have the deep rose-red petal color, the greenish yellow petal color of this specimen is considered just a color variant of *L. modesta*.

Lennea modesta is very similar to *L. melanocarpa* and is distinguished primarily by the leaf and inflorescence rachis characteristics given in the key. Also, the style of *L. modesta* seems to be somewhat longer (5–6 mm) compared to that of *L. melanocarpa* (3.5–4.5 mm); however, this appears to be the only floral difference between the two species. The close relationship of *L. modesta* and *L. melanocarpa*, which together are more distantly related to *L. viridiflora*, is suggested, in part, by the glabrate leaves and inflorescences of the former two and the more densely sericeous to tomentose leaves and inflorescences of *L. viridiflora*. One specimen of *L. modesta* (*Góngora 591*, MEXU) has distinctly tomentose inflorescence rachises suggestive of *L. viridiflora*, though in every other character and in geographical distribution it is clearly a specimen of *L. modesta*.

4. Hybosema Harms, Repert. Spec. Nov. Regni Veg. 19: 66. 1923.—TYPE: *Hybosema ehrenbergii* (Schlechtendal) Harms.

Single- to multistemmed trees, and large to small shrubs. Stems erect, branches alternate; bark smooth to superficially furrowed, more deeply furrowed in the oldest specimens. Roots often superficial and producing leafy sprouts. Leaves alternate, pulvinate, imparipinnate; leaflets opposite, narrowly elliptic to elliptic, rounded, caducous from the more persistent rachis during dry periods or with age, adaxial surface mostly glabrate, mostly without finely scattered patchy tannin deposits, abaxial surface mostly strigose to sericeous, rarely glabrate, with finely scattered patchy tannin deposits; leaf rachis canaliculate, the groove without a dense indumentum, stipels absent to barely visible; stipules conspicuous. Inflorescences of pedunculate lax axillary racemes, borne in axils of coeval leaves and not clustered at branch ends, or borne on short shoots in axils of undeveloped leaves, with a synchronous (or nearly so) maturation of flowers on a single rachis, each with mostly 20–70 nodes, stipitate glands absent. Flowers ebracteolate, with a well-developed hypanthium; floral bracts conspicuous, caducous to sometimes persistent; pedicels articulated with the calyx, straight and ascending during anthesis and seed set, often persisting after floral abortion. Calyx narrowly cylindrical, generally attenuate at base, glabrate, rarely strigose, stipitate glands absent, with five unevenly spaced lobes shorter than the tube, the entire to bifid vexillary and the trifid carinal lobes mostly triangular and not overlapping at anthesis. Standard orbicular, emarginate, reflexed at most 90° at anthesis, dorsal surface glabrous, subequal in length to the keel petals, with a single prominent linear callosity along the midrib in the region of the nectar-guide, claw abruptly contracted from the blade, inflexed auricles present; wing petals sculptured, barely projecting beyond the keel at anthesis, and loosely attached to it by a fold in the basal auricle, the distal ends glabrous, the blades perpendicular, or nearly so, to the plane of the keel; keel valvately connate distally to nearly the tip, apex blunt. Staminal tube pseudomonadelphous, with basal fenestrae; anthers included in the keel tip, basally attached to the filaments, the broad connective reddish brown; intrastaminal nectariferous disk absent. Pollen spheroidal, tricolporate, mesocolpium rugulate-perforate, polar regions smooth. Ovary stipitate, the stipe as long as the calyx tube, glabrous, eglandular, with 5–7 or 13–16 ovules; styles glabrous, the base thin and imperceptibly confluent with the distal portion of the ovary, not becoming displaced to the carinal side during fruit maturation; stigma glabrous, terminal on a distally tapered style. Legume subsessile, glabrate, eglandular, inner epidermis of the endocarp of adjacent valves spongy and abutting between seeds, forming somewhat uniform seed compartments, the valves thickly woody at maturity, elastically and explosively dehiscent, the dorsal and ventral margins generally parallel to subsinuous in fruit. Seeds transversely oriented, lenticular, not constricted in the middle, accumulating canavanine, uniformly light to dark brown, without dark mottling, hilum apical and attached to a very short funiculus, testa firmly adherent. Seedlings with thickened, somewhat foliar, epigeal cotyledons, single eophyll usually uni- to trifoliolate, alternate, metaphylls many-foliolate and alternate. Base chromosome number unknown.

Hybosema traditionally has been relegated to synonymy under *Gliricidia* in accordance with Rydberg's (1924b) treatment in the North American Flora. No doubt, *Hybosema* and *Gliricidia* are very closely related, but any recent common ancestry that might be unique to these two genera is also, according to the evidence, shared by the West Indian genus *Poitea*. Thus, a grouping comprising just *Hybosema* and *Gliricidia* is paraphyletic with respect to *Poitea*. The picture that emerges now is that *Hybosema*, *Poitea*, and *Gliricidia* form a tight-knit monophyletic subgroup of Robinieae. Each of the three genera, as now circumscribed, represents one of the three major lineages within this subgroup. The genus *Lennea* is the putative sister group to these genera (see phylogeny). *Hybosema* is marked by inflorescences with many crowded flowers and standard petals bearing a single callus from the middle of the nectar-guide; like *Lennea*, *Hybosema* is marked by flowers with pseudomonadelphous staminal tubes. *Poitea* is very similar to *Hybosema* in having flowers with bilobed cylindrical calyces, but differs by the many characters listed in the key to the genera, including flowers with diadelphous staminal tubes. *Gliricidia* stands apart from both of these genera by having truncated campanulate calyces and standard petals that are fully reflexed at anthesis (although one highly derived species of *Poitea*, *P. paucifolia*, has this latter character). Since *Poitea* is recognized as distinct from *Gliricidia* and ranked at the generic level, so must also *Hybosema* be likewise distinguished and ranked.

Key to the Species of Hybosema

1. Single- to multistemmed trees 5–16 m tall, 5–28 cm in diameter at base; petals brick red at anthesis and aging dark purple to black; ovary with 13–16 ovules; mature legumes (8–) 10–17 cm long, (13–) 15–22 mm wide, with 5–12 seeds; larger leaves 18–35 cm long; leaflets (13–) 19–27 per leaf, elliptic, the larger 3.5–7.0 cm long, 10–35 mm wide; stipels absent; inflorescences produced on short shoots in the axils of underdeveloped leaves or at leafless nodes. 4.1. *H. robustum*.

1. Treelets to shrubs 0.5–3 m tall, 0.5–3.0 cm in diameter at base; petals reddish pink at anthesis and aging the same; ovary with 5–7 ovules; mature legumes 4.0–8.5 cm long, 9–14 mm wide, with 2–5 seeds; larger leaves 9–21 cm long; leaflets (11–) 15–23 per leaf, narrowly elliptic to

elliptic, the larger 1.4–3.5 cm long, 7–17 mm wide; stipels barely conspicuous; inflorescences produced on long shoots in the axils of coeval leaves. 4.2. *H. ehrenbergii.*

4.1. Hybosema robustum M. Sousa & Lavin, Anales Inst. Biol. Univ. Nac. México, Ser. Bot. 63: 143. 1992.—Type: Mexico. Chiapas: Mirador Manos que Imploran, Cañón del Sumidero, camino a Chicoasén, Mpio. San Fernando, 13 Dec 1987, *Martínez S. & Reyes G. 22047* (holotype: MEXU!; isotypes: BM! MEXU! MO!).

Single- to multistemmed trees 5–16 m tall. Stems erect to ascending, mostly single, 5–28 cm in diameter at base, with or without major branches from near base; bark light grayish brown, mottled with large whitish patches (other than the distinct lenticels), minutely furrowed toward the base on older stems; branches glabrate to sericeous with whitish trichomes. The larger leaves 18–35 cm long; leaflets (13–) 19–27 per leaf, not distally accrescent, elliptic, the larger 3.5–7.0 cm long, 10–35 mm wide, adaxial surface strigose, usually without darker tannin deposits, abaxial surface glabrate to sericeous, often with darker tannins deposited in patches or uniformly over the entire surface; stipules 2.5–3.5 mm long, readily caducous; stipels absent. Inflorescences borne on short shoots in axils of underdeveloped leaves or at leafless nodes, rachises 4–10 cm long, stipitate, the stipe 1–3 cm long, strigose to sericeous, with 20–50 nodes, internodes mostly 0.1–3.0 (–4) mm long; floral bracts 1.1–2.2 mm long, linear-lanceolate, sericeous, variably persistent to caducous; pedicels 3.5–5.0 mm long at anthesis, glabrate. Calyx cylindrical, attenuate at base, generally glabrate but pilose at the distal margin, tube 4–5 mm long, the lobes 0.5–1.0 mm long, the vexillary lobe bifid, the carinal trifid. Standard brick red, aging dark purple, blade reflexed 90° at anthesis, 11–13 mm long, 11–14 mm wide, orbicular, emarginate, with a yellowish nectar-guide; wing petals 11–12 mm long, parallel to the plane of the keel, brick red; keel petals 10.5–12.0 mm long, adnate for much of the distal end, straight, with a blunt tip, brick red. Style 3–4 mm long at anthesis; ovary tomentose with fine whitish to brownish trichomes, with 13–16 ovules. Legume (8–) 10–17 cm long, (13–) 15–22 mm wide, strongly laterally compressed, light brown, mostly glabrate after dehiscence, but commonly villous before; seeds 5–12, lenticular, (7–) 8–10 mm in diameter, uniformly light brown, aging dark brown. Figs. 40–50.

Phenology. Collected in flower from late February through April, with mature fruits in March, August, September, and December.

Distribution (Fig. 51). Mexico; lowland seasonally deciduous primary forests, on calcic substrates of canyon sides in Aguacero and Sumidero Canyons, Chiapas (the northwestern region of the central depression); 600–850 m. Although of primary forests, *H. robustum* occasionally inhabits disturbed sites, such as very steep rocky slopes, especially on exposed sites. Co-occurring dominant genera often include *Acacia*, *Aeschynomene*, *Agave*, *Alvaradoa*, *Bursera*, *Cnidoscolis*, *Comocladia*, *Guazuma*, *Leucaena*, *Lonchocarpus*, *Lysiloma*, *Moringa*, *Senna*, and genera of Bignoniaceae.

Additional Specimens Examined. **Mexico.** Chiapas: Mpio. Chicoasén: a 33 km al NE de Tuxtla Gutiérrez, camino a Chicoasén, *Martínez S. et al. 20091* (MEXU). Mpio. San Fernando: Mirador for Chicoasén Dam, along rd from Tuxtla Gutiérrez to Chicoasén Dam, *Breedlove 39954* (MEXU, MO);

FIGS. 40–50. *Hybosema robustum*. 40. Flowering branch. 41. Fruiting branch. 42. Leaf. 43. Seedling. 44. Flower bud. 45. Calyx. 46. Androecium. 47. Gynoecium. 48. Standard petal. 49. Wing petal. 50. Keel petal.

Mirador Manos que Imploran, off the main rd running N from Tuxtla Gutiérrez to Chicoasén, about 6 km from Chicoasén, *Hughes 1186* (FHO), *1488* (FHO, MEXU, MONT); slopes of the Sumidero Canyon, close to the rd running N from Tuxtla Gutiérrez towards Chicoasén, about 10 km NE of San Fernando, *Hughes & Lewis 1291* (MEXU, MONT); 30 km from Tuxtla Gutiérrez towards Chicoasén, *Lavin & Delgado 8207* (MONT); Mirador Manos que Imploran, a 10 km al WSW de Chicoasén, *M. Sousa et al. 13212* (MEXU). Mpio. Ocozocoautla: Parque Nacional Aguacero, limestone cliffs above

FIG. 51. Distribution of *Hybosema* in Mexico and Central America.

Río La Venta, *Lavin & Delgado 8201* (MEXU, MONT); Cascada El Aguacero en el Río La Venta, *Martínez S. & Reyes 22023* (MEXU).

That *H. robustum* is one of the common tree species in the forests of the canyon sides of Aguacero and Sumidero makes this recent discovery exceptional (Sousa & Lavin 1992). Notably, *H. robustum* is one of the very few species of robust trees in the tribe Robinieae that form single trunks, a trait otherwise known only in *Lennea*, *Hebestigma*, and *Sesbania*. Its petal color of brick red that ages dark purple (superficially almost black) is unknown in other members of the tribe and is reminiscent of the petal color found in the species of *Macroptilium* (Benth.) Urb. (tribe Phaseoleae) and in two species of *Brachysema* R. Br. (Mirbelieae; M. Crisp, pers. comm.). This large tree species, which inhabits the dense forests of Aguacero and Sumidero, is often found growing erect, but can assume a horizontal scandent growth form even with trunks that measure up to 28 cm in diameter.

The circumscription of *Hybosema* and *Gliricidia* presented in this treatment may appear problematic, especially if one compares the leaves of *H. robustum* to the very similar leaves of *G. brenningii*. The similarity is due to the large leaf size and to the similar number of elliptic rounded leaflets that are not distally accrescent. Such leaf variation is not necessarily considered as evidence of close relationship, because all genera of the *Gliricidia*-group, except *Hybosema*, have leaves with two distinctly variable types: one in which the leaflets are acuminate and distally accrescent (e.g., *Lennea viridiflora* var. *novogaliciensis*, *Poitea campa-*

nilla, and *Gliricidia sepium*) and another in which the leaflets are rounded and not distally accrescent (e.g., *Lennea viridiflora* var. *viridiflora*, *Poitea galegoides*, and *Gliricidia brenningii*).

Hybosema robustum is, upon superficial inspection, very unlike its sister species, *H. ehrenbergii*, although both have very similar flowers with bilobed cylindrical calyces, pseudomonadelphous staminal tubes, and standard petals with a single callus along the midrib in the middle of the nectar-guide. They differ in that *H. robustum*, a large tree with profuse foliage, produces inflorescences at essentially leafless nodes, whereas *H. ehrenbergii*, a multistemmed shrub or treelet, bears inflorescences in the axils of its small leaves. These differences are especially pronounced in disturbed habitats, because slashing and burning of native pine-oak woodlands maintains *H. ehrenbergii* in a small shrub growth-form with stems rarely surpassing 1 m in height.

4.2. Hybosema ehrenbergii (Schlechtendal) Harms, Repert. Spec. Nov. Regni Veg. 19: 66. 1923. *Robinia ehrenbergii* Schlechtendal, Linnaea 12: 303. 1838. *Gliricidia ehrenbergii* (Schlechtendal) Rydberg. N. Amer. Fl. 24: 239. 1924.—TYPE: MEXICO. In solo calcareo boream versus ab aquis calidis pr. Grande, Jul 1831, *Ehrenberg 645* (holotype: HAL!, photo: MONT!; isotypes: B, destroyed, HAL!, fragment of B isotype: F!, photos of B isotype: F! G! GH! MEXU! MICH! MO! NY! TEX!).

Gliricidia guatemalensis Micheli, Bull. Herb. Boissier 2: 442, pl. 10. 1894.—TYPE: GUATEMALA. Quiché: prope Sacabajá, 1500 m, Mar 1892, *Heyde & Lux in J. D. Smith 3315* (holotype: G; isotypes: GH!, photos: MEXU! NY! US!).

Gliricidia meistophylla (J. D. Smith) Pittier, Contrib. U.S. Natl. Herb. 20: 86. 1917. *Lonchocarpus meistophylla* J. D. Smith, Bot. Gaz. (Crawfordsville) 56: 55. 1913.—TYPE: GUATEMALA. Baja Verapaz: in apricis ad Cuesta de Quililhá prope Purulhá, 1400 m, 21 Apr 1905, *Pittier 141* (holotype: US!, photos: MEXU! NY!).

Shrubs to occasionally treelets, 0.5–3 m tall. Stems erect to ascending, 0.5–3.0 cm in diameter, with major branches from near base; bark usually dark grayish to reddish brown, often mottled with distinct whitish lenticels, minutely furrowed toward the base on older stems; branches glabrate to sparsely strigose with whitish trichomes. The larger leaves 9–21 cm long; leaflets (11–) 15–23 per leaf, not distally accrescent, narrowly elliptic to elliptic, the larger 1.4–3.5 cm long, 7–17 mm wide, adaxial surface glabrous to sparsely strigose, usually without darker tannin deposits, abaxial surface glabrate to sericeous, often with darker tannins deposited in patches or uniformly over the entire surface; stipules 3–8 mm long, readily caducous (rarely persistent); stipels barely conspicuous. Inflorescences borne on long shoots in axils of coeval leaves, rachises 6.0–10.5 cm long, stipitate, the stipe 0.5–2.5 cm long, sparsely strigose to sericeous, with 25–70 nodes, internodes mostly 0.1–3.0 mm long; floral bracts 2.0–2.5 mm long, ligulate to linear-oblanceolate, densely sericeous, readily caducous; pedicels 3.5–4.0 mm long at anthesis, sparsely (rarely densely) strigose. Calyx cylindrical, attenuate at base, most often glabrate (rarely strigose) but pilose at the distal margin, tube 3.5–5.0 (–6.0) mm long, the lobes 0.5–1.5 mm long, the vexillary lobe bifid to entire, the carinal trifid.

Standard reddish pink, aging same, nectar-guide greenish, blade reflexed almost 90° at anthesis, 11.5–16.0 mm long, 12–16 mm wide, orbicular, emarginate, with a yellowish nectar-guide; wing petals 11.0–15.5 mm long, nearly perpendicular to the plane of the keel, reddish pink; keel petals 10.0–15.5 mm long, adnate for much of the distal end, straight, with a blunt tip, reddish pink. Style 3.0–4.5 mm long at anthesis; ovary glabrous, with 5–7 ovules. Legume 4.0–8.5 cm long, 9–14 mm wide, strongly laterally compressed, dark brown, glabrous; seeds 2–5, lenticular, 5–7 mm in diameter, uniformly light brown, aging dark brown. Figs. 52–59.

Phenology. Flowering from January through November, but the predominant season occurs from February through May or the first part of June, and this uniformly throughout Mexico and Central America; fruits persisting potentially all year round, but predominantly from March through November. Anthesis is reached by mid-day to afternoon and the brilliant masses of reddish pink flowers are very conspicuous in spite of the well-developed foliage.

Distribution (Fig. 51). Eastern and southern Mexico (Tamaulipas south to Chiapas), Guatemala, and southern Honduras (Comayagua, Intibucá, and Morazán); mostly in pine-oak woodlands, particularly at cliff edges, roadsides, and other occasionally disturbed sites in partially closed understory; (550–) 600–1700 (–2400) m. Co-occurring dominant genera usually include *Acacia*, *Juniperus*, *Liquidambar*, *Pinus*, and *Quercus*.

ADDITIONAL SPECIMENS EXAMINED. **Mexico.** CHIAPAS: Aguacatengo, Mpio. Venustiano Carranza, *Breedlove 9660* (F, MICH, NY, TEX); 6 mi S of Bochil along rd to Soyalo, Mpio. Bochil, *Breedlove 10226* (TEX); Río Hondo, 4 mi N of Jitotol on rd to Pueblo Nuevo Solistahuacán, Mpio. Jitotol, *Breedlove 12033* (MEXU, NY, TEX, US); Río Hondo, 6.5 km N of Jitotol along rd to Pichucalco, Mpio. Jitotol, *Breedlove & Thorne 21426* (MEXU, MICH, MO, NY); deep shaded canyon with pine-oak, at Rancho Ch'a ha', 5–6 km W of Teopisca, Mpio. Totolapa, *Breedlove 22899* (MEXU); SE of Cerro Baúl on the border with Oaxaca, 16 km NW of Rizo de Oro along rd to Colonia Figueroa, Mpio. Cintalapa, *Breedlove 24703* (MEXU, MICH, MO, NY); Rancho Ch'a ha' 5–6 km W of Teopisca, Mpio. Totolapa, *Breedlove 26203* (MEXU, MICH, MO); 6–8 km W of Jitotol, Mpio. Jitotol, *Breedlove 26375* (MEXU); 17.5 km SE of San Cristóbal along Hwy 190, Mpio. San Cristóbal de Las Casas, *Breedlove 34358* (MEXU); 6–10 km NNE of La Soledad from Las Margaritas to Campo Alegre, Mpio. La Independencia, *Breedlove 34764* (MEXU); San Lucas Zapotal, 2–4 km from Hwy 190, Mpio. San Cristóbal de las Casas, *Breedlove 37343* (MEXU); along Hwy 190 in the Zinacantán paraje of Muctajoc, Mpio. Ixtapa, *Breedlove 39746* (MEXU, MO); on SW slope of Muk'ta vits near Hwy 190, Mpio. San Cristóbal Las Casas, *Breedlove 41303* (MEXU, MO); near paraje Hok' Ch'enlom, Mpio. Zinacantán, *Breedlove 50139* (MEXU, NY); Salida Tunel 4–7 km W of San Cristóbal Las Casas, *Breedlove 50275* (MEXU, NY); along Río Hondo, 6.5 km N of Jitotol along rd to Pichucalco, Mpio. Jitotol, *Breedlove & Smith 32729* (MEXU, MO, NY, TEX); 17 km SE of San Cristóbal along Hwy 190, *Breedlove 34358* (MICH, MO); 6–10 km NNE of La Soledad from Las Margaritas to Campo Alegre, Mpio. La Independencia, *Breedlove 34764* (MO); 20 km E de San Cristóbal, *Cabrera & H. de Cabrera 2869* (MEXU); 16 km al SW de San Cristóbal de las Casas por el Camino a Zacualpan, *Cabrera & H. de Cabrera 3769* (MEXU, MO); locality unknown, *Ghiesbreght 865* (GH); 70 km W de San Cristóbal a Tuxtla, *Hernández & Shilom Ton 551* (MEXU); from Zinacantán Center to San Lucas at Vitstik, Mpio. Zinacantán, *Laughlin 753* (NY, US); Tenochtitlán, *MacDougall 7942* (MEXU); V. Carranza, *Matuda 15958* (F, MEXU, MICH); 3 km SE de Comitán, *Medrano & Hiriart s.n.* (MEXU); cerca Teopisca, encinares, *Miranda 2670* (MEXU); cerros al NW de Comitán, *Miranda 5077* (MEXU); Tenochtitlán, *Miranda 7942* (MEXU); 5 km NW de Teopisca, Mpio. Teopisca, *Sousa et al. 6705* (MEXU); 33 km al N de la desv. a Pichucalco, carr. San Cristóbal de las Casas–Villahermosa, *Téllez et al. 649* (MEXU); 33 km al N de la desv. a Pichucalco, carr. San Cristóbal de las Casas–Villahermosa, *Téllez et al. 650* (MEXU); 5 km al NE de El Rancho la Gloria, 14 km NE de Rizo de Oro, hacia la Col. Rodolfo Figueroa, *Torres C. &* Martínez S. 4822 (MEXU).—GUERRERO: 17 km de Chilapa por la carr. a Tlapa, Mpio. Chilapa, *Contreras 792* (MEXU); Mpio. Chichihualco, 4 km del Palmar hacia Chichihualco, 2 km al S de

FIGS. 52–59. *Hybosema ehrenbergii*. 52. Branch with inflorescences and infructescence. 53. Seedling. 54. Flower. 55. Calyx. 56. Standard petal. 57. Wing petal. 58. Keel petal. 59. Androecium.

Atlixtax, *Delgado & García 1068* (F, MEXU, NY); adelante de Alotenango, entre Tixtla y Chilapa, *Germán & Téllez 956* (MEXU); 1 km adelante de Tlapa, *López F. 179* (MEXU); 20 km al W de Chilpancingo, camino a Omiltemi, Mpio. Chilpancingo de los Bravos, *Martínez & Téllez 204* (MEXU, MO); Tixtla Chilapa, *Miranda 3982* (MEXU); 5 km W de Amojileca, camino Chilpancingo–Amojileca, *Rodríguez & Martínez 38* (MEXU).—HIDALGO: Hwy 85, 56 mi S of Tamazunchale, *Ahshapanek 262* (TEX); 3 mi N of Jacala, *Barkley 17M061* (F, TEX, UC); Puerto de La Zorra, Mpio. Jacala, *Cano P. 47* (MEXU); 6 km al N del Zoquital, Mpio. Atotonilco el Grande, *Hernández M. et al. 6029* (MEXU); Cuesta Colorada, 12 km NE de Jacala, Mpio. Jacala, *Hernández M. & Tenorio L. 6957* (MEXU, MO); 28 mi N of Pachuca, 5 mi N of Antotonilco on Hwy 105, *Hess & Byrne 4656* (MEXU); NE from Jacala towards Tamazunchale about 9 km from Jacala, *Hughes et al. 1588* (MONT); near Jacala, *Kenoyer s.n.*

(MO); Coñecita, *Kenoyer 678* (F); about 12 mi N of Jacala, *Manning & Manning 53585a* (GH); Puerto de la Zorra, Mpio. Jacala, Km 284 on hwy N of Jacala, *Moore 1665* (GH), *Moore 2668* (GH), *Moore 5031* (MEXU), *Moore & Wood 3640* (MEXU); mountains near Tula, *Pringle 13868* (GH, MO); 6.5 km S de El Alamo, carr. a Jacala, *Torres 3176* (MEXU); bei Pinalito oberhalb Jacala, *Troll 456* (G).—México: Dtto. Temascaltepec, Tenayac, *Hinton 5928* (GH); Dtto. Temascaltepec, Puñón, *Hinton 7983* (F, GH, MEXU, NY).—Oaxaca: Mpio de Sola de Vega, 20 km de Sola de Vega, *Delgado S. et al. 638* (F, MO); Dtto. Centro, Mpio. Huayapan, Sierra de San Felipe on trail from Estación Microondas Corral de Piedras, *Gereau & Saynes V. 2139* (NY); 12 km SW of Sola de Vega toward Puerto Escondido, *Hughes 385* (MEXU); to San Baltazar Guelavila, about 3 km E of the main rd from Tehuantepec to Oaxaca, 65 km SE of Oaxaca, *Hughes 1498* (MONT); ESE from the main Tehuantepec to Oaxaca rd towards San Baltazar Guelavila, 4 km from Guelavila, *Hughes 1726* (MONT); 15 mi N of Juchetango bridge on Hwy 131 to Oaxaca, *Lavin & Luckow 5819* (MICH, MO, NY, TEX); Cerro Lazo, Sola de Vega, *MacDougall s.n.* (NY); Cerro de Sola, Sola de Vega, *MacDougall 550.S* (NY); El Llano, 1 km al NE de Taraguntín, 14 km al NE de Sola de Vega, *Sousa et al. 6300* (MEXU, MICH); El Vado, Dtto. Ejutla, *Sousa et al. 8283* (MEXU, UC); 14 km SW de Sola de Vega, *Sousa 8337* (MEXU, UC); 9 km al S de Sola de Vega, Dtto. Sola de Vega, *Sousa & Basurto 9971* (MEXU); 11 km al S de Sola de Vega, Dtto. Sola de Vega, *Sousa & Basurto 9973* (MEXU); 16 km al WSW de San Pedro y San Pablo Ayutla, Dtto. Mixtepec, *Sousa et al. 10050* (MEXU); El Vado, en el extremo NW del Dtto. Ejutla, *Sousa et al. 10475* (MEXU); 10 km SW de Sola de Vega, *Sousa 10504* (MEXU); 3 km al W del entronque a San Sebastián con la carr. a Oaxaca–Pto. Escondido, Dtto. Sola de Vega, *Torres et al. 433* (MO).—Tamaulipas: Rancho del Cielo area, near La Joya, *Richardson 1308* (TEX).

Guatemala. Baja Verapaz: NE de Salamá, *Molina & Molina 12355* (NY).—Chiquimula: 15 km from Quetzaltepeque toward Esquipulas, *Hughes 1458, 1459* (MEXU, MONT); SE from Quetzaltepeque towards Esquipulas on the NW slopes of Montaña La Ruda about 15 km from Quetzaltepeque, *Hughes 1529* (MONT).—Huehuetenango: W from Colotenango towards Cuilco about 1 km W of the small town of Ixtahuacán, valley of the Río Cuilco, *Hughes 1468* (MONT); NE from Sacapales towards Cunen and Uspatan, ±10 km from Sacapales, *Hughes 1475* (MEXU, MONT); 3 km W of Ixtahuacán, *Hughes et al. 1688* (MONT); E from Huehuetenango towards Aguacatán, about 18 km from Huehuetenango, *Hughes et al. 1692* (MONT); Cumbre Papal, between Cuilco and Izmoquí, *Steyermark 50926, 50927* (F).—Quiché: Quiché, Aguilar 762 (F). **El Salvador.** La Libertad: above Zaragosa, *Williams & Molina 15043* (F).—Santa Ana: Hacienda San Miguel near Metapán, between Río San Miguel and summit of Cerro El Pinal, *Carlson 758* (F, UC). **Honduras.** Comayagua: above Siguatepeque, *Williams & Molina R. 12556* (F, GH); cerca de la Villa San Antonio de Flores, *Molina R. 8283* (F); near El Achote, hills above the plains of Siguatepeque, *Yuncker et al. 5834* (F, GH, MICH, MO, NY).—Intibucá: Quebrada El Huise, 9 km al E de La Esperanza, *Molina R. & Molina 14003* (F, NY); Quebrada Lejarsis, km 9–11 on rd La Esperanza–Marcala, *Molina R. & Molina 24284* (F, NY).—Morazán: Cerro Uyuca, 8 km W of El Zamorano, *Davidse & Pohl 2101* (MEXU); along Hwy 5, 13 mi S of Tegucigalpa, *Dziekanowski et al. 3408* (MO, NY, WIS); alrededores Nueva Tatumbla, 15 km E de Tegucigalpa, *Ferrera B. 150* (NY); near San Antonio, *Molina R. 333* (F, GH, MEXU); Valle de Angeles, near San Antonio, *Molina R. 390* (F, GH, MEXU, MO); Mt. Uyuca, entre Hoya Grande y San Juan Rancho, Río Yeguare, *Molina R. 1765* (F); Montaña Lepaterique, entre Las Tablas y Lepaterique, cerca de Quebrada Honda, *Molina R. 7829* (F); Valle Encantado y La Labranza de Mt. Uyuca, *Molina R. 8520* (F); campus de la E.A.P., El Zamorano, *Molina R. 13626* (F, NY); entre Zambrano y La Piramide, *Molina R. 14244* (F); entre Cuesta de Los Muertos y Monte Obscuro, Cordillera Azaculpa, *Molina R. 14518* (F); cerca del Cementerio de San Antonio de Oriente, *Molina R. & Molina 14131* (F, NY); between La Vuelta del Zope and San Juan del Rancho, *Molina R. & Molina 25825* (F, MO, NY); on bank of Lepaterique River, 5 km to Lepaterique, *Molina R. & Molina 31497* (F, MO); Piedra Herrada, 15 km SE de Tegucigalpa, *Ortega U. 168* (NY); near Joya Grande, on rd from El Zamorano to Suyapa, *Standley & Molina R. 4481* (F, NY); Camino Viejo between Suyapa and Tegucigalpa, *Standley 14226* (F); Cerro de Uyuca, between Hoya Grande and Valle Encantando, *Standley 15227* (UC); San Juan del Rancho, N of Cerro de Uyuca, *Standley 15467* (F, GH); toward San Antonio de Oriente, region of El Jicarito, above El Zamorano, *Standley 22545* (F, GH); El Quebracho, Cerro de Uyuca, *Standley 23750* (F, GH, NY); Chauite, *Valerio R. 905* (F); Río Yeguare, *Valerio R. 2113* (F); vicinity of Hoya Grande, *Williams & Molina R. 10169* (F, GH, MICH, MO, TEX, UC); near Hoya Grande, *Williams & Molina R. 10556* (F, GH); entre Uyuca y El Empalme at Tatumbla, *Williams s.n.* (F, NY, TEX); above Hoya Grande, *Williams & Molina R. 11658* (F); Hoya Grande, *Williams & Molina R. 11965* (F, MICH, MO); above Hoya Grande, *Williams & Molina R. 12658* (GH, MO).—El Paraíso: puente de Guinope, Río Yeguare, *Molina R. 3340* (F, GH); Barranco de Manzaragua, Río

Yeguare, *Molina R. 7583* (F); entre Galeras y Las Casitas, *Molina R. 10772* (F); entre Cuesta de Galeras y Las Casitas, *Molina R. 13727* (F, NY, TEX); Quebrada Honda, Montaña Tapahuasca, 3 km a San Lucas, *Molina R. 14230* (F, NY, TEX); Tapahuasca, between Manzaragua and San Lucas, *Molina R. 18751* (F, NY, US); Yuscaran Rd, 2 km S of Hwy 4 intersection, *Pohl 12603* (F); 2 mi W of Guinope, *Williams & Molina R. 10320* (F).

The leaves and leaflets of *H. ehrenbergii* enlarge significantly during anthesis and especially after fruit set, such that leaflets collected with flowering or fruiting specimens may be smaller than those realized later in the growing season. In spite of this, even the largest leaf and leaflet sizes of *H. ehrenbergii* do not reach those of *H. robustum*. In this latter species, the larger leaves of a specimen measure 18–27 cm long, and the largest leaflets measure 3.5–7.0 cm long by 10–35 mm wide. In comparison, the larger leaves of *H. ehrenbergii* measure 9–17 cm long, and the largest leaflets measure (even after fruiting season) 1.4–3.5 cm long by 7–17 mm wide. Thus, leaf and leaflet length provide readily useful distinguishing characteristics for the two species of *Hybosema*.

Intraspecific morphological variation is notably correlated with geographical distribution in *H. ehrenbergii*. In Honduras, specifically in the departments of Morazán and El Paraíso (the easternmost portion of the distribution of this species), the undersurfaces of the leaflets are distinctly glabrate, even on the newly expanding leaves, whereas in specimens collected in Depto. Comayagua and westward in Guatemala and Mexico the undersurfaces of the leaflets are distinctly sericeous. In any geographical region, the lower surfaces of the leaflets of *H. ehrenbergii* may be glabrate with age, so these indumentum differences are most conspicuous on developing or newly expanded leaves (i.e., those co-occurring with flowers). Corresponding with size difference is leaflet number, which tends to be more uniform toward the larger end of the variation scale in Honduras (19–23 leaflets per leaf) and toward the smaller end or at least more variable northward (15–23 leaflets per leaf). In Hidalgo, Mexico, leaflets commonly number 15 per leaf, although some individuals have leaves with only 11 leaflets. Also, petals tend toward the larger end of the variation scale in Honduras and toward the smaller end northward. Although the last two characters, leaflet number and petal length, display a gradual transition, the change in indumentum is abrupt and suggests either a founder event or some limited barrier to migration between southwestern Honduras and northward regions of the range

One additional character making a gradual transition from south to north is the length of the calyx tube. In southern Mexico (Oaxaca and Chiapas) and Central America, the tube measures mostly 3.5–5.0 mm long, but in more northern regions of Mexico (Guerrero, Hidalgo, and Tamaulipas), the calyx tube often measures 5.5–6.0 mm long. This transition of calyx tube length from small in the south to large in the north is in direct contrast to that of petal length (i.e., largest in the south and smaller to the north).

5. Poitea Ventenat, Mém. Math. Phys. Inst. Nat. France 1: 16. (March) 1807. *Poitea* sect. *Eupoitea* Urban, Symb. antill. 2: 291. 1900.—Type: *Poitea galegoides* Ventenat.

Single- and multitrunked trees and shrubs. Roots commonly superficial and often producing leafy sprouts. Stems erect to ascending or splayed laterally, branches alternate; bark smooth to superficially or deeply furrowed. Leaves alternate, pulvinate, imparipinnate or paripinnate by reduction of the terminal leaflet to a seta; leaflets opposite, narrowly to broadly elliptic, retuse to rounded, caducous from the more persistent rachis during the onset of the dry season or with age, adaxial surface glabrous to puberulous or sericeous, with or without finely scattered patchy tannin deposits, abaxial surface glabrate to densely sericeous or tomentose, usually with scattered patchy tannin deposits; leaf rachis canaliculate, the groove without a dense indumentum, stipels present; stipule-pairs adnate, broadly attached to the stem and base of petiole, herbaceous, at least the basal portion persistent. Inflorescences of subsessile to pedunculate, pendulous to lax, axillary racemes to solitary flowers, in the axils of coeval leaves or at leafless nodes of older stems, with a synchronous maturation of flowers on few-flowered rachises, or with a sequential acropetal maturation on elongate many-flowered rachises, each rachis ranging from 1–60 nodes, stipitate glands absent. Flowers ebracteolate, rarely bracteolate, with a well-developed hypanthium; floral bracts conspicuous, at least the basal portion persistent; pedicels mostly articulated, or sometimes confluent with the calyx, straight and ascending during anthesis and seed set, persisting on the rachis or falling with the aborted flower. Calyx campanulate, generally attenuate at base, glabrate to sericeous, stipitate glands absent, with two prominent lobes mostly shorter than the tube, the vexillary lobe bifid, the carinal lobe trifid, the ultimate lobes triangular if at all conspicuous, not overlapping at anthesis. Standard orbicular to oblong, emarginate, reflexed from 90–180° at anthesis in bee-pollinated species, not at all reflexed in bird-pollinated species, dorsal surface glabrous, shorter than the keel petals, with two small callosities about the midrib in the middle of the nectar-guide, claw gradually contracted from the blade, inflexed auricles absent; wing petals sculptured, shorter than the keel at anthesis, and loosely attached to it by a fold in the basal auricle or by a boss-and-socket-joint, the distal ends glabrous, the blades parallel or nearly so to the plane of the keel; keel tips completely free and fused only a short distance along the dorsal side, or valvately connate distally to nearly the tip, apex blunt in the former case and acute to strongly rostrate in the latter. Staminal tube mostly diadelphous, rarely pseudomonadelphous, with basal fenestrae; anthers projecting beyond or included in the keel tip, basally attached to the filaments, the broad connective reddish brown; intrastaminal nectariferous disk absent. Pollen spheroidal, tricolporate, mesocolpium rugulate-perforate to smooth, polar regions smooth. Ovary stipitate, the stipe generally as long or longer than the calyx tube, glabrous, eglandular, with 15–30 ovules; style glabrous, the base thin and imperceptibly confluent with the distal portion of the ovary, not becoming displaced to the carinal side during fruit maturation; stigma ciliate to glabrous, terminal on a distally tapered style, or subterminal on a distally introrse, subclavate style. Legumes stipitate, glabrous, eglandular, inner epidermis of the endocarp of adjacent valves smooth and not abutting, not adherent and not forming seed compartments, the valves thinly woody at maturity, elastically and explosively dehiscent, the dorsal and ventral margins generally parallel in fruit, except where tandem seed abortion occurs. Seeds transversely oriented, thinly lenticular, not constricted in the middle, accumulating canavanine, uniformly light to dark brown, without dark mottling, hilum apical and attached to a very short funiculus, testa firmly adherent. Seedlings with thick and fleshy, or thin and foliar, epigeal cotyledons, eophylls opposite and unifoliolate, metaphylls and mature leaves alternate and 5–many-foliolate. Base chromosome number $x = 8, 10$ (Goldblatt 1981).

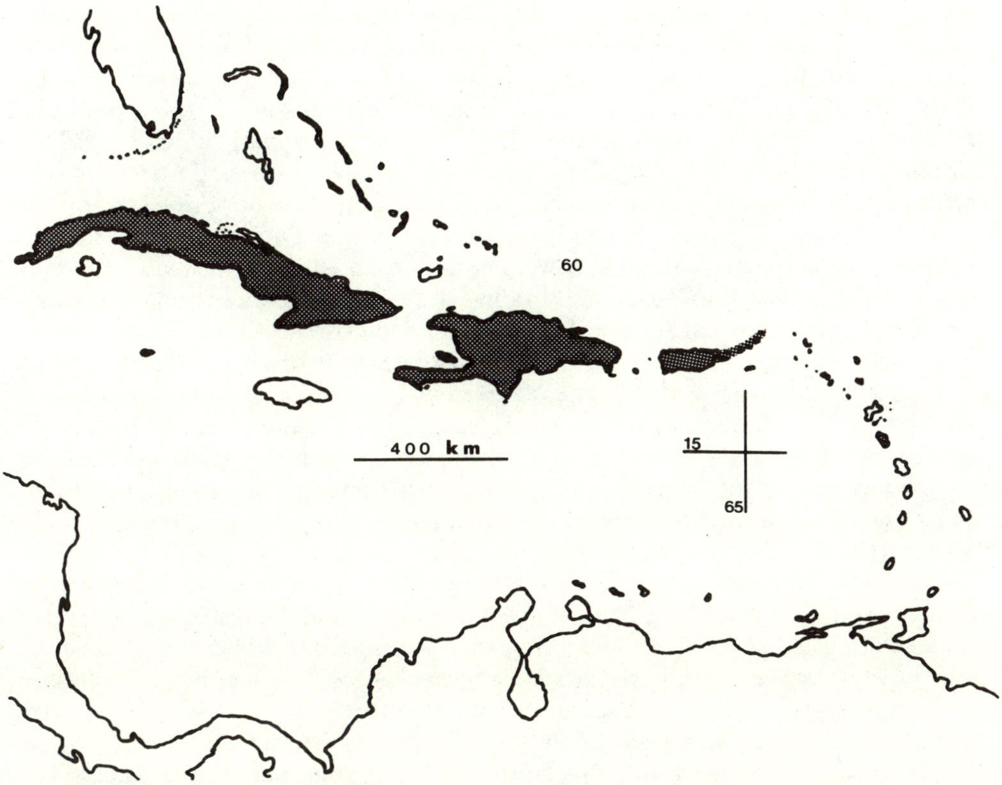

FIG. 60. Distribution, indicated by shading, of the genus *Poitea* in the Greater Antilles and on the island of Dominica.

Poitea comprises 12 species confined to the Greater Antilles and the island of Dominica (Fig. 60). It is easily distinguished from other members of Robinieae by its persistent connate stipules, which are distichously arranged on short shoots. The flowers of this genus are distinctive in that the keel petals are uniformly longer than the standard and wing petals. Also, the mature legumes of *Poitea* stand in contrast to those of other genera in the tribe in that they are thinly woody and have no spongy growth of the inner epidermis; the valves dehisce rapidly and very explosively, such that it is nearly impossible to find mature seed persisting on the mother plant for any length of time. In the *Gliricidia*-group, *Poitea* is distinguished by its inflorescences; flowers are borne on a long rachis and reach anthesis in a sequential acropetal manner, as in *Coursetia* (Lavin 1988), or they are borne on small rachis of 1–3 nodes, which is the presumed ancestral condition (Lavin 1993). The close relatives of *Poitea* (e.g., *Lennea*, *Hybosema*, and *Gliricidia*) have inflorescence rachises in which all of the flowers reach anthesis synchronously.

Potential economic uses of *Poitea* are those characteristic of much of the rest of the tribe. In particular, most of the species have very showy blossoms and would make attractive ornamentals (they are used for this purpose to a very limited extent in the Caribbean Basin). Also, all of the species profusely root-sprout and often inhabit very unstable slopes; thus they have potential in soil stabilization projects.

The broad circumscription of the genus *Poitea* reflects the very close relationship of the constituent species. Five previously recognized genera are now included in *Poitea*. These include *Sauvallella* Rydb. (from western Cuba), *Sabinea* DC. (from Puerto Rico and Dominica), and *Bembicidium* Rydb., *Notodon* Urb., and *Corynella* DC. (a collection of species with small bee-pollinated flowers from Cuba, Hispaniola, and Puerto Rico). Lavin (1993) argued that *Poitea* should be broadly circumscribed, because genera such as *Bembicidium*, *Notodon*, and *Corynella* were artificial (e.g., *Bembecidium* was based on a single immature flowering specimen of *Corynella gracilis* Griseb.). Second, the two major lineages of *Poitea*, the *Poitea galegoides*-alliance (with tubular flowers borne on an elongate rachis, and imparipinnate leaves) and the *Poitea florida*-alliance (with papilionaceous flowers borne on short shoots, and paripinnate leaves) are joined by the Cuban *Poitea immarginata* (C. Wright) Lavin (with an intermediate tubular-papilionaceous flower borne on short shoots, and imparipinnate leaves). Third, from a phylogenetic perspective, it is our intention to emphasize relationships among species by broadly circumscribing genera rather than maximizing the number of segregate genera.

A recent taxonomic recognition of the two principal subgroups, the *Poitea galegoides*-alliance and the *Poitea florida*-alliance, makes biogeographical sense with regard to the vicariance model of Caribbean biogeography put forth by Rosen (1976, 1985). Though some of the distribution of *Poitea* in the Antilles can be explained by chance dispersal, much of it is best explained by dispersal directed by island drift. That is, the *P. galegoides*-alliance is centered in westernmost Cuba and southwestern and northern Hispaniola, whereas the *P. florida*-alliance is centered in eastern Cuba, northern Hispaniola, and Puerto Rico (see Lavin 1993). It is hypothesized that these subsets of land masses (e.g., western Cuba and southwestern Hispaniola, and eastern Cuba, northern Hispaniola, and Puerto Rico) were connected before and during the beginning of the late Eocene tectonic movements that relegated them to their present day positions and configurations to the east of their Central American origin. With one exception, the species of *Poitea* are endemic to a single island in the Caribbean, and the genus is notably absent from recently emergent islands, such as Jamaica, St. Croix, and most of the Lesser Antilles. Such attributes, as enumerated by Page and Lydeard (1994), render *Poitea* an ideal test group for Caribbean vicariance hypotheses.

The following synopsis provides a classification of the species of *Poitea* based on a cladistic analysis of morphological and chloroplast DNA data (Fig. 48 in Lavin 1993).

Synopsis of the Genus Poitea

I. The *Poitea galegoides*-alliance. Tubular or tubular papilionaceous flowers; wing petals either assuming the function of the standard or highly reduced; flowers borne on pendent to lax raceme rachises, or on short shoots; imparipinnate leaves.

Poitea sect. **Poitea.**

Vilmorinia DC., Prod. 2: 239. 1825. *Poitea* sect. *Vilmorinia* (DC.) Urban, Symb. antill. 2: 291. 1900.—Type: *Vilmorinia multiflora* (Swartz) DC. [=*Poitea multiflora* (Swartz) Urban].

Flowers tubular, borne on pendent to lax raceme rachises.

5.1. Poitea galegoides Ventenat.

5.1a. Poitea galegoides var. **galegoides.** Widespread on Hispaniola.

5.1b. Poitea galegoides var. **stenophylla** Ekman ex Lavin. Southcentral Dominican Republic

5.2. Poitea longiflora Urban. Southcentral Dominican Republic.

5.3. Poitea campanilla DC. Central Dominican Republic.

5.4. Poitea multiflora (Swartz) Urban. Haiti and adjacent Dominican Republic.

5.5. Poitea glyciphylla (Poiret) Urban. Southcental Hispaniola.

Poitea sect. **Sauvallella** (Rydberg) Lavin, comb. nov. *Sauvallella* Rydberg, Amer. J. Bot. 11: 480. 1924.—TYPE: *Poitea immarginata* (C. Wright) Lavin.

Cajalbania Urban, Symb. antill. 9: 449. 1928.—TYPE: *Cajalbania immarginata* (C. Wright) Urban [=*Poitea immarginata* (C. Wright) Lavin].

Flowers papilionaceous-tubular, borne on short shoots.

5.6. Poitea immarginata (C. Wright) Lavin. Westernmost Cuba.

II. The *Poitea florida*-alliance. Papilionaceous flowers; wing petals not assuming the function of the standard and not highly reduced; flowers borne on short shoots; paripinnate leaves.

Poitea sect. **Sabinea** (DC.) Lavin, comb. nov. *Sabinea* DC., Ann. Sci. Nat. (Paris) 4: 92. 1825. *Sabinea* sect. *Eusabinea* Urban, Symb. antill. 1: 324. 1899.—TYPE: *Poitea florida* (Vahl) Lavin.

Sabinea sect. *Sabineopsis* Urban, Symb. antill. 1: 323. 1899.—TYPE: *Sabinea punicea* Urban [=*Poitea punicea* (Urban) Lavin].

Trees and shrubs; large bird-pollinated flowers with distally free keel petals that are rounded at the tip; bilobed calyx with a tendency to become truncate. (Figs. 61–62.)

5.7. Poitea florida (Vahl) Lavin. Puerto Rico and the Virgin Islands.

5.8. Poitea carinalis (Grisebach) Lavin. Dominica.

5.9. Poitea punicea (Urban) Lavin. Puerto Rico.

Poitea sect. **Corynella** (DC.) Lavin, comb. nov. *Corynella* DC., Ann. Sci. Nat. (Paris) 4: 93. 1825. *Corynitis* Sprengel, Syst. veg. 4: 263. 1827, nom. superfl.—TYPE: *Poitea dubia* (Poiret) Lavin.

Toxotropis Turczaninow, Bull. Soc. Imp. Naturalistes Moscou 4: 506. 1846.—TYPE: *Toxotropis jaegeri* Turczaninow [=*Poitea dubia* (Lamarck) Lavin].

Notodon Urban, Symb. antill. 1: 324. 1899.—TYPE: *Notodon gracilis* (Grisebach) Urban [=*Poitea gracilis* (Grisebach) Lavin].

Bembicidium Rydberg, Mem. Torrey Bot. Club 16: 68. 1920.—TYPE: *Bembicidium cubense* Rydberg [=*Poitea gracilis* (Grisebach) Lavin].

Shrubs; small bee-pollinated flowers with distally fused keel petals that are acute to rostrate at the tip; calyx distinctly bilobed.

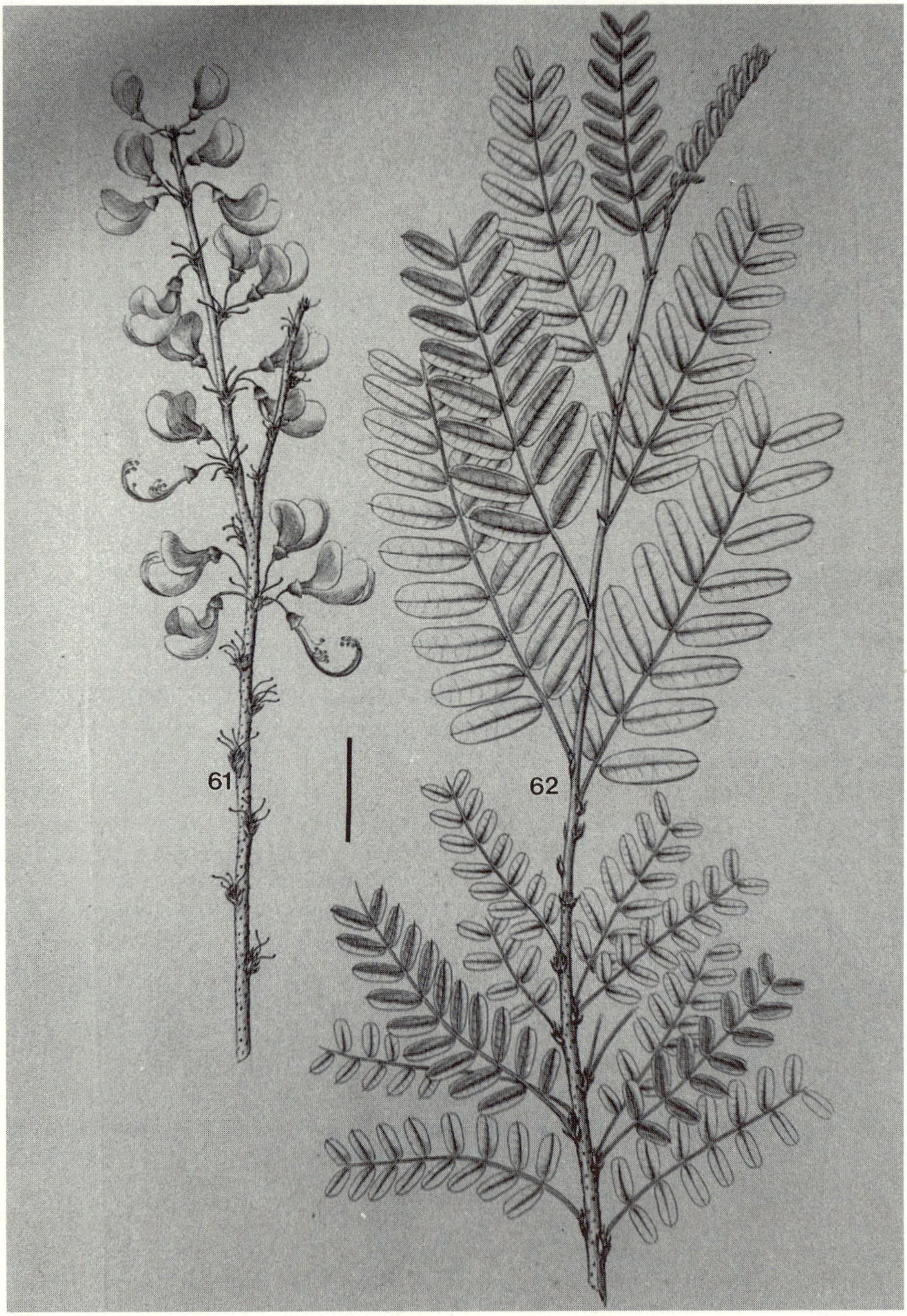

FIGS. 61–62. *Poitea florida*. Reproduced from Vahl, Symb. bot. 3: 89, t. 70. 1794. 61. Flowering branch. 62. Leafy branch. (bar = 20 mm).

5.10. Poitea paucifolia (DC.) Lavin. Hispaniola and Puerto Rico.
5.11. Poitea dubia (Poiret) Lavin. Haiti and Dominican Republic.
5.12. Poitea gracilis (Grisebach) Lavin. Cuba.

6. Gliricidia H. B. K., Nov. gen. sp. 6: 393. 1823.—TYPE: *Gliricidia sepium* (Jacquin) Steudel.

Yucaratonia Burkart, Darwiniana 15: 523. 1969.—TYPE: *Yucaratonia brenningii* (Harms) Burkart [=*Gliricidia brenningii* (Harms) Lavin].

Single- and multitrunked trees and occasionally shrubs. Roots commonly superficial and often producing leafy sprouts. Stems erect to ascending or splayed laterally, branches alternate, rarely subopposite; bark smooth to superficially furrowed. Leaves alternate to occasionally subopposite, pulvinate, imparipinnate; leaflets generally opposite, occasionally alternate, narrowly to broadly elliptic, rounded to acuminate, sometimes cuspidate, caducous from the more persistent rachis during dry periods or with age, adaxial surface glabrate to sericeous, usually without finely scattered patchy tannin deposits, abaxial surface glabrate to densely sericeous, usually with finely scattered or large patchy tannin deposits; leaf rachis canaliculate, the groove without a dense indumentum, stipels absent or inconspicuous; stipules free, narrowly to broadly attached to the stem adjacent to the petiole, herbaceous, readily caducous. Inflorescences subsessile to pedunculate, of pendent or lax to ascending axillary racemes, in the axils of coeval leaves or at leafless nodes of older stems, with mostly synchronous maturation of flowers on a single rachis, each rachis with mostly 25–100 nodes, stipitate glands absent. Flowers ebracteolate, with a well-developed hypanthium; floral bracts well developed and caducous, or minute and persistent; pedicels articulated with the calyx, straight and ascending during anthesis and seed set, often persisting on the rachis after flower abortion. Calyx campanulate, attenuate at base, glabrate to strigose, stipitate glands absent, distally truncated, with five inconspicuous more or less evenly spaced lobes, lobes not overlapping at anthesis. Standard orbicular, emarginate, reflexed nearly 180° at anthesis, dorsal surface usually glabrous, sometimes villous, subequal in length to the keel petals, with two prominent callosities on either side of the midrib in the region of the nectar-guide, claw abruptly contracted from the blade, inflexed auricles present; wing petals sculptured, commonly projecting beyond the keel at anthesis, and loosely attached to it by a fold in the basal auricle, the distal ends glabrous, the blades parallel to perpendicular to the plane of the keel; keel valvately connate distally to nearly the tip, apex blunt. Staminal tube diadelphous, with basal fenestrae; anthers included in the keel tip, basally attached to the filaments, the broad connective reddish brown; intrastaminal nectariferous disk present in *G. brenningii*, absent in *G. maculata* and *G. sepium*. Pollen spheroidal, tricolporate, mesocolpium rugulate-perforate, polar regions smooth. Ovary stipitate, the stipe nearly as long as the calyx tube, glabrous, eglandular, with 7–15 ovules; style glabrous, the base thin and imperceptibly confluent with the distal portion of the ovary, not becoming displaced to the carinal side during fruit maturation; stigma ciliate, terminal on a distally tapered style. Legumes stipitate, glabrous, eglandular, inner epidermis of the endocarp of adjacent valves spongy and abutting between seeds, forming irregular seed compartments, the valves woody at maturity, elastically and explosively dehiscent, the dorsal and ventral margins generally parallel in fruit. Seeds transversely oriented, lenticular, not constricted in the middle, accumulating canavanine, uniformly light to dark brown, without dark mottling, hilum apical and attached to a very short funiculus, testa firmly adherent. Seedlings with foliar epigeal (hypogeal in *G. brenningii*) cotyledons, 1–several eophylls alternate and 1-foliolate, metaphylls and mature leaves alternate and 5–many-foliolate. Base chromosome number x = 10, 11.

Gliricidia is here circumscribed to exclude *G. ehrenbergii* [=*Hybosema ehrenbergii*] and to include *G. brenningii* (based on *Yucaratonia brenningii*). Though inconsistent with previous taxonomies (e.g., Urban 1900; Harms 1923; Rydberg 1924b; Burkart 1969), the present classification is based on a cladistic analysis of morphological and chloroplast DNA evidence (see phylogeny section; Lavin et al. 1991). Synapomorphies marking the monophyly of the constituent species of *Gliricidia* include flowers with campanulate truncate calyces that persist well into fruit maturation and fully reflexed standard petals. Furthermore, the species of *Gliricidia* are not most closely related to those now in *Hybosema,* but may be most closely related to those placed in the Antillean genus *Poitea*. For example, *Poitea* and *Gliricidia* both have flowers with diadelphous staminal tubes and seedlings with foliar epigeal cotyledons. *Hybosema* and more distant relatives, such as *Lennea*, have flowers with pseudomonadelphous staminal tubes and seedlings with hypogeal or thickened, somewhat foliar cotyledons. At the least, *Hybosema*, *Poitea*, and *Gliricidia* represent a trichotomously branching lineage (cf. Figs. 1–6); combining *Hybosema* with *Gliricidia* would result in a paraphyletic group with respect to *Poitea*.

The flowers of the Ecuadorian *G. brenningii*, which have whitish petals and are borne on long-pedunculate axillary racemes, are very similar to those of *G. maculata* from the Yucatan Peninsula. The inclusion of this species in *Gliricidia*, though at odds with most traditional treatments (e.g., Burkart 1969; Rydberg 1924b) is nevertheless consistent with that of Spruce (in Urban 1900). Though Harms (1923) rejected the placement of *G. brenningii* into *Gliricidia* as then circumscribed, he did state that its inclusion in *Sesbania*, then as *S. brenningii*, was not completely certain, especially since he was dealing with very few, preserved specimens. Harms certainly was not explicit about his reasoning for excluding *S. brenningii* from *Gliricidia*.

Key to the Species of Gliricidia

1. Leaflets 33–41 per leaf, rounded and not distally accrescent, 7–12 (–19) mm wide, with few tannin patches in just the very center of the abaxial surface of the leaflet; stipules 4–7 mm long; stipels 0–1 mm long; inflorescence rachis (10–) 16–35 cm long, pendent, internodes 2–6 mm long; floral bracts 2.5–3.0 mm long, generally readily caducous; wing petals 16–18 mm long, whitish; ovary borne on a gynoecial disk, with 12–15 ovules; legume 16–31 cm long; Ecuador and adjacent Peru. 6.1. *G. brenningii*.
1. Leaflets 7–25 per leaf, rounded to acuminate or cuspidate, distally accrescent or not, 16–48 mm wide, with large conspicuous tannin patches covering much of the abaxial surface of the leaflet; stipules 0.5–2.0 mm long; stipels absent; inflorescence rachis 2.5–23 cm long, ascending to lax or pendent, internodes 0.1–4.0 mm long; floral bracts 0.5–2.0 mm long, generally persistent; wing petals 15–23 (–27) mm long, whitish or pinkish; gynoecial disk absent, ovary with 7–12 ovules; legume (6–) 8–17 (–23) cm long; Mexico and Central America, introduced in northern South America but uncommon in Ecuador and Peru.
    2. Leaflets coriaceous, rounded, rarely acuminate, 7–15 (–19) per leaf, rarely distally accrescent; inflorescence rachis (4–) 6–23 cm long, pendent; calyx 5.0–6.5 (–7.0) mm long,

the attenuate base 0.5–2.0 mm long, strigose to sericeous, commonly granuliferous; pedicels strigose to sericeous, often granuliferous; petals predominantly whitish, occasionally pinkish, the wing petals 18–23 (–27) mm long; legume 13–16 (–21) mm wide; seed 5.5–9.0 mm in diameter; Mexico in the Yucatan Peninsula and uncommon in Chiapas, also in Belize and Guatemala (Petén). 6.2. *G. maculata.*

2. Leaflets thin, acuminate to sometimes cuspidate, rarely rounded, (7–) 13–21 (–25) per leaf, slightly to conspicuously distally accrescent; inflorescence rachis 2.5–15 (–17) cm long, ascending to lax; calyx (5.5–) 6.0–9.0 mm long, the attenuate base (1–) 2–4 mm long, glabrous to sparsely strigose, rarely granuliferous; pedicels glabrous to sometimes sparsely strigose, rarely granuliferous; petals predominantly light pink to pinkish, rarely whitish, the wing petals 15–23 mm long; legume 14–22 mm wide; seed 8.5–11.5 mm in diameter; widespread in seasonally dry forests of Mexico and Central America, but uncommon in the native vegetation of the Yucatan Peninsula, Belize, and Petén, Guatemala; introduced widely in northern South America, the Caribbean, and throughout tropical Africa and Asia. 6.3. *G. sepium.*

6.1. Gliricidia brenningii (Harms) Lavin, comb. nov. *Sesbania brenningii* Harms, Repert. Spec. Nov. Regni Veg. 19: 68. 1923. *Yucaratonia brenningii* (Harms) Burkart, Darwiniana 15: 526. 1969.—TYPE: ECUADOR. Guayaquil, an Wegrändern in der Nähe des Río Salado, Jan 1896, *Brenning 238* (holotype: B, destroyed, fragment: F!, photos: F! NY!).

Single- to multistemmed trees 3–18 m tall. Stems erect, 3–20 cm in diameter at base, with or without major branches from near base; bark yellowish brown to whitish, minutely furrowed toward the base on older stems; branches glabrate to sericeous (and then mostly in occasional patches) with whitish trichomes. The larger leaves 15–35 cm long; leaflets 33–41 per leaf, thin, not distally accrescent, elliptic, apex rounded, the larger 20–35 (–55) mm long, 7–12 (–19) mm wide, (at maturity) adaxial surface glabrate to puberulous, without darker tannin deposits, abaxial surface strigose to sericeous, usually with few purplish tannins deposited in very small scattered patches in just the very center; stipules linear-lanceolate, 4–7 mm long, very readily caducous; stipels 0–1 mm long. Inflorescence rachises (10–) 16–35 cm long, pendent, stipitate, the stipe 1.3–2.5 cm long, puberulous, with 25–50 nodes, internodes 2–6 mm long; floral bracts 2.5–3.0 mm long, linear-lanceolate, sericeous, generally readily caducous, but occasionally persisting through anthesis; pedicels 5–9 mm long at anthesis, glabrous. Calyx campanulate, distinctly attenuate at base for 0.2–1.0 mm, glabrate to puberulous, tube 6.5–7.0 mm long, the five lobes 0.1–0.3 mm long, more-or-less evenly spaced. Standard mostly whitish, aging same, nectar-guide a broad yellowish patch, blade reflexed about 180° at anthesis, 16–18 mm long, 16–18 mm wide, orbicular, emarginate; wing petals 16–18 mm long, perpendicular to the plane of the keel, whitish; keel petals 15–17 mm long, adnate for much of the distal end, straight, with a blunt tip, whitish. Style 4–5 mm long at anthesis; ovary glabrous, with 12–15 ovules, borne on a gynoecial disk. Legume 16–31 cm long, 17–21 mm wide, strongly laterally compressed, dark brown, glabrous; seeds 5–15, lenticular, 9–11 mm in diameter, uniformly light brown, aging dark brown. Fig. 63.

Phenology. The predominant flowering season occurs, depending on local rains, from January through April, sporadically later; woody fruits persist from May through October, but probably bear mature seed only during September and October (C. Hughes, pers. comm.).

FIG. 63. *Gliricidia brenningii*. Inflorescence of *C. E. Hughes 1009* (FHO). Photograph by Colin E. Hughes. The raceme in foreground in 22 cm long.

Distribution (Fig. 64). Pacific slope of Ecuador and immediately adjacent Peru; lowland seasonally deciduous and evergreen forests and thorn scrub, and commonly as part of living fence systems or cultivated as shade over coffee crop; 0–820 m. Co-occurring predominant species include members of *Guazuma*, *Inga*, and *Mutingia*.

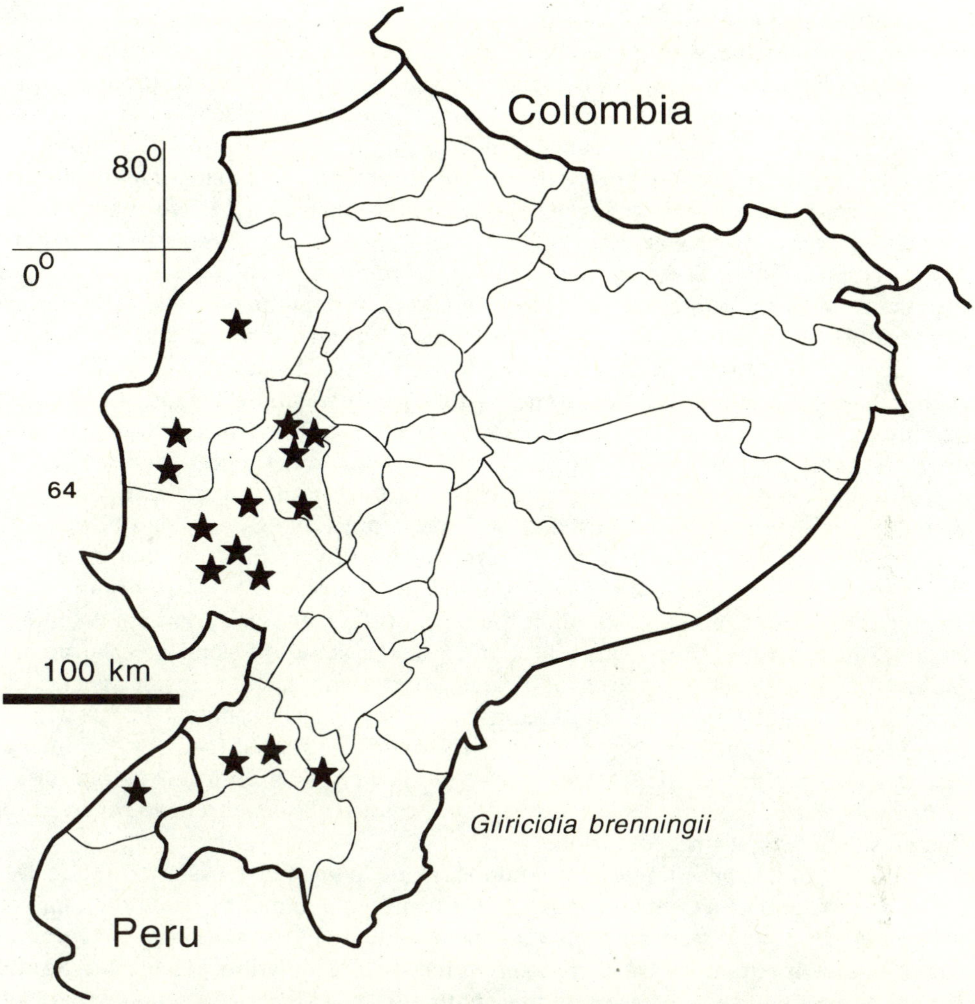

FIG. 64. Distribution of *Gliricidia brenningii* in Ecuador and adjacent Tumbes, Peru.

ADDITIONAL SPECIMENS EXAMINED. **Ecuador.** GUAYAS: area of Miguel Wagner's Hacienda, 10 km N of Guayaquil, *Dodson & Thien 1788* (NY); Guayaquil, Urdesa, Río Guayas, *Gilmartin 830* (MO); near Pedro Carbo, *Haught 3020* (F); rd from Palestina to Vinces about 10 km NE of Palestina, *Hughes 1009* (FHO); along rd W of Guayaquil, *Little 6570* (F, UC); sylvis prope Guayaquil, *Mille 811* (F); prope Guayaquil, *Mille 889* (F); Guayaquil, *Valverde 596* (MO).—LOJA: 10 km S of Portovelo, rd towards Loja, S side of Río Ambacas, *Hughes 1015* (FHO).—MANABÍ: about 25 km N of Chone, rd towards El Carmen and Santo Domingo del los Colorados, *Hughes 993* (FHO); San Pablo, between Sancan and Santa Ana about 35 km S of Portoviejo, *Hughes 1193* (FHO); Montalvo, on the main rd S from Jipijapa towards Pajan and Guayaquil about 19 km S of Jipijapa, *Hughes 1194* (FHO).—EL ORO: W of Palmales close to the rd towards Quebrada Seca about 15 km SE of Arenillas, *Hughes 1012* (FHO); Hacienda Ingenio to Piedras over Cordillera Marcabeli, *Little 6683* (F, UC).—LOS RIOS: Jauneche Forest, Jauneche, Km 70 Quevado–Palenque via Mocachi, Cantón Vinces, *Dodson et al. 7462* (MO); 22–30 km S of Empalme on rd to Guayaquil, near border with Guayas, *Gentry 9657* (MO); N from Vinces towards Palenque about 7 km S of Palenque, *Hughes 1010* (FHO); Estación Experimental, property of Corporación Ecuatoriana de Fomento, *Little 6444* (UC). **Peru.** TUMBES: along the Pampas de Hospital to El Caucho rd, about 11 km from El Caucho, Bosque Nacional de Tumbes cerca El Caucho o entre El Caucho Y Pampas de Hospital, *Simpson & Schunke 516* (F).

The intrastaminal disk of *G. brenningii* is remarkable in that it is otherwise unknown in the tribe Robinieae. With this character and the truncated calyx, *G. brenningii* distinctly converges on species of *Lonchocarpus* (tribe Millettieae), and superficially appears to be, in this regard, one of the earliest offshoots in the tribe. However, a greater body of evidence points to a close relationship of *G. brenningii* within *Gliricidia* and the *Gliricidia*-group. The truncated calyx that persists with the maturing fruit and the fully reflexed standard petals form a distinctive combination found in the tribe Robinieae otherwise only in *G. sepium* and *G. maculata*. Furthermore, the position of *G. brenningii* within the *Gliricidia*-group is well supported by the occurrence of stipitate ovaries, thin cotyledons, and tannin patches on the lower surfaces of the leaflets. In the phylogeny section, *Gliricidia* was coded as a multistate taxon for the presence or absence of nectaries from a gynoecial disk (#11). As such, *Gliricidia* was consistently grouped with *Hybosema* and *Poitea* in all the most parsimonious reconstructions, and no alternative relationship of *Gliricidia* to genera possessing the gynoecial disk (e.g., *Lonchocarpus*) was suggested.

Given its distribution in South America, away from the main diversity of Robinieae, *G. brenningii* is most likely to be confused with species of *Lonchocarpus*. This genus is distinguished from *G. brenningii* (and in fact from all Robinieae) by its flowers borne in pairs at the tip of a peduncle. Furthermore, the combination of bracteoles subtending the base of the calyx, pseudomonadelphous staminal tube, and sericeous abaxial surface of the standard petal are characters uniformly present in all species of *Lonchocarpus* and lacking in all species of Robinieae (though *Olneya*, from the Sonoran Desert, has the standard petal with a tomentose abaxial surface). Since many of the close relatives of the tribe Robinieae (especially those in the tribe Millettieae) have flowers with an intrastaminal disk, *G. brenningii* can be regarded as the only species of Robinieae to have reverted to this presumably ancestral condition.

The use of *G. brenningii* as a component of living fence systems, shade for coffee plants, and as a control agent for burrowing mammals (its common name is *yucaratón*, "the plant that kills mice") is nearly identical to those of *G. sepium*, a native of North America. *Gliricidia sepium* has been widely introduced into South America in spite of the existence on this continent of a close native relative that can serve similar purposes. Collections of *G. sepium* are few from Ecuador where *G. brenningii* is native, and this species probably remains relatively unknown. The native distribution of *G. brenningii* is tentatively considered to be Ecuador and adjacent northern Peru, even though nearly all collections come from cultivation, especially as part of living fence systems. Specimens indicated to be taken from native habitat are very few and are generally from trees in native secondary vegetation along streams.

6.2. Gliricidia maculata (H. B. K.) Steudel, Nom. Bot., ed. 2, 1: 688. 1840. *Robinia maculata* H. B. K., Nov. gen. sp. 6: 393. 1823. *Lonchocarpus maculatus* (H. B. K.) DC., Prodr. 2: 260. 1825. *Gliricidia sepium* f. *maculata* (H. B. K.) Urban, Symb. antill. 2: 289. 1900.—TYPE: MEXICO. Campeche: crecit prope Campeche, *Humboldt & Bonpland s.n.* (holotype: P-HBK, microfiche IDC 6209. 161:II.7!).

Single- to multistemmed trees, also shrubby, (1–) 2–13 m tall. Stems erect, 3–30 (–38) cm in diameter at base, with or without major branches from near base;

bark light brown to gray, minutely to not at all furrowed toward the base on older stems; branches glabrate to densely sericeous with brownish to whitish trichomes especially at the distal branch ends. The larger leaves 13–29 cm long; leaflets 7–15 (–19) per leaf, coriaceous, rarely distally accrescent, elliptic to broadly elliptic, rarely narrowly elliptic, rounded, rarely acuminate, the larger 30–75 mm long, 16–48 mm wide, (at maturity) adaxial surface glabrous, usually without dark tanniniferous patches, abaxial surface glabrate to strigose, usually with purplish tanniniferous patches evenly scattered over the leaflet lamina; stipules triangular, 0.5–2.0 mm long, readily caducous; stipels absent. Inflorescence rachises (4–) 6–23 cm long, pendent, subsessile to pedicellate, the stipe 1–35 mm long, sparsely strigose to sericeous or tomentose, often sparsely to densely granuliferous, with 20–90 nodes, internodes 0.1–4.0 mm long; floral bracts 0.5–1.3 mm long, triangular to ligulate, tomentose usually over the entire abaxial surface, generally persistent; pedicels 4–12 mm long at anthesis, strigose to usually sericeous, often granuliferous. Calyx campanulate, distinctly attenuate at base for 0.5–2.0 mm, strigose to sericeous, granuliferous, tube (including the long-attenuate base) 5.0–6.5 (–7.0) mm long, the five lobes 0.1–0.2 mm long, more-or-less evenly spaced. Standard predominantly whitish or occasionally predominantly pinkish, aging mostly a whitish mottled-brown or faint purple, nectar-guide a large central yellowish patch, blade reflexed about 180° at anthesis, 18–23 (–27) mm long, 18–23 (–27) mm wide, orbicular, emarginate; wing petals 18–23 (–27) mm long, perpendicular to the plane of the keel, usually whitish; keel petals 17–21 (–25) mm long, adnate for much of the distal end, straight, with a blunt tip, usually whitish but with distal ends tinged pinkish. Style 4–6 mm long at anthesis; ovary glabrous, with 9–12 ovules, gynoecial disk absent. Legume (6–) 8–17 (–18.5) cm long, 13–16 (–21) mm wide, strongly laterally compressed, light to dark reddish brown, glabrous; seeds 3–8, lenticular, 5.5–9.0 mm in diameter, uniformly light brown, aging dark brown. Figs. 65–66.

Phenology. Flowering occurs primarily from November through the following March and occasionally into April, similar to the flowering period of the *G. sepium*, and may be dependent on local rains; mature fruits can persist throughout much of the year, February through November.

Distribution (Fig. 70). Native in Mexico, primarily in the Yucatan Peninsula but sporadically south into Chiapas, and also in Belize and adjacent Guatemala (Petén); seasonally deciduous forests and thorn scrub, occasionally in lowland rain forest (Chiapas: Laguna Miramar); 0–400 m (–950 m in Chiapas). Co-occurring predominant species in native regions include species of *Bursera*, *Caesalpinia*, *Cedrela*, *Chloroleucon*, *Cordia*, *Crescentia*, *Ficus*, *Guazuma*, *Haematoxylum*, *Heliocarpus*, *Ipomoea*, *Karwinskia*, *Leucaena*, *Lysiloma*, *Manilkara*, *Mimosa*, *Pachira*, *Piscidia*, *Senna*, *Tabebuia*, and *Talisia*.

Representative Specimens. **Mexico.** Campeche: Isla del Carmen, *Barlow 7/5C* (MEXU); 500 m N de Zpujil, rumbo a Zohlaguna, *Cabrera & H. de Cabrera 2151* (MEXU); Campo Experimental Forestal Tropical "El Tormento" Km 5 carr. Escárcega a Candelaria, *Chavelas P. ES195* (MEXU), *Chavelas P. ES844* (MEXU, MICH); Isla del Carmen, Mpio. Carmen, *Flores 9797* (MEXU); 70 km SW of Francisco Escárcega towards Villahermosa, *Hughes 678* (MEXU); 20 km S of El Quebrache on main rd from Villahermosa to Francisco Escárcega, *Hughes 944* (MEXU); 196 km NE of Villahermosa towards Francisco Escárcega, *Hughes 946* (MEXU); Tuxpeña, *Lundell 1272* (WIS), *Lundell 1356* (F, GH, MICH, MO, NY, UC, WIS); 36 km al NE de Campeche camino a Mérida, cerca de la torre de Microondas de San Antonio Bubola, Mpio. Tenabo, *Martínez S. et al. 3006* (MEXU, MO); Isla del Carmen, *Sauer 7/5* (GH, WIS).—Chiapas: La Cima, 15 km WNW of Ocozocoautla, *Breedlove 29005*

FIGS. 65–69. *Gliricidia maculata* and *G. sepium*. 65–66. *G. maculata*. 65. Flowering branch. 66. Leaf. 67–69. *G. sepium*. 67. Flowering branch. 68. Leaf. 69. Mature legume. (65, 67, bar = 3 cm; 66, 68, 69, bar = 4 cm).

FIG. 70. Native distribution of *Gliricidia maculata* in Mexico and Central America.

(MEXU, MO, NY); 1 km N of Ocozocoautla, *Breedlove & Thorne 30414* [specimen on the left] (MEXU); western side of Laguna Miramar, E of San Quintín, Mpio. Las Margaritas, *Breedlove 33292* (MEXU, MO, NY); en Campamento COFOLASA a 24 km al SE de Crucero Corozal, camino a Boca Lacantum, Mpio. Ocosingo, *Martínez S. 11343* (MEXU); en Ejido Roberto Bárrios a 60 km al S de Boca Lacantum, camino a Chajul, Mpio. Ocosingo, *Martínez S. 18361* (MEXU); near jct of Río Perlas and Río Jataté at San Quintín and near Laguna Miramar, *Sohns 1712* (MICH, UC).—Quintana Roo: Cobá, *Barrera et al. 450* (MEXU); 1 km S de Akumal, carr. Cancún–Tulum, *Cabrera & Ibarra 1152* (MEXU); 8 km S de San José, rumbo a Tomás Garrido, *Cabrera 1252* (MEXU); parque Chancanab, Isla de Cozumel, *Cabrera et al. 11171* (MEXU); 4 km N de San Miguel de Cozumel, *Cabrera & H. de Cabrera 12992* (MEXU); Pamul, 12 km S de playa del Carmen, Mpio. Carrillo Puerto, *Durán & Olmsted 798* (MEXU); 5 km W de Vigía Chico, Mpio. Carrillo Puerto, *Durán & Olmsted 814* (MEXU); 6 km S of Playa del Carmen, *Hughes 488* (MEXU); a few km N of Puerto Morelos, *Hughes 489* (MEXU); Cozumel, center of island, *Millspaugh 1553* (F); Tulum–Cobá, 10 km antes de Cobá, *Moreno 493* (MEXU); Mpio. Benito Juárez, crucero Puerto Morelos, *Palma G. 85–40* (MEXU); Isla de Cozumel, E de la laguna de Chancanab, *Pérez & Ramos 1577* (MEXU, MO); 7 km SW of Puerto Morelos, carr. a Playa del Carmen, *Sousa et al. 11053* (MEXU, MO); 22 km W de Puerto Morelos, *Sousa et al. 12331* (MEXU); 6 km W de Cancún, sobre carr. a Mérida, *Téllez & Cabrera 1320* (MEXU, MO); Zona Arqueológica de Cobá, *Téllez & Cabrera 1381* (MO); 18 km NE de F. Carrillo Puerto, hacia Vigía Chico, *Téllez & Cabrera 1452* (MEXU); 4 km E de Ucum, *Téllez & Cabrera 1733* (MO); 23 km S de Chiquila, *Téllez & Cabrera 1857* (MEXU, MO); 12 km S del Ingenio Alvaro Obregón, *Téllez &* Cabrera 1986 (MEXU); Puerto Morelos, *Uitzil et al. 01* (MEXU).—Tabasco: sobre la carretera no. 20 a 3 km de la W, Mpio. Tenosique, *Calzada & Arellano 2225* (F, MEXU, TEX); Mpio. Tenosique, Km 24 carr. de Tenosique hacia Nuevo México, *Magaña & Zamudio 774* (MEXU).—Yucatán: Yokdzonot Presentados, 60 km del Mpio. Tizimín, *Aguilar et al. 193* (MEXU); Rancho San Mateo, 5 km E de Ticacalteyub, Mpio. Yaxcabá, *Arias & Vara 232* (MEXU); Chankom, *Bequaert 85* (GH); 15 km N of Chunchucmil, *Darwin 2455* (MEXU, MO, NY); Hunucma, *Enriquez 298* (MEXU); Progreso, *Flores 5*

(F); Xanaba, *Gaumer 448* (WIS); Kancabonot, *Gaumer 23603* (F); Yucatán, *Gaumer 24108* (F); 2 km N of Piste close to rd towards Dzitas, *Hughes 500* (MEXU); 25 km N of Valladolid, *Hughes 939* (MEXU); 35 km W of Tizimin, *Hughes & Lewis 1281, 1282* (MEXU, MONT); Sisal, *Matuda et al. 38314* (MEXU); Hac. San José, *Seler & Seler 3883* (F, GH); Yucatán, *Steggerda 30C* (F); carr. Pixoy a Valladolid Km 2, Mpio. Valladolid, *Ucán E. 708* (F); Ejido de Chankom a Zanlah, Mpio. Chankom, *Ucán E. 725* (F); Zona Arquelógica de Chichen Itza rumbo al Cenote Sagrado, Mpio. Tinum, *Ucán E. 777* (F); vicinity of grottos above Calcehtok, *White & Mott 64* (F).

Belize. BELIZE: Maskall, *Arvigo et al. 377* (MEXU); Cornhouse Creek, *Bartlett 11305* (F, MICH); Salt Creek Estate, mile 18 on Northern hwy, *Dieckman 158* (TEX); 3 mi W of Belize, *Dwyer S–69* (MO), *Dwyer et al. 21* (MO); 7 mi W of Belize, *Dwyer et al. 23* (MO); 12.5 mi NW of Belize, Northern hwy, *Dwyer 10701* (MO); mile 11, Northern hwy, *Dwyer 10934* (MO); Maskall Pine Ridge, *Gentle 1138* (F, GH, MEXU, MO, NY, WIS); 1.5 mi E of Hattieville on the coastal pine savanna, *Huston 0564* (MO); Manatee Lagoon, *Peck 402* (GH, NY); along Western hwy at mile 26, *Proctor 30292* (TEX); 8 km W del Tropical Park, sobre la carretera a Burrel Boom de Hattieville, *Ramamoorthy et al. 3640* (MEXU).—COROZAL: Cerros Maya Ruins, Lowry's Bight, *Crane 289* (MO, TEX); vicinity of Little Belize, *Davidse & Brant 32585* (MEXU); 1931–1932, *Gentle 401* (F, MICH); 3 km N of Corozal towards Mexican border at Santa Elena, *Hughes 943* (MEXU).—EL CAYO: Vaca, *Gentle 2529* (F, GH, MEXU, MICH).—ORANGE WALK: Indian Church, *Arnason & Lambert 17162* (MO).—STANN CREEK: vicinity of Cabbage Haul, Fire Lookout, *Proctor 35764* (MO).—TOLEDO: lower part of Richardson Creek, affluent of Bladen Branch, Maya Mtns, *Davidse & Brant 31939* (MEXU); Punta Gorda, *Kinloch 56* (F). **Guatemala.** PETÉN: Santa Cruz, *Bartlett 12374A* (MICH); Tikal, on temple in "H" group of ruins, *Contreras 486* (F, MEXU, MO, TEX); Tikal National Park, 1.5 km on Uaxactun Rd, *Contreras 7671* (F, MEXU, MO, TEX); 15 km S of Tikal, N of Lago Petén Itza, *Hughes & Lewis 1276* (MEXU, MONT); La Libertad, *Lundell 2179* (MICH), *Lundell 2420* (MEXU, MICH), *Lundell 2852* (MICH), *Lundell 2885* (F, MICH); Tikal, bordering airfield, *Lundell 15709* (F, MEXU, MO, TEX), *Lundell 15710* (F, MEXU, TEX); orillando el camino para Remate, 60 km de Santa Elena, Parque Nacional Tikal, *Tún Ortiz 790* (F, MICH, US); 2 km de Santa Elena, *Tún Ortiz 888* (F, MO, NY); en orillando el camino para Pactún, Km 80, *Tún Ortiz 976* (F, TEX, UC); San Luis, *Ventur 213* (F).

The differences between *G. maculata* and *G. sepium* are many, though often subtle. The distribution of *G. maculata* is centered in the Yucatan Peninsula, including Belize and northern Guatemala; it also occurs occasionally southward in Chiapas, Mexico. Unlike *G. sepium*, populations of *G. maculata* often comprise scattered individuals often in undisturbed forests. Like *G. sepium*, *G. maculata* is brought into cultivation but only in this region, where it is used for living fence systems and in construction or occasionally as an ornamental. Notably, no evidence suggests a use of the leaves of this species for livestock fodder, and much label data on herbarium specimens state that the leaves are poisonous. No specimen assigned to *G. maculata* is known to have been brought into cultivation elsewhere in Central America or Mexico, let alone in the Caribbean, South America, or the Paleotropics. In contrast, the native range of *G. sepium* extends from Pacific coastal Mexico into Central America, and populations most often comprise dense stands in occasionally disturbed areas, such as arroyos, roadsides, and steep hillsides; most commonly, *G. sepium* is found in or escaped from cultivation.

There appear to be two absolute differences between *G. maculata* and *G. sepium* from Mexico and Central and South America. The former has leaflets that are coriaceous and flowers with a short-attenuate hypanthium that is strigose to sericeous, like the adjacent pedicel, with a density of 9–48 trichomes per 0.25 mm^2. In contrast, *G. sepium* has leaflets that are thinly textured and flowers with a long-attenuate hypanthium that is glabrous to sparsely strigose, like the adjacent pedicel, with a density of 0–3 trichomes per 0.25 mm^2. There are also other characters that show high frequency differences, which, when taken together, can lead to the successful determination of a given specimen. The following eight characters were scored in 189 herbarium specimens (i.e., individual collections) of *G. maculata* and *G. sepium*, and analyzed in a phenetic (average taxonomic distance) and cluster (UPGMA) analysis using NTSYS-pc (version 1.80; Rohlf 1993). The resulting phenogram (too large to be figured, but available from the authors upon request) had a cophenetic correlation coefficient of 0.69668. The result was that all specimens could be assigned unequivocally to either *G. maculata* or *G. sepium* based on a minimum of five of the following eight characters (e.g., when data was missing on a herbarium specimen).

1. Leaflet apices of mature leaves. One out of 92 herbarium specimens of *G. sepium* with mature leaflets showed rounded apices, whereas eight out of 83 herbarium specimen of *G. maculata* with mature leaflets showed acuminate apices. Thus, rounded leaflet apices are a fairly good indicator of *G. maculata*, but acuminate leaflet apices are a somewhat less accurate indicator of *G. sepium*. Leaflets of young expanding leaves of *G. sepium*, especially those first emerging during or after flowering, have a tendency to have rounded apices, whereas those of mature leaves seldom do; thus only mature leaves were scored in this study.

2. Leaflet texture. Leaflets of *G. sepium* are uniformly thin, whereas those of *G. maculata* are uniformly coriaceous. The thin leaflets of *G. sepium* are greenish and the finest (quaternary and finer) vascular branches are distinct on the adaxial surface; often, in dried herbarium specimens, the leaflets are cracked along these finest veins. In contrast, the adaxial leaflet surfaces of *G. maculata* are grayish green and very rough, and the fine veins are not distinct, nor do they direct cracks on upper leaflet surfaces of dried herbarium specimens.

3. Petal color. Two out of 97 herbarium specimen of *G. sepium* with flowers showed white petals, whereas 10 out of 63 herbarium specimen of *G. maculata* with flowers showed pink petals. Thus, whitish petals are a fairly good indicator of *G. maculata*, but pinkish petals do not distinguish as well *G. sepium* from *G. maculata*. The standard petals of dried herbarium specimens differ between the two species; in *G. maculata*, the back of the standard tends to be distinctly finely mottled, brownish red on a brownish white background, and that of *G. sepium* tends to be uniformly light to reddish brown, and mottling, if it is present, tends to be inconspicuous.

Though white petal color appears in a very high frequency in *G. maculata*, herbarium specimens taken from Petén, Guatemala, suggest that in *G. maculata*, the two color-morphs co-exist in a single populations (*Kinloch 56*, *Lundell 15709*, *15710*). Also, individuals assigned to *G. maculata* from Chiapas are all recorded to have flowers with pinkish petals, and the lowland rain forest populations at Laguna Miramar are recorded to be large trees that have the largest petals (up to 27 mm long). Though they are otherwise very similar to the Yucatan populations, they are so different from the main body of *G. maculata*, that they may represent populations that have been long isolated from the rest of the species.

4. Indumentum of the pedicel and adjacent calyx base. *Gliricidia sepium* has distinctly glabrous to sparsely strigose pedicels and adjacent calyx base. Trichomes in this region range in density from 0 to 3 trichomes per 0.25 mm^2 (mean = 0.7810, S.D. = 0.9804, N = 105). In contrast, trichome densities of these same structures in *G. maculata* range from 9 to 48 trichomes per 0.25 mm^2 (mean = 22.7403, S.D. = 8.0629, N = 77). These differences are statistically significant ($p < 0.001$) and also readily useful for distinguishing the two species. The basal cell of each trichome

tends to persist in *G. maculata*, often giving the pedicel and attenuate portion of the calyx tube a granuliferous appearance; rarely do such trichomes bases persist in *G. sepium*.

5. Length of the inflorescence rachis. *Gliricidia sepium* tends to have shorter inflorescences rachises, which range in length from 2.5 to 17 cm (mean = 8.0381, S.D. = 2.9529, N = 105). Those of *G. maculata* range from 5.0 to 23.0 cm (mean = 13.0658, S.D. = 4.0070, N = 79). Though the difference in the means are statistically significant ($p < 0.001$), overlap is often too great to be useful, and use of this character in diagnosis has to be in conjunction with other attributes. Actually, inflorescence length may underlie a more qualitative trait, for according to label data and personal observation, it appears that the flowering racemes of *G. maculata* are always pendent and those of *G. sepium* are always lax to ascending. This qualitative difference between these two species may be absolute between *G. maculata* and *G. sepium*, and rachis length may be only an incidental quantitative measure.

6. The length of the calyx tube in *G. sepium* ranges from 5.5 to 9.0 mm (mean = 7.0243, S.D. = 0.8783, N = 103), whereas that of *G. maculata* ranges from 5.0 to 7.5 mm (mean = 5.9867, S.D. = 0.5512, N = 75). Again, these differences are significant ($p < 0.001$), but only lengths greater than 7.5 mm can be used confidently to place a specimen into *G. sepium*, whereas lengths less than 7.5 mm necessitate the use of other traits to facilitate identification. One aspect of the calyx that renders it more useful for discrimination is that the generally longer calyx tube of *G. sepium* has a more pronounced attenuate base, (1–) 2–4 mm long; that of *G. maculata* is 0.5–2.0 mm long.

7. Leaflet number per leaf. This is a somewhat useful character for distinguishing these two species. In *G. sepium*, leaflet number per leaf ranges from 5 to 25 (mean = 15.3516, S.D. = 3.3940, N = 91), whereas in *G. maculata* it ranges from 7 to 19 (mean = 12.7971, S.D. = 2.3300, N = 69). Though the differences are significant ($p < 0.001$), this character, by itself, would be useful only in assigning a specimen with 21–25 leaflets per leaf to *G. sepium*.

8. Accrescence of leaflets. The leaflets of a given leaf in *G. sepium* are nearly always distally accrescent, whereas those of *G. maculata* are not. As a measure of this, the ratio of the length of the terminal leaflet to that of the most adjacent lateral leaflet in *G. sepium* ranges from 0.87 to 1.46 (mean = 1.0775, S.D. = 0.1114, N = 88). In contrast the same ratio in *G. maculata* ranges from 0.74 to 1.05 (mean = 0.9257, S.D. = 0.0678, N = 69). Although there is overlap in these ratios, the means are significantly different ($p < 0.001$).

Characters of the mature legume and seed were not used in the phenetic analysis, because they were not present on a high percentage of the herbarium specimens that were analyzed. The legume of *G. maculata* is generally narrower, 13–16 (–21) mm wide, than that of *G. sepium*, which is 14–22 mm wide. Also, the mature seed of *G. maculata* is correspondingly smaller, measuring 5.5–9.0 mm in diameter, whereas that of *G. sepium* is larger, 8.5–11.5 mm in diameter. Though not absolute, these characters combined with those above should lead to successful determination of a specimen.

The distinction of *G. maculata* from *G. sepium* is in line with recent work on population variation (C. Hughes and T. Simons, pers. comm.) and chloroplast DNA restriction site analysis (Lavin et al. 1991). Morphologically, the two species are subtly distinct, whereas genetically, they are very different. Though Simons (unpubl., see herbarium specimen *Simons 1* at FHO and MONT) has shown that these two species are interfertile, F_1 hybrids between *G. sepium* (the male parent) and *G. maculata* (the female parent) are relatively stunted. After about two years of growth, the hybrids attain an average height of less than 1 m; in contrast, non-hybrid lines attain heights of several meters during such a time period. The hybrids do produce flowers, and the petals are predominantly pink (mostly like *G. sepium*) but arranged on long pendent raceme rachises (like *G. maculata*). The leaves are most like those of *G. sepium* (thin and acuminate), whereas the indumentum on the pedicel and calyx is like that of *G. maculata* (strigose to sericeous, and granuliferous).

The degree of natural introgressive hybridization between *G. maculata* and *G. sepium* is uncertain, and it could be argued that such hybridization is the cause that so few fixed differences are found between these two species; however, retention of ancestral polymorphisms is an equally likely explanation. If gene flow via introgressive hybridization were occurring to even a small degree, it would probably not be possible to assign 100% of the herbarium specimens unequivocally into one or the other species, as has been possible in this analysis.

Around Ocozocoautla, Chiapas, Mexico, both *G. maculata* and *G. sepium* co-occur. One herbarium sheet, *Breedlove & Thorne 30414* (MEXU) contains a mixed collection of the two species. As noted, the petal color of *G. maculata* in Chiapas is uniformly pinkish. But even in this part of Chiapas, the thick-textured leaflets with rounded apex and not distally accrescent, and flowers on long racemes and with a calyx tube that has a very short attenuate base distinctly covered with hairs readily separates these *G. maculata* specimens from *G. sepium*. All collections of *G. maculata* from Chiapas come from native forest habitat, whereas more than half of those of *G. sepium* come from weedy roadside or locally introduced populations (as inferred from label data).

At the least, if a given specimen of *Gliricidia* from North America consists of just flowering material (i.e., without leaves) and the flowers are pinkish, it can be assigned to *G. maculata* if all flowers are borne along a raceme rachis that is commonly at least 10 cm long, and if the calyx tube is short, usually less than 7 mm (with the attenuate calyx base strigose to sericeous and usually less than 2 mm long).

6.3. Gliricidia sepium (Jacquin) Steudel, Nom. Bot., ed. 2, 1: 688. 1841. *Robinia sepium* Jacquin, Enum. syst. pl. 28. 1760 [Select. stirp. amer. hist. 211, t. 179, f. 101. 1763.]. *Lonchocarpus sepium* (Jacquin) DC., Prodr. 2: 260. 1825.—TYPE: COLOMBIA. Cartagena, *N. J. Jacquin s.n.* (holotype: not located).

Robinia rosea Miller, Gard. dict., ed. 8, no. 4. 1768. *Lonchocarpus roseus* (Miller) DC., Prodr. 2: 260. 1825.—TYPE: MEXICO. Campeche, in Herbarium Sloane 292, f. 64, *Houstoun s.n.* (holotype: BM, photos: BM! MONT!).

Robinia variegata Schlechtendal, Linnaea 12: 301. 1838.—TYPE: MEXICO. Veracruz: Actopan, Apr, *Schiede & Deppe s.n.* (holotype: HAL!, photo: MONT!; isotype: HAL!).

Millettia luzonensis A. Gray, U.S. Expl. Exped., Phan. 1: 456. 1854.—TYPE: THE PHILIPPINES. Luzon, province of Laguna, Los Baños, 1838–1842, *Wilkes s.n.* (holotype: US!).

Gliricidia maculata var. *multijuga* Micheli in J. D. Smith, Bot. Gaz. (Crawfordsville) 20: 284. 1895.—TYPE: GUATEMALA. Santa Rosa: Santa Rosa, 3000 ft, Apr 1892, *Heyde & Lux 3296* (holotype: US!).

Gliricidia lambii Fernald, Bot. Gaz. (Crawfordsville) 20: 533. 1895.—TYPE: MEXICO. Sinaloa: Río Rosario, Rosario, 12 Jan 1895, *Lamb 451* (holotype: GH!; isotypes: NY! DS! UC!, photo of DS isotype: MEXU!).

Single- to multistemmed trees, rarely shrubby, 2–15 m tall. Stems erect, 5–30 cm in diameter at base, with or without major branches from near base; bark grayish brown to whitish, minutely to deeply furrowed toward the base on older stems; branches glabrate to sparsely or densely sericeous with mostly whitish, sometimes brownish, trichomes especially at the distal branch ends. The larger leaves (15–) 19–30 (–35) cm long; leaflets (7–) 13–21 (–25) per leaf, thin, slightly to conspicuously distally accrescent, narrowly elliptic to elliptic, rarely broadly elliptic, acuminate to sometimes cuspidate, rarely rounded, the larger 44–83 mm long, (17–) 21–42 (–48) mm wide, (at maturity) adaxial surface glabrous to strigose, usually without dark tanniniferous patches, abaxial surface glabrate to strigose, usually with purplish tanniniferous patches concentrated toward the center of the leaflet lamina or scattered evenly throughout; stipules triangular, 0.5–2.0 mm long, readily caducous; stipels absent. Inflorescence rachises 2.5–15.0 (–17) cm long, ascending to lax, subsessile, the stipe 1–8 mm long, glabrate to sparsely strigose, rarely tomentose or granuliferous, with 30–100 nodes, internodes 0.1–2.0 (–4) mm long; floral bracts 0.8–1.2 (–2.0) mm long, triangular, rarely ligulate, tomentose usually over the entire abaxial surface, or at least on the distal portion, generally persistent; pedicels 5–11 (–15) mm long at anthesis, glabrous to sometimes sparsely strigose, rarely granuliferous. Calyx campanulate, distinctly attenuate at base for (1–) 2–4 mm, glabrous to sparsely strigose, rarely granuliferous, tube (including the long-attenuate base) (5.5–) 6–9 mm long, the five lobes 0.1–0.2 mm long, more-or-less evenly spaced. Standard light-pink to pinkish, rarely entirely whitish, aging mostly a whitish discolored brown or faint purple, nectar-guide a broad light to deep yellowish patch, blade reflexed about 180° at anthesis, 15–23 mm long, 15–23 mm wide, orbicular, emarginate; wing petals 15–23 mm long, perpendicular to the plane of the keel, usually pinkish, rarely entirely whitish; keel petals 15–22 mm long, adnate for much of the distal end, straight, with a blunt tip, usually pinkish, rarely entirely whitish. Style 4–5 mm long at anthesis; ovary glabrous, with 7–12 ovules, gynoecial disk absent. Legume 10–17 (–23) cm long, 14–22 mm wide, strongly laterally compressed, light to dark reddish brown, glabrous; seeds 3–10, lenticular, 8.5–11.5 mm in diameter, uniformly light brown, aging dark brown. Figs. 67–69, 71–82.

Phenology. Both in native and introduced populations, flowering potentially occurs anytime of the year; however, in native populations (Pacific coastal Mexico and Central America) the predominant flowering season, depending on local rains, is from November through March; in a population of full-grown trees, mature fruits can persist throughout the year, even in native stands.

Distribution (Fig. 83). Native to much of the southern two-thirds of Mexico and to Central America, especially along the Pacific slope, but introduced widely in tropical regions throughout the world; seasonally deciduous forests and thorn scrub, usually in lowlands, and commonly as part of living fence systems or cultivated for

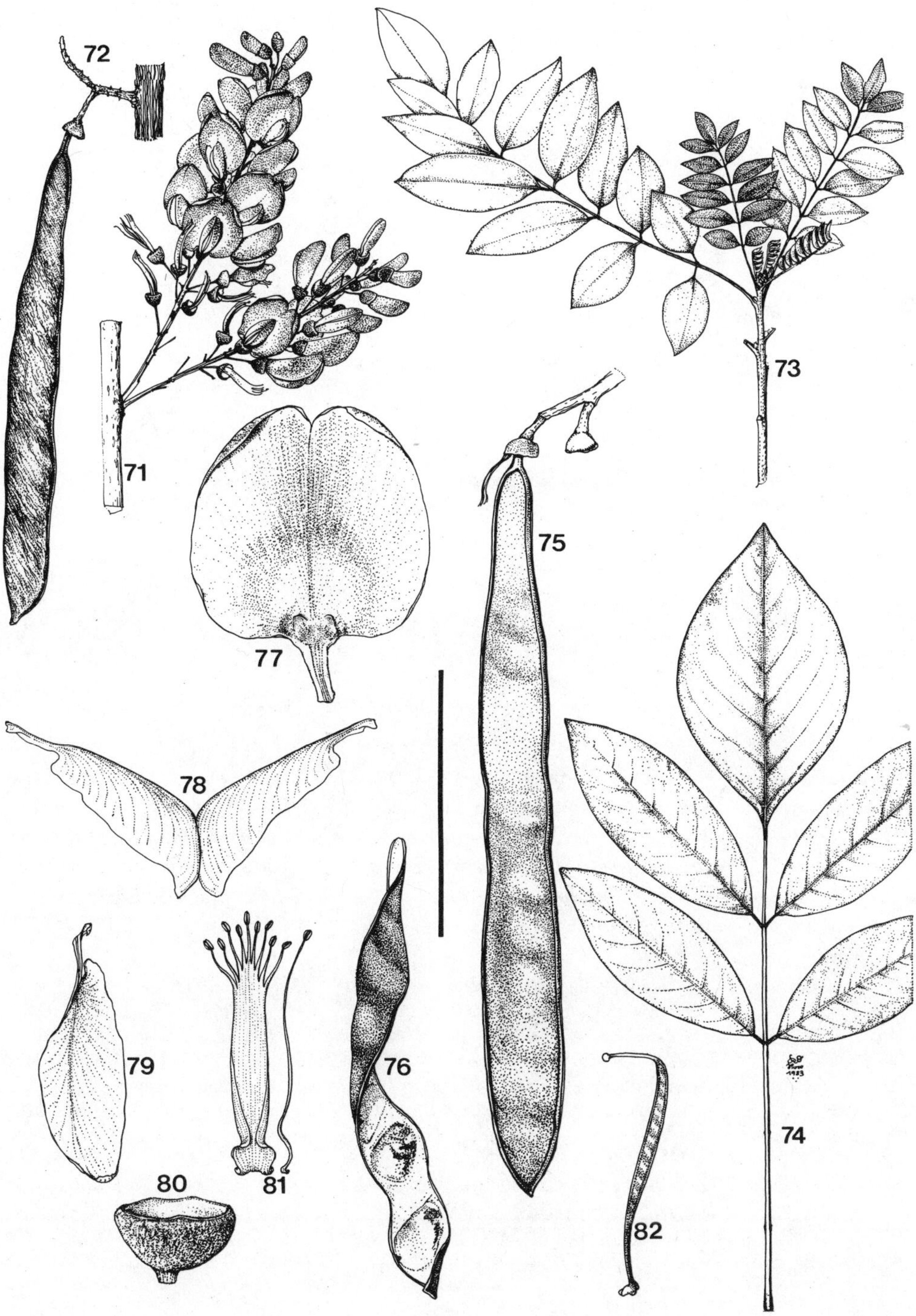

FIGS. 71–82. *Gliricidia sepium*. 71. Flowering branch. 72. Infructescence. 73. Leafy branch. 74. Leaf. 75. Mature legume. 76. Dehisced legume valve. 77. Standard petal. 78. Keel petals. 79. Wing petal. 80. Calyx. 81. Androecium. 82. Gynoecium. Reproduced from vol. 5 of *Flora Novo-Galiciana* by Rogers McVaugh. (71, 72, 76, bar = 8 cm; 73, bar = 10 cm; 74, 75, bar = 6 cm; 77–82, bar = 20 mm).

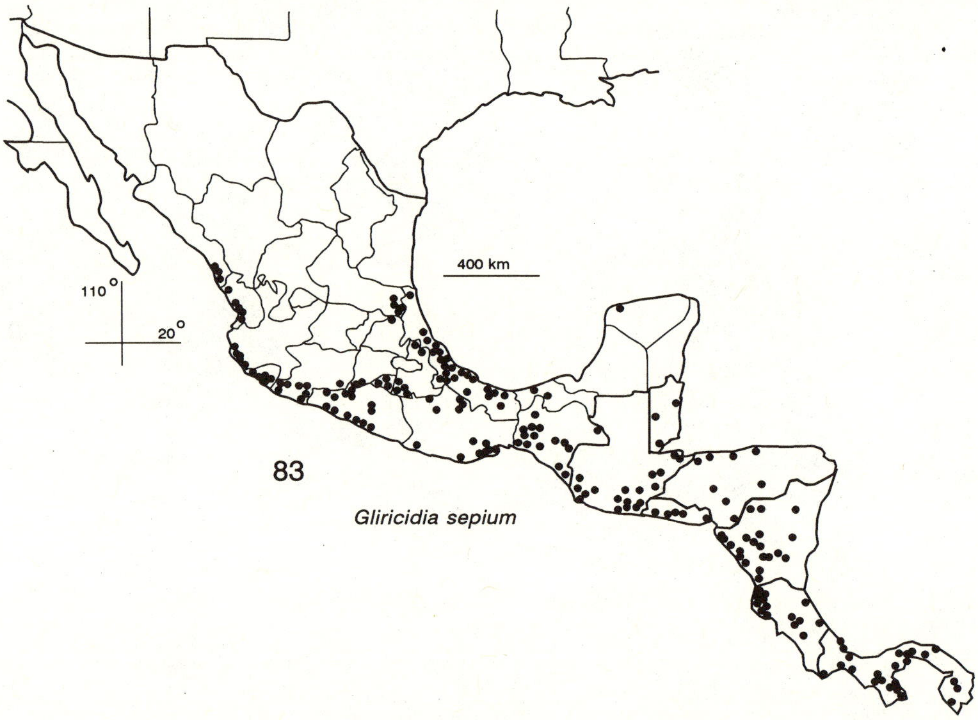

FIG. 83. Probable native distribution of *Gliricidia sepium* in Mexico and Central America.

livestock fodder or as shade over crops; 0–1200 (–1500) m. Co-occurring predominant genera in native regions include *Acacia*, *Bauhinia*, *Bursera*, *Brosimum*, *Byrsonima*, *Caesalpinia*, *Cedrela*, *Cephalocereus*, *Cochlospermum*, *Combretum*, *Comocladia*, *Cordia*, *Crataeva*, *Cyrtocarpa*, *Dalbergia*, *Guazuma*, *Haematoxylum*, *Hura*, *Ipomoea*, *Juliania*, *Lonchocarpus* , *Lysiloma*, *Poulsenia*, *Pseudosmodingium*, *Pterocarpus*, *Rhacoma*, *Senna*, *Solanum*, *Tabebuia*, and *Trichilia*.

Representative Specimens. **Cuba.** Camagüey: Camagüey, *Shafer 1180* (NY).—Cienfuegos: Cieneguita, *Combs 736* (F, GH, NY, US); Soledad, *Jack s.n.* (MICH); Limones, *Singleton 50* (GH), *Singleton 67* (GH).—La Habana: Santiago de las Vegas, *Abarea 4321* (NY); Playa de Mariano, *Britton & Wilson 4528* (NY); Santiago de las Vegas, *Cook 4* (NY), *van Hermann 525* (F, NY); Buena Vista, *Shafer 546* (NY).—Holguín: Nicaro, *Lavin 7148* (MONT); valley of Río Matamoros S of Holguín, *Shafer 1397* (NY, US).—Isla de la Juventud [Isla de Pinos]: Nueva Gerona, *Curtiss 437* (F, GH, MO, NY); E of Presedio, *Killip 44537* (US).—Pinar del Río: Cayajabos, near Artemisa, *Ames & Leavitt s.n.* (GH); Rangel, *Lavin 7147* (MONT).—Sancti Spíritus: Sancti Spíritus, *Shafer 12123* (NY).—Santiago de Cuba: Santiago, *Hamilton 114* (NY); Santa Ana, *Hamilton 126* (NY); Santiago, *Havard 81* (NY), *Havard 154* (NY); Santiago, San Juan Hill, *Millspaugh 1055* (F); Santiago, *Underwood & Earle 97* (NY). **Jamaica.** Roaring River, *Britton 2477* (NY); Hanover Parish, Middlesex, *Clarkson & Kress 75–269* (GH); St. Andrew Parish, Mona Hotel, *Cooley & Proctor 10982* (GH); St. Ann Parish, Middlesex County, *Hunnewell 11007* (GH); St. Mary's Parish, Gray's Inn, *Orcutt 4290* (UC); Clarendon Parish, near Kellits, *Proctor 27778* (MICH, TEX); near Kingston, *Rehder s.n.* (GH); 3 mi W of Port Antonio, *Wight 191* (F, NY); St. Ann Parish, 8 mi W of Ocho Rios, *Yunker 17775* (F, MICH, NY). **Haiti.** Nord: vicinity of Pilate, *Leonard 9694* (F).—Nord-Ouest: Massif du Nord, Port-de-Paix, *Ekman 9846* (GH).—Ouest: Port au Prince, *Holdridge 1105* (MICH, NY); vicinity of Port au Prince, *Leonard 2812* (NY). **Dominican Republic.** La Altagracia: Higüey, *Ekman 12762* (NY); 21 km E of Higüey, *Zanoni* *et al. 18956* (NY).—Distrito Nacional: Ciudad Trujillo, *Allard 18936* (US); Jardín Botánico, Sto. Domingo, *Liogier 21030* (NY); Santo Domingo, *Millspaugh 783* (F).—La Romana: La Uvita, S of Cumayasa, *Mejía & Zanoni 9371* (NY).—San Cristóbal: 7.7 km N of San Cristóbal, *Lavin & Zanoni 8025* (MONT); 14 km N of San Cristóbal, *Mejía et al. 10350* (NY).—San Juan: 12 km W of San Juan, *Lavin et al. 8029* (MONT).—Santiago Rodríguez: Monción, *Valeur 341* (MICH), *Valeur 596* (F, MICH, MO, UC). **Puerto Rico.** San Juan, *Britton & Britton 7961* (NY); near Cerro La Badera, *Duke 7323* (MO); 5 km from Bargain Town, *Evans 54* (GH); S of Sabana Grande near Machucha, *Lavin & Lavin 8013* (MONT); Km 8.5 N of Coamo, *Lavin & Lavin 8016* (MONT); Experiment Station, U.S.D.A., Mayagüez, *Moore 3672* (GH); Río Piedras, *Otero 381* (MO); Río Piedras, Station Grounds, *Otero M–83* (MICH); Hotel El Conquistador, Fajardo, *Wagner 514* (GH). **St. Croix.** La Grange rd, Frederiksted, *Hewit 406* (MICH); Tamarind Reef Hotel grounds, *Nelson s.n.* (MO); Areria's Hope, *Thompson 644*, *1093* (GH). **St. John.** Cannel Bay, *Nelson s.n.* (MO). **Tortola.** Pasea Estate, *D'Arcy s.n.* (MO). **Dominica.** Clarke Hall, Layou River Valley, *Chambers 2705* (GH, MEXU); Bleuheiu, *Cooper 127* (NY); Imperial Rd segment from Mt. Joy to abandoned L'Imprevue Estate, *Hodge & Hodge 1255* (GH); Mahaut, *Hodge 1308* (GH); N of Prince Rupert Bay, West Cabri, *Smith 10326* (NY). **Martinique.** 1 km W of le Diamant, *Kimber 1328* (WIS); Trois Ilets, Pointe du Bout, *Simon 80–447* (WIS); Trois Ilets, *Stehlé 2942* (NY). **St. Lucia.** Along pathways and fencerows, *Benz & Pierre 1157* (WIS); Summersdale, Casteries, *Pierre 72* (GH). **St. Vincent.** Waldfragment, *Gilli s.n.* (W). **Grenadine Islands.** Port Elizabeth, Bequia, *Sauer & Sauer 4269* (WIS). **Aruba.** Near Seroe Pretoe, *Stoffers 1682* (GH). **Curaçao.** Groot Kwarkien, *Arnoldo-Broeders 3879* (GH); Pescadero Bay, *Britton & Shafer 3132* (NY); Hofie Abau, *Curran & Haman 182* (GH, NY). **Tobago.** Botanic Station, *Broadway s.n.* (F, NY); Scarborough, *Broadway s.n.* (F, GH, MO, NY), *Broadway 2433* (F, MO). **Trinidad.** Saddle Rd, *Britton & Hazen 165* (GH, NY); Port of Spain, Harris Square, *Broadway 5552* (MO, UC); St. Clair, *Broadway s.n.* (MO, UC); Port of Spain, Harris Square, *Broadway 6294* (F, MO, UC); Aripa Savannah at edge of abandoned Waller Air Force Base, *Harriman 17668* (NY); near Arima, *Howard 10388* (GH).

U.S.A. Florida: Dade Co., Coconut Grove, *Moldenke 5808* (NY); Monroe Co., Big Coppitt Key, *Killip 43405* (F); Monroe Co., Georgia St., Key West, *Isely 7836* (GH). **Mexico.** Chiapas: 1 km N of Ocozocoautla, *Breedlove 23892* (MEXU, MO); 10–12 km SW of Colonia Agronomos Mexicanos, Mpio. Villa Corzo, *Breedlove 23925* (MEXU, MO); 6.5 km W of Tuxtla Gutiérrez along Hwy 190, *Breedlove & McClintock 23453* (MEXU, NY); 3 km N of Ocozocoautla, *Breedlove & McClintock 23576* (MEXU); 15 mi S of La Trinitaria, *Breedlove & Raven 8435* (F, WIS); 1 km N of Ocozocoautla, *Breedlove & Thorne 30414* [specimen on the right] (MEXU); about 22 mi SE of Comitán, *Carlson 1936* (F); E from Cintalapa de Figueroa towards Ocozocoautla, *Hughes 1678* (MONT); 12 mi S of Tonala, *Johnson 206–68* (MO); Escuintla, *Matuda 15* (GH, MEXU, MICH); Centro Arqueológico Bonampak, Mpio. Ocosingo, *Meave et al. B–245* (MEXU); Finca Exp. Tapachula, *Ross s.n.* (UC); along Hwy 195 near Monte Bonito, Mpio. Arriaga, *Shilom Ton 3382* (MEXU, NY); along rd from Acala to Pugiltik, Mpio. Venustiano Carranza, *Shilom Ton 3715* (NY, WIS); along Río Cuztepeques near Finca Cuztepeques, Mpio. Angel Albino Corzo, *Shilom Ton 3899* (TEX, WIS); Copainalá, *Wonderly 6* (MICH).—Colima: 4 mi S of Cihautlan, *Clarke et al. 2029–1* (MICH); 20 mi S of Colima on main hwy, *Alava & Cook 1694* (MICH, UC); between Cuyutlán and Colima, *Gilly et al. 33* (MICH, NY); 12 mi S of Colima, *Johnson 387–93*, *397–73* (MO); 11 mi SSW of Colima on Manzanillo rd, *McVaugh & Koelz 1050* (MEXU, MICH); 19–21 km SSW of Colima, *McVaugh 26186* (MICH); 11 mi SSW of Colima, *McVaugh 22944* (MICH); Armería, *Palmer 1279* (GH, MO, NY, UC); 5–10 mi N of Tecomán along hwy, *Thompson & Fields 329* (TEX).—Guerrero: 21.2 mi N of Chilpancingo, *Anderson & Laskowski 4470* (GH, MICH, NY); Acapulco, *Andrews 580* (NY); Playa Hornita, near Acapulco, *Barkley 14141* (MO); orillas de la carr. Zihuatenejo–Ixtla, *Germán et al. 218* (MEXU, MO, NY, UC); Rosario, Dtto. Mina (Coyuca), *Hinton 5869* (F, GH); Vallecitos, Dtto. Montes de Oca, *Hinton 10230* (GH, NY); Placeres, Dtto. Mina, *Hinton 11246* (GH); Icacos, Acapulco, *Huerta s.n.* (MICH); San Felipe, *Langlassé 710* (GH); 24 mi W of Tecpan, *Lavin et al. 5355a* (MONT); 1 km N de El Aguacate, *Martínez S. & Morales 3390* (MEXU); 3 km al E de Guayameo, Mpio. Zirandaro, *Martínez S. & Soto N. 3606* (MEXU, MO); 35 km NNW of Chilpancingo, *McVaugh 22184* (MICH); Temisco, Dtto. Adama, N of Río Balsas, *Mexia 8888* (F, GH, MO, NY, UC); Acapulco and vicinity, *Palmer 360* (F, GH, MO, NY, UC); Pié del La Cuesta, carr. a Zihuatanejo, *Pennington & Sarukhán K. 9464* (GH, MEXU, NY).—Hidalgo: Taman, *Ahshapanek 186* (TEX); 64 km NE de Jacala, carr. a Tamazunchale, 10 km SW del limite con SLP, *Zárate 318* (MEXU).—Jalisco: 10 km S of Chamela, *Hughes 622* (MEXU); 3 km NE of Nacastillo, *Iltis & Nee 1549* (MEXU, MICH, WIS); 18 mi N of Barra de Navidad, *Johnson 23–73* (MO); Chamela, Estación de Investigación, *Magallanes 955* (MEXU, MO); 9–11 km E of Chamela, *McVaugh 25176* (MEXU, MICH, NY);

Chamela, *Pérez J. 114* (MO).—México: San Lucas del Maíz, *Hinton 3331* (MEXU, MO); San Antonio, Tlatlaya, *Matuda et al. 27960* (MEXU, NY).—Michoacán: Hacienda Coahuayula, *Emrick 99* (F); banks of Río Ostula, Mpio. Aquila, *Hill 11* (MICH); Huetamo–Mal Paso, Dtto. Huetamo, *Hinton 5713* (F, GH); Coalcomán, Dtto. Coalcomán, *Hinton 10230, 11246* (GH, MICH, NY); Huizontla, *Hinton 12581* (F, GH, MEXU, MICH, NY); Aquililla, *Hinton 15741* (NY); Palmitas, *Hinton 15922* (MICH, NY); Aquila, *Hinton 16234* (MICH, NY, TEX, UC), *Hinton 16288* (MICH, NY, UC), *Hinton 16289* (NY, TEX); 35 km W of Playa Azul, *Hughes 630* (MEXU); 66 mi S of Nueva Italia, *Johnson 639–79* (MO); 45–48 km S of Arteaga, 12–15 km N of Playa Azul, *McVaugh 22594* (MEXU, MICH).—Morelos: Oaxtepec, *Converse 33* (UC); entronque a Xochitepec, *Flores 1* (MEXU); 12 km S de Jolotepec, *González Q. s.n.* (MEXU); Yautepec, camino rumbo a Cuautla, *Palacios s.n.* (MICH).—Nayarit: along rd to Santiago, 0.5 mi W of hwy 15, *Alava & Cook 1519* (MICH); 1 mi S of Rosamorada, *Johnson 275–68* (MO); 43 mi SW of Compostela, *Johnson 465–73* (MO); Acaponeta, *Jones 23005* (NY, WIS); San Blas, *Lehto 24252* (NY); 13 mi SE of turn-off to San Blas, *Mahler & Thieret 5881* (NY); San Blas, *Mahler & Thieret 5886* (NY); 1 mi N of Mecatán, *McVaugh 14900* (MICH); 6–12 km NE of Miramar, *McVaugh 23568* (MICH); vicinity of Acaponeta, *Rose et al. 14291* (NY).—Oaxaca: Santiago Astata, Mpio. Tehuantepec, *Calzada & Elizondo 8486* (F, MEXU); Dtto. Juchitán, Vista Hermosa al N de La Ventosa, *Cedillo T. 545* (MEXU, NY); Estación San Antonio, Dtto. Teotitlán, *Conzatti 2116* (F, MICH); along hwy 185, 4 mi N of jct of Hwy 190 and 185, *Croat 39993* (MO); Salina Cruz, *Deam s.n.* (GH); Dtto. Juchitán, Mpio. Niltepec, Marita, *Delgado et al. 964* (F, MEXU, MO); La Soledad, *Ernst 2562* (MEXU, MICH); 5 km W of San Mateo del Mar about 20 km E of Salina Cruz, *Hughes 545* (MEXU); Ixtepec–Tlacotepec, *Kelly 967* (UC); La Ventosa, 7 km E of Salina Cruz, *King 1283* (TEX); beach at Salina Cruz, *King 2458* (MICH, TEX); Bahía Ventosa, *King 2466* (MICH, TEX); Chivela, *Mell 2245* (NY); between San Carlos and San Bartolo, *Nelson 2570* (NY); Salina Cruz, *Orcutt 3453* (F, GH, MO); Ejido Benito Juárez, Tuxtepec, *Pennington & Sarukhán K. 9438* (MEXU, NY); La Ventosa Beach, 6 mi E of Salina Cruz, *Reznicek & Gregory 299* (MICH, NY); base of Cerro El Palacio, Chiltepec, *Schultes & Reko 578* (GH); Yautepec, *Seler & Seler 1674* (GH); Cuicatlán, *Smith 453* (GH); 2 km S de Pinotepa Nacional, rumbo a la Lagartera y Barra de Oro, *Sousa et al. 5515* (MEXU, US); La Ventosa, Dtto. Juchitán, *Sousa et al. 9150* (MEXU, UC); 2 km W of Tuxtepec, *Téllez et al. 400* (MEXU, MO); 31 km N de Juchitán, carr. a Matias Romero, *Tenorio L. et al. 3430* (MEXU, MO); Dtto. Juchitán, 10.4 km S de El Barrio, carr. a Almoloya, Mpio. El Barrio, *Torres et al. 9755* (MEXU); Ubero, *Williams 9119, 9184* (F); Almoloya, Isthmus of Tehuantepec, *Williams 9776* (F).—Puebla: 41 km al N de Mecapalapa, *Basurto & Durán 562* (MEXU); Cuauhtapanaloyan, Cuetzalan, *Basurto & Patrón 53* (MEXU); Mpio. Jolalpan, 4 km SW de Huachinantla, *Guízar 1924* (MEXU); Los Amates, between Izúcar de Matamoros and Acatlán, *Hughes 700* (MEXU); 54 mi N of Huajuapan along Hwy 190, *Lavin et al. 4623* (MO, NY, TEX); Rancho Los Amates, 20 km SE of Izúcar de Matamoros, *McVaugh 22487* (MICH); 20 km SE of Izúcar de Matamoros, *McVaugh 22495* (MICH); NW of Acatlán, *Walker 74H06* (NY).—San Luis Potosí: Hwy 85, 5 mi N of Tamazunchale, *Ahshapanek 304* (TEX); Tancuime, Mpio. Aquismón, *Alcorn 2898* (TEX); 50 mi S of Valles, *Barkley 17M133* (F, TEX); Mpio. San Martín, Terrerias, *Calzada et al. 2577* (WIS); Tancanhuitz, *Rzedowski 7481* (MEXU, MICH).—Sinaloa: 8 mi SE of Escuinapa, *Elias 45912* (MO); 15 mi from Rosario on rd to Matatán, *Kimnach 718B* (MEXU); El Monte, Los Labrados, *Mexia 931* (GH, MO, NY, UC); Mpio. Mazatlán, Bellavista, *Ortega 6831* (F); cerco del Barro, al N de Agua Caliente de Zevada, *Pérez 4* (MEXU, TEX); Agua Caliente de Zevada, camino a Los Gatos, al N de Sinaloa de Leyba, *Shapiro 288* (MEXU, NY).—Tabasco: near Municipal Hospital, NW of Villahermosa, *Barlow 2/3* (GH, MICH, WIS); Km 10 de la carr. de la Villa Tecolutilla hacia el Campo Tupilco, Mpio. Comalcalco, *Magaña & Suárez 640* (MEXU).—Veracruz: between Veracruz and Alvarado, *Ahshapanek 660* (TEX); Rancho el Nanchal, 20 km NE de Alvarado frente de Playa Camaronera, *Calzada et al. 4394* (F); J. de Ferrer, *Castillo C. et al. 1923* (F); Ojo de Agua, Mpio. Espinal, *Cortés 145* (MEXU, MO); 40 mi N of Chachalaca, 10 mi S of Punta Delgado, *Davis & Kincaid 55–11* (TEX); 3 km SW de La Mancha, carr. Cardel–Nautla, Mpio. Actopan, *Dorantes 34* (F, GH); Plán del Río, Mpio. Emiliano Zapata, *Dorantes L. 479* (F, MEXU); falda de Cerro Azul, *Dorantes L. et al. 1118* (MEXU, F); Mpio. Emiliano Zapata, Km 370 between Cerro Gordo and Plán del Río, *Gilly 29* (MICH, NY); Mpio. La Antigua, between Puente Naciónal and José Cardel, *Gilly et al. 104* (MICH, NY); city of Veracruz, *Greenman 94* (F); camino a San Cristóbal, *Gutiérrez B.* (F); near Jalapa, *Halsted s.n.* (NY); cerca del Puente Nacional, Rinconada, *Hernández M. & Dorantes 1803* (F, MEXU, NY); 1 mi E of Zempoala, *Higgins 2575* (MICH); 80 km N of Veracruz, *Hughes 681* (MEXU); 56 mi E of Orizaba on Hwy 150, *Hunsaker s.n.* (TEX); Morro de la Mancha Biological Station, 25 km N of Cardel, *Judziewicz et al. 3481* (WIS); Ejido Lázaro Cárdenas, 35 km NE de la carr. Catemaco–Montepío, *Ibarra M. 487* (MO,

NY); 34 mi NW of Veracruz, *Johnson 81–68* (MO); 4 mi S of Acayucan, *Johnson 118–68* (MO); 1 mi E of Cuitláhuac, *King 2377* (MICH, TEX); 8 mi S of Martínez de la Torre, *Lavin & Luckow 7149* (MONT); 9 mi S of Veracruz, *Lloyd & Mueller 3679* (UC); Texas, Mpio. Cosamaloapan, *Martínez C. 1867* (F, GH, MEXU, TEX); Orizaba, *Müller 85* (NY); Río Tepango, 3 km SW of jct with Hwy 180 in Santiago Tuxtla, *Nee & Taylor 26474* (F); Río Papaloapan, 2 km NE of Tlacotalpan, *Nee & Taylor 26552* (F); 16 km SE of Cuitláhuac, Mpio. Cuichapa, *Nee & Taylor 29353* (F, NY); 4 km W of Tenenexpan, Mpio. Soledad de Doblado, *Nee & Taylor 29631* (F, NY); near Campamento La Laguna, Mpio. Hidalgotitlán, *Nee 29986* (F, NY); Lagunetas de Nistamalapa, *Nevling & Gómez-Pompa 883* (MEXU), *884* (GH); Laguna de la Mancha, Mpio. Actopan, *Novelo 355* (NY); adelante de Actopan, Km 43 carr. La Concha–Actopan, *Ortega O. 231* (F, MEXU); vicinity of Pánuco, *Palmer 360* (F, GH, MO, NY); Salto de Eyipantla, *Ponce C. 102* (F, MEXU); Isla Agaltepec, Catemaco, *Ponce C. & Alvarez del C. 294* (MEXU, TEX); Remudadero, *Purpus 8836* (GH, MO, NY, UC); 10 km ESE de Cuitláhuac, Km 373 Córdoba–Veracruz hwy, *Romero R. 80* (GH); Barra La Mancha, *Sauer & Gade 3060* (WIS); Las Cruces, rd to Acayucan, *Schubert & Rojas 1863* (GH); La Victoria, orilla de La Laguna Catemaco, *Sousa 2888* (MEXU, F); Playa Azul, N de La Laguna de Catemaco, *Sousa s.n.* (F); 21 mi S of Veracruz, *Taylor & Taylor 7352* (NY); Cerro Monte de Oro, *Vázquez Y. 621* (MEXU, F); camino a Atoyac, Tepeatlaxco, *Velázquez L. 277* (F, MEXU); El Carrizal, Mpio. Dos Rios, *Ventura A. 3224* (MICH, TEX); Campo Experimental de Hule, El Palmar, Zongolica, *Vera S. 2833* (MICH); Plán del Río, Mpio. Emiliano Zapata, *Villanueva G. 70* (NY, WIS), *Villanueva G. 79* (WIS), *Villanueva G. 86* (NY, WIS); Fortuño, Coatzacoalcos River, *Williams 9083* (F), *Williams 9119* (MICH); San Lorenzo Tenochtitlán, *Wing 46* (GH).—YUCATÁN: Progreso, *Flores s.n.* (F).

Belize. BELIZE: Belize City, Fabers Rd, *Dieckman 140* (MO); 1.5 mi W of city limits on Western hwy, *Liesner & Dwyer 1546* (MEXU, MO).—EL CAYO: El Cayo, *Bartlett 12955* (F, GH, MICH, NY, WIS); El Cayo and vicinity, *Chanek 113* (F, MICH); W of Spanish Lookout, *Dieckman 140–79* (MO).—TOLEDO: Punta Gorda, *Cosminsky 213* (F). **Guatemala**. CHIQUIMULA: 2–3 km E of Vado Hondo, *Hughes 437* (MEXU); eastern portions of Verapaz and Chiquimula, *Watson 495* (GH).—ESCUINTLA: Escuintla, *Ruano 868* (F), *Smith 2813* (GH); near San José, *Standley 64026* (F).—GUATEMALA: 21 mi SE of Guatemala, *Johnson 148–68* (MEXU, MO); Laguna Amatitlán, *Smith 2314* (GH).—IZABAL: Santo Tomás La Unión, *Marshall 81* (GH); Bay of Santo Tomás, between Escobas and Santo Tomás, *Steyermark 39219* (F).—JALAPA: La Laguna, Volcán Jumay and Jalapa, *Steyermark 32275* (F).—JUTIAPA: 6 km W of Jutiapa, *Iltis s.n.* (WIS); 20 mi SW of Jutiapa, *Johnson 147–68* (MO); near Jutiapa, *Standley 60498* (F); El Barril, E of Jutiapa, *Standley 75805* (F).—QUETZALTENANGO: between Coatepeque and turnoff to Colomba, *Croat & Hannon 63385* (MEXU, MO, NY).—RETALHULEU: 7 km W of Cuyotenango, *Hughes 430* (MEXU); Champerico, *Standley 66617* (F), *Standley 87562* (F).—SAN MARCOS: 13 mi SW of Tumbador, *White 5273* (MICH).—SANTA ROSA: vicinity of Chiquimulilla, *Standley 79274* (F); Volcán Tecuamburro, N of Chiquimulilla, *Steyermark 33130* (F, GH).—ZACAPA: Gualán, *Deam 292, 295* (GH, MICH, NY); 7 km W of El Lobo on Hwy 9, *Harmon & Fuentes 1844* (MO, TEX); vicinity of Río Hondo, *Steyermark 29372* (F). **El Salvador.** LA LIBERTAD: Finca Paraíso, S of Santa Tecla, *Carlson 110* (F, UC); 11 mi N of La Libertad, *Johnson 835–79* (MO).—SAN MIGUEL: 13 km S of Lake Olomega, *Tucker 862* (F, MICH, UC).—SAN SALVADOR: San Salvador, *Calderón 109* (GH, NY); vicinity of San Salvador, *Renson 72* (NY), *Standley 19263* (GH); from San Martín to Laguna de Ilopango, *Standley 22601* (GH); vicinity of San Salvador, *Standley 23133* (GH).—SAN VINCENTE: vicinity of San Vicente, *Standley & Padilla V. 3496* (F).—SONSONATE: vicinity of Acajutla, *Standley 21897* (GH).—LA UNIÓN: La Unión, *Beetle 26255* (UC). **Honduras.** ATLÁNTIDA: vicinity of Tela, *Standley 53630* (F, GH); plains near Ceiba, *Yunker et al. 8814* (GH, MICH, NY).—CHOLUTECA: vicinity of San Marcos de Colón, *Standley 15712* (F).—COLÓN: Jericho, 3.5 km E of Trujillo, *Saunders et al. 770* (F, MO, NY, TEX).—COMAYAGUA: Siguatepeque, *Edwards P557* (F, GH, UC); vicinity of Siguatepeque, *Standley & Chacón P. 6373* (F).—CORTÉS: San Pedro Sula, *Yale Forestry School 579762* (F).—MORAZÁN: Tegucigalpa, *Gillis 9054* (GH, TEX); Zamorano, *Valerio R. 2299* (F), *Williams & Molina R. 11283* (F, GH, MEXU, MICH); San Francisco, *Williams & Molina R. 12212* (F, GH, MO).—OLANCHO: vincinity of Juticalpa, *Standley 17685* (F). **Nicaragua.** BOACO: 1 km W of Boaco, *Atwood 1421* (MO); rd toward Camoapa, *Atwood 5403* (MO); 2.4 km N of Río Las Canas, *Stevens 5851* (MEXU, MO).—CARAZO: vicinity of Jinotepe, *Standley 8450* (F).—CHINANDEGA: Chinandega, *Baker 190* (GH, MICH, NY, UC); vicinity of Chichigalpa, *Standley 11439* (F).—CHONTALES: Hwy 7 between Boaco cutoff and Acoyapa, *Bunting & Licht 694* (F); between Santo Tomás and Villa Somozo (Zelaya) near limits of Deptos. Chontales and Zelaya, *Bunting & Licht 1122* (F); Hda. Corpus, W of Juigalpa, *Stevens 22872* (MO).—GRANADA: Isla Zapatera, Ensenada Los Chiqueros, entre Sta. María y Ladera, *Sandino 1835*

(MO); Isla Zapatera, camino entre La Habana y Cerro El Llano, *Sandino 1950* (MO).—JINOTEGA: al NE de Wiwilí, camino entre El Carmen y Wamblán, *Araquistain & Castro 1861* (MEXU, MO); al NE de Wiwilí, camino entre El Carmen y Wamblán, *Araquistain & Moreno 1499* (MEXU, MO); camino entre Wiwilí y El Carmen, *Araquistain & Moreno 1602* (MO).—LEÓN: Puerto Somoza, 0.6 km N of rd to Miramar, *Stevens 6227* (MEXU, MO).—MANAGUA: Península de Chiltepe, 9 km al SE de Mateare, *Araquistain & Moreno 1328* (MEXU, MO); NE del Volcán Mombacho, los caminos a Sta. Isabel y a Cutirre, *Moreno 2629* (MO); Entrada Silvio Mayorga (La Ceiba), Km 48 carr. a Masachapa, *Moreno 3504* (MO); El Ventarrón, camino a finca Las Cichillas, *Sandino 296* (MO); Escuela Nacional de Agricultura y Ganadería, 12 km E of Managua, *Seymour 2231* (GH, MO); 2.3 km from Hwy 12, Sierra de Managua, *Stevens 5306* (MEXU, NY); 5 km NNW of Hwy 12, Sierra de Mateare, *Stevens 6195* (MEXU, MO); Km 15 on carr. Panamericana Norte between Managua and Tipitapa, *Vincelli 109* (MEXU, MO).—MATAGALPA: Cerro Apante, *Moreno 130* (MO); lower W slope of Cerro Musun, Quebrada El Jobo, *Stevens 12019* (MEXU, MO); Mataguás to Río Blanco, 6.6 km SE of Río Paiwas bridge, *Stevens 12263* (MEXU, MO).—NUEVA SEGOVIA: Ciudad Antigua, Valle Rancho Grande, *Moreno 5796* (MO).—RIVAS: Hacienda La Flor, 18 km S de San Juan del Sur, *Araquistain & Moreno 1280* (MEXU, MO); Isla de Ometepe, Volcán Masearas, *Stevens 6622* (MO); Isla de Ometepe, E of Cuatro Esquinas, *Stevens 6645* (MO).—ZELAYA: vicinity of La Luz–Siuna, *Bunting & Licht 634* (F, GH, NY, US); San Antonio, *Englesing 134* (F); Mpio. Siuna, Río El Bambú abajo, *Ortiz 1686* (MO); 44 km W del Recreo, límites entre Zelaya y Chontales, *Sousa et al. 13012* (MEXU); vicinity of Wani including Río Uli, *Stevens 7240* (MEXU, MO); Cerro Waylawas, *Stevens 7430* (MO). **Costa Rica.** CARTAGO: Villa Quesada, *Carpenter 478* (F); along rd from Turrialba N to Pavones and Siquirres, *Lems 5056* (NY).—GUANACASTE: Finca La Pacífica, vicinity of Cañas, *Daubenmire 469* (F); 4 km E of Cañas, *Davidse & Pohl 1434* (MO, WIS); roadsides S of Liberia, *Dodge & Thomas 6263a* (MO); Port Parker, Salinas Bay, *Elmore E7* (MICH); 28 km N of Cañas, *Frankie 327* (F, MO); 5 mi N of Cañas, *Frankie 353* (F, MO); Finca La Pacífica, along Panamerican Hwy just NW of Cañas, *Gentry 324* (MO, WIS); Comelco Ranch, 7 km NW of Bagaces, *Heithaus 34* (MO); Finca Jenny, at S border of Santa Rosa National Park, *Hill 17619* (NY); Tamarindo, *Hunter 561* (WIS); Hda. Palo Verde, *Keeler 188* (UC); 5 km NW of Bagaces, *Keeler 169* (MO); Santa Rosa National Park, *Liesner 2311* (MO); between Manilkara forest and fork to Naranjo Beach, *Liesner 4296* (MO); 2 km S of Finca La Pacífica, *Mathias 114* (F); Comelco Property near Bagaces, *Opler 510* (F); Nicoya, *Pittier 13588* (GH); Playa del Coco, *Plowman 3012* (GH); Parque Nacional Santa Rosa, Llano Jicacal, *Sousa et al. 12689* (MEXU, MO); Playa El Hachal, en la Bahía de Santa Elena, *Sousa et al. 12729* (MEXU, MO); 4–6 km S of La Cruz, *Weston et al. 4348* (UC); 5 km S of La Cruz, *Williams & Williams 24510* (F, NY).—HEREDIA: Río Virilla, hwy to Heredia, *Allen 566* (F, MO); Heredia, *Brenes 6* (F).—LIMÓN: between Limón and Puerto Viejo, *Baker & Burger 166* (F); 29 km W of Tortuguero, *Davidson et al. 6816* (F).—PUNTARENAS: Punta Morales, *Gómez L. 9977* (F).—SAN JOSÉ: Guadelupe pr. San José, *Porsch 742* (W); E of San Isidro del General, *Stork 4560* (MICH, UC). **Panama.** BOCAS DEL TORO: Almirante, along rd to the Bomba, *Blum 1349* (MO); region of Almirante, *Cooper 528* (F); Bocas del Toro, *Gordon 78* (MO); Bocas del Toro Island Airport, *Lewis et al. 779* (MO); Isla Colón, *von Wedel 505* (GH, MO); vicinity of Chiriquí Lagoon, *von Wedel 1198* (GH, MO); vicinity of Chiriquí Lagoon, Water Valley, *von Wedel 2747* (GH, MO).—CANAL ZONE: near mouth of Río Chagres, *Allen 871* (F, GH, MO); near Paraíso, *Croat 13953* (MO, UC); vicinity of Ancón, *Piper 5133* (US); along shore of Gatún Lake, *Schmalzel 481* (MO); summit, rd passing Navy Radio Station, *Stern & Chambers 157* (MO, NY); Fort Sherman near point, *Tyson & Blum 3781* (MO); summit, *Woodworth & Vestal 477* (F, GH).—CHIRIQUÍ: Quebrada Mérida, 6 km S of center of Puerto Armuelles, *Busey 697* (MO); 1.6 mi W of Puerto Armuelles, *Croat 21927* (MO, NY, TEX, WIS); from Boquete to 3 mi N, *Lewis et al. 393* (MO, UC); 2 km NE of Remedios, *Nee 10124* (MO, WIS).—COCLÉ: Penonome and vicinity, *Williams 341* (NY).—DARIÉN: Patiño, *Duke 10526* (NY); vicinity of El Real, *Stern et al. 289* (GH, MO, UC); 3 km SW of Jaqué, *Sytsma & D'Arcy 3466* (MEXU, MO).—HERRERA: 4 mi S of Los Pozos, *Tyson 2701* (MO); París de Parita, *Carrasquilla 173* (MO); 5 mi S of Pese, *Tyson et al. 2886* (MO); outskirts of Chitré, *Stern et al. 1230* (MO).—PANAMÁ: roadside near Panamá Viejo, *Dodge 17517* (GH, MO); Taboga Island, *Dwyer 2816* (MO), *Dwyer 2872* (MO), *Dwyer 4467* (F, MO); between Chepo and Río Bayano, *Porter et al. 5176* (MO).—SAN BLAS: Comarca, Soskatupu, *Duke 10179* (GH, MO).—LOS SANTOS: southern tip of the Azuero Peninsula, 3 km E of Pedasí, *Hughes 790* (MEXU); El Ejido de Los Santos, *Lao 88* (MO); rd from Macaracas to Chitre hwy, *Tyson 3151* (MO).—VERAGUAS: vicinity of Santiago, *Allen 1077* (MO); vicinity of Sona, *Allen 4261* (MO); Cañazas, *Tyson 3608* (MO), *Tyson 3626* (MO); 2 mi S of Cañazas, *Tyson 3723* (MO).

Colombia. ANTIOQUIA: between Punta Las Barcas and Turbo, *Duke 9744* (NY); 1 km W of Turbo,

Feddema 2039 (MICH); 2 km W of Bolombolo, Cauca Valley, *Gentry & Rentería 23688* (MO); vicinity of Medellín, *Toro 240* (NY).—ATLÁNTICO: along rd from Malambo to Sabanagrande, *Dugand 282* (F); Barranquilla, *Dugand 5257* (W); cerca de Barranquilla, *Dugand 5611* (W).—BOLÍVAR: 25 km NE of Cartagena, *Hughes 781* (MEXU); Cartagena, *Schotts 13* (NY).—BOYACÁ: 130 mi N of Bogota, El Humbo, Guadua Negra, *Lawrance 638* (F, GH, MO, UC).—CALDAS: 14–21 km N of La Dorada on rd to San Miguel, *Gentry et al. 18156* (MO); Quebrada Yeguas, 20 km N of Honda, *Gentry et al. 18191* (MO).—CHOCÓ: Mpio. Riosucio, Parque Natural Nacional Los Katyos, Río Peye, *León 24* (MO).—CUNDINAMARCA: cercanías de Arbeláez, *Cuatrecasas 6507* (F); 3 km NW of Guaduas, *Gentry et al. 18059* (F, MO, NY); Girardot, roadside to Melgar, *Soejarto 364* (GH).—HUILA: about 5 km N of Villavieja, Río Magdalena, *Mason 13803* (GH, MO, UC).—LA GUAJIRA: Barrancas, at Puerto Arturo on the Río Rancheria, *Bunch et al. 0487* (MO).—MAGDALENA: Los Cocos, *Daniel 353* (WIS); Santa Marta, *Espina 14* (F), *Smith 15* (F, GH, MICH, NY, UC, WIS).—SANTANDER: Mesa de los Santos, *Killip & Smith 15437* (GH, NY); 2 km along rd W of Quebrada Mensuli, 1 km S of Floridablanca, *Langenheim 3278* (UC).—TOLIMA: Espinal, *Pennell & Rusby 176* (NY).—VALLE DEL CAUCA: Mpio. Tuluá, Mateguadua, Jardín Botánico, *Cuatrecasas 27492* (NY); Cali, *Forero & Hernández 1609* (MO); Cauca Valley, near Yotoco, N of Cali, *Gentry et al. 54072* (MO); Río Meléndez, charco El Aguacate, *Willard 7* (F). **Ecuador.** GUAYAS: 2–4 km W of Bucay, *Gentry 12306* (MO).—LOS RIOS: 1 km W from Los Angeles, *Dodson & Gentry 6480* (MO). **Venezuela.** ARAGUA: between Caracas and Puerto Cabello, *Rose 21843* (GH).—CARABOBO: 8 km W of Mariaria on rd to Valencia, *Hughes 779* (MEXU); vicinity of Valencia, *Pittier 8710* (GH, NY).—DISTRITO FEDERAL: Caracas Botanical Garden, *Berry 3511* (MO).—GUÁRICO: 4 km E of Ortiz on rd to El Sombrero, *Gentry 10314* (MO).—MÉRIDA: near Barinitas, *Breteler 3158* (NY).—YARACUY: Yaracuy, *Curran 258* (NY). **Guyana.** Cultivated in Georgetown Botanical Garden, *Persaud 1744* (NY). **Brazil.** BAHIA: Mpio. Ilhéus, área do CEPEC, Km 22 rodovia Ilhéus–Itabuna, *Mori 10302* (NY).—MATO GROSSO: main square of Chapada dos Guimarães, *Prance et al. 19303* (NY).—RIO DE JANEIRO: Jacarepaguá, Estrada da Curicica em frente ao Hospital da Curicica, *Cominote 66* (NY); Ilha do Governador, Aeroporto do Galeáo, *Pabst 5412* (NY).—SÃO PAULO: cultivada em São Paulo, no Jardim Botánico, *Hoehne 43078* (NY); Arvore, Fazenda Santa Elisa, Campinas, *Hoehne 994* (F); Fazenda Santa Elisa, Campinas, *Houk 31* (GH, MICH, UC).

Senegal. Dakar, Harin, *Jacques-Georges 203, 204* (MO). **Nigeria.** OYO: Ibadan, Polytechnic campus, *Pilz 1913* (F); Ibadan, Garden Quarter 680, Jericho, *Wit 36* (MO). **South Africa.** NATAL: Mbazwana, littoral dunes, *Balsinhas 3248, 3258* (MO). **Madagascar.** TAMATAVE [TOAMASINA]: Hwy 5 from Fenerive to Maroanseta, *Croat 32542* (MO).

India. Yercad, Salem, *Matthew 26861* (GH); Hassan, Mysore, about 5 mi from Hassan, *Saldanha 12853* (MO); Hassan, Mysore, Mallipatna–Arkalgud rd, *Saldanha 16326* (MO). **Sri Lanka.** Ruhuna National Park, Hambantota–Tissa rd, *Comanor 838* (GH, MO, NY). **Malaysia.** Petaling, Selangor, Shah Alam, *Worthington 12415* (NY). **The Philippines.** Palawan, *Bermejos 39771* (NY); E. Mindoro, 5 km SW of Puerto Galera, *Burley 208* (F); Bauang, Prov. Union, Luzon, *Fénix 166* (F, GH, MO); Mt. Pinatubo, Zambales Prov., Luzon, *Fox 2099–4* (MICH); Guimaras Island, *Gammill 294* (NY); Bohol, *McGregor 42885* (NY); Manila, *Merrill 289* (MICH, NY, UC); Antipolo, Rizal Prov., Luzon, *Merrill 613* (F, GH, MO, NY, W); Clapan, Mindoro, *Merrill 894* (GH, NY); Mindoro, *Merritt 61686–2* (MO); Lamao River, Mt. Mariveles, Prov. Bataan, Luzon, *Meyer 28667–1* (F, NY); Bontoc subprov., Luzon, *Vanoverbergh 3404* (F); Manila, *Vidal 2570* (GH); Los Baños, Prov. Laguna, Luzon, *Williams 3072* (NY). **Marianas Islands.** GUAM. Piti, *Evans 1524* (F, GH, NY); Asan, *Fosberg 35490* (NY). **Society Islands.** Moorea, Papetoai village, *Smith 132* (GH, MO, NY, UC); Uturoa, *Florence 3826* (NY).

Gliricidia sepium is one of the more widely introduced, economically important species of Leguminosae. *Madricacao, madriado, mataratón*, and *cacahuananche* are some of the very widely used common names of *G. sepium*; such widespread use of multiple common names, undoubtedly due to the relatively high economic value of *G. sepium,* is otherwise unknown for the species of the tribe Robinieae. Hughes (1987) and Soto and Sousa (in press) enumerate the many economic applications of this species. The wood is used in construction, leaves are used for medicinal purposes, cattle fodder, and shade in coffee or cacao plantations, and the tree itself is grown for living fence systems. Label data suggest some unusual economic uses, such as in the manufacture of candy from the flowers. On the label of one specimen

(*Moore 3672*, GH) from the Puerto Rico Experiment Station (U.S.D.A., Mayagüez, Puerto Rico), use as a "fish poison" is suggested. This is otherwise seldom reported from the tribe Robinieae, but is very common in such related tribes as Millettieae, where the production of isoflavonoids is common (Gomes et al. 1981; Geesink 1984).

Differences between *G. sepium* and the very similar *G. maculata* are enumerated in the discussion of the latter. Leaflet traits most diagnostic of *G. maculata* will rarely occur in specimens of *G. sepium*, and when they do, it is mostly from a few Caribbean collections. For example, the leaflets are rounded in a specimen from Aruba (*Stoffers 1682*), Curaçao (*Arnoldo-Broeders 3879*), Haiti (*Leonard 9694*), and St. Croix (*Nelson 20–2–1974*). No doubt, the leaves of these specimens of *G. sepium* approach those of *G. maculata,* because they are all young developing leaves produced during the flowering season (see discussion of *G. maculata*). Also, white petals, generally diagnostic of *G. maculata*, occur rarely in Chiapas populations of *G. sepium* (*Breedlove 23925*), where they co-occur with the much more common pink-flowered individuals (*Hughes et al. 1678*). Although seemingly confounding, the simultaneous use of the most or all of the eight characters listed in the discussion of *G. maculata* should unequivocally provide an accurate determination of a specimen in question.

Specimens of *G. sepium* occasionally display the subopposite branching pattern that is otherwise distinctive of the Cuban *Hebestigma*. Furthermore, some of these same specimens also show the *Hebestigma*-like inflorescence production: racemes clustered at the tip of a stem and in the axils of aborted or highly reduced leaves. It appears that in these same specimens, the arrested vegetative development during raceme production is released following flowering. Such specimens are superficially very similar to *Hebestigma cubense*, but they are readily distinguished as *G. sepium* (in addition to the obvious leaf and mature legume characters—see key to genera of Robinieae) by the glabrate to sparsely strigose pedicels and calyces; in *Hebestigma*, these structures are tomentose.

7. Robinia L., Sp. pl. 722. 1753.—TYPE: *Robinia pseudoacacia* L.

Trees and large shrubs. Roots superficial and spreading, stout and woody, sprouting leafy shoots. Stems commonly scandent, but if erect, then crooked, branches alternate; bark thick and furrowed. Leaves alternate, pulvinate, imparipinnate; leaflets mostly opposite, sometimes alternate at maturity, orbicular or narrowly to widely elliptic, rounded or retuse, caducous from the more persistent rachis during the onset of winter or during age, adaxial surface glabrate to sericeous, usually without finely scattered tannin deposits, abaxial surface strigose to densely sericeous, with or without finely scattered tannin deposits; leaf rachis canaliculate, the groove without a dense indumentum, stipels present; stipules free, mostly spinescent and persistent. Inflorescences of pendent to lax, pedunculate axillary racemes, in axils of coeval leaves, with a nearly synchronous maturation and with many flowers reaching anthesis simultaneously on a single rachis, the rachis with 10–30 or more nodes, strigose to tomentose, often with stipitate or sessile glands. Flowers ebracteolate, with a well-developed hypanthium; floral bracts broadly ovate to linear-lanceolate, readily caducous; pedicels confluent with the calyx, straight and ascending during anthesis and seed set, falling with the aborted flower. Calyx cylindrical, attenuate at base, strigose to sericeous, sometimes stipitate-glandular or with sessile glands, with five prominent lobes equalling the length of the tube or longer or slightly shorter, the two vexillary ones often fused at the base, the lobes narrowly triangular, margins not overlapping at anthesis. Standard orbicular to oblong, emarginate, reflexed about 90° at anthesis, dorsal surface glabrate, subequal to or longer than the keel petals, with two prominent callosities about the midrib in the middle of the nectar-guide, claw abruptly contracted from the blade, inflexed auricles present; wing petals sculptured, commonly protruding beyond the keel at anthesis and loosely attached to it by a fold in the basal auricle, the distal ends glabrate, the blades usually nearly perpendicular to the plane of the keel; keel petals valvately connate distally nearly to the tip, apex blunt. Staminal tube usually pseudomonadelphous, with basal fenestrae; anthers included in the keel tip, basally attached to the filaments, the broad connective reddish brown; intrastaminal nectariferous disk absent. Pollen spheroidal, tricolporate, mesocolpium rugulate-perforate, polar regions smooth. Ovary subsessile, the stipe much shorter than the calyx tube, glabrate to stipitate-glandular, with 10–15 ovules; style with a pollen brush uniformly surrounding the distal half of the style, the style base thickened and becoming displaced to the carinal side during fruit maturation; stigma ciliate, capitate and terminating the distally tapered style. Legumes subsessile, glabrate or with conspicuously enlarged stipitate glands, inner epidermis of the endocarp of adjacent valves thin and not abutting, not adherent between seeds and not forming uniform seed compartments, the valves very thinly woody at maturity, dehiscent but not explosively so, the dorsal and ventral margins generally parallel in fruit, except where tandem seed abortion occurs. Seeds obliquely oriented, lenticular at maturity, not constricted in the middle, accumulating canavanine, brownish but mottled with dark purple blotches, hilum distinctly lateral and attached to a relatively long, curved funiculus, testa firmly adherent. Seedling with epigeal germination, cotyledons orbicular and foliar, eophyll unifoliolate, orbicular, alternate to the first metaphyll, metaphylls with intermediate leaflet number relative to the eophyll and mature leaves. Base chromosome number $x = 10$ (Isely & Peabody 1984). Figs. 84–87.

Robinia stands apart from the rest of the tribe by its elongate funiculus, which is attached to a laterally positioned obliquely notched hilum, and the thinly woody legume valves, which are tardily elastically dehiscent. Glandular indumenta and stipitate glands are found commonly within the *Robinia*-group and are not therefore diagnostic of *Robinia* (see phylogeny). *Robinia* is also distinctive in that it has a strictly woody habit and occurs only in cool-temperate regions (Figs. 88–89); only one additional herbaceous species of Robinieae is restricted to cool-temperate regions, *Peteria thompsonae* of the west-central United States.

In spite of these distinctions, *Robinia* has never been explicitly separated from close relatives. Isely and Peabody (1984) provided a preliminary account of the species and varieties, but avoided any discussion of what formally constitutes the genus and how it is related to other genera of Papilionoideae. Earlier characterizations of the genus were likewise vague, such that many woody species of unrelated Papilionoideae could be accommodated within *Robinia*. For example, Herendeen et al. (1992) comment that the name *Robinia* has been used uncritically for fossil species of Leguminosae. Undoubtedly, this was due to the lack of explicit criteria for defining the genus.

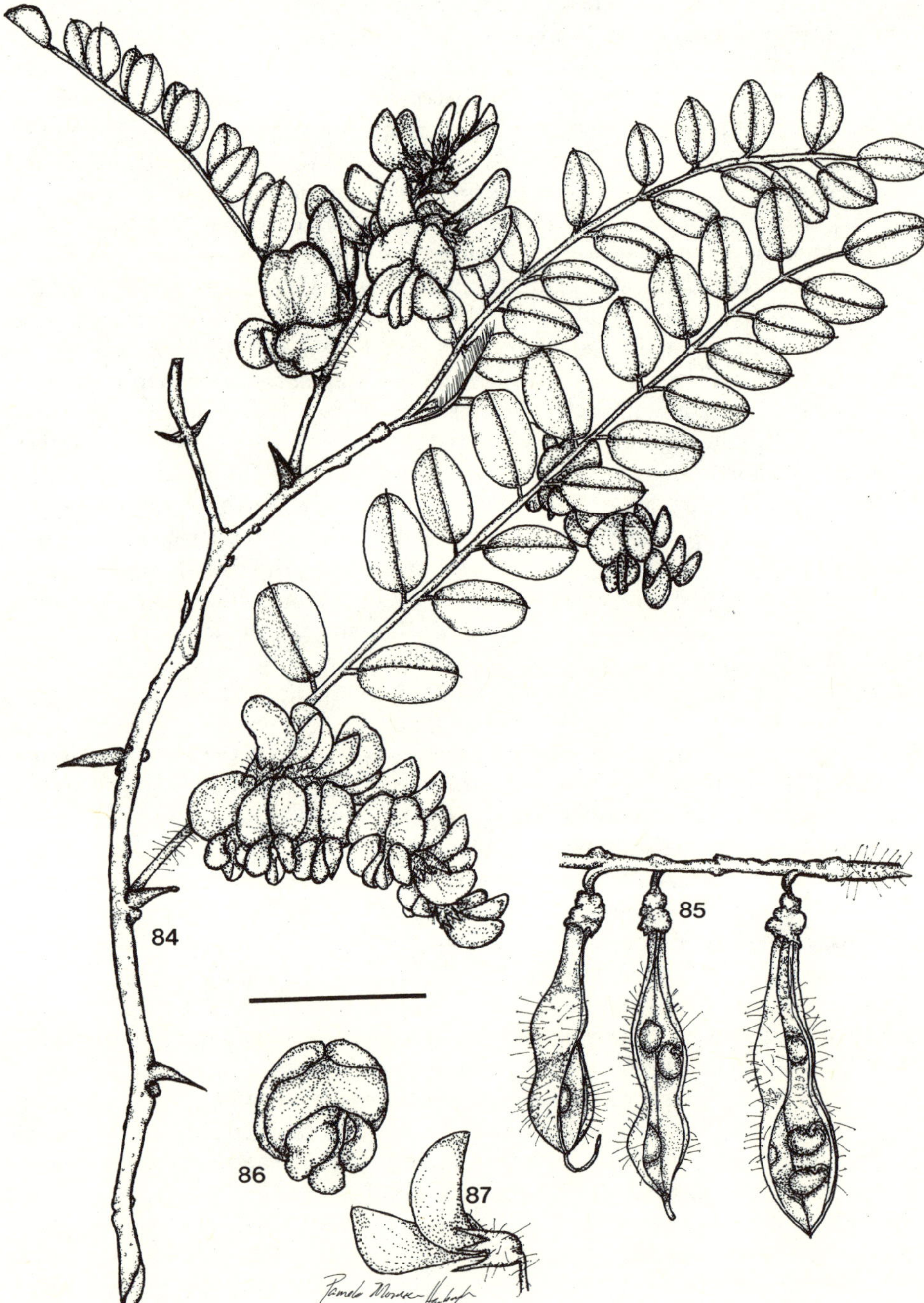

FIGS. 84–87. *Robinia neomexicana*. 84. Flowering branch. 85. Infructescence. 86. Front view of flower. 87. Side view of flower. (84, bar = 3 cm; 85–87, bar = 2 cm).

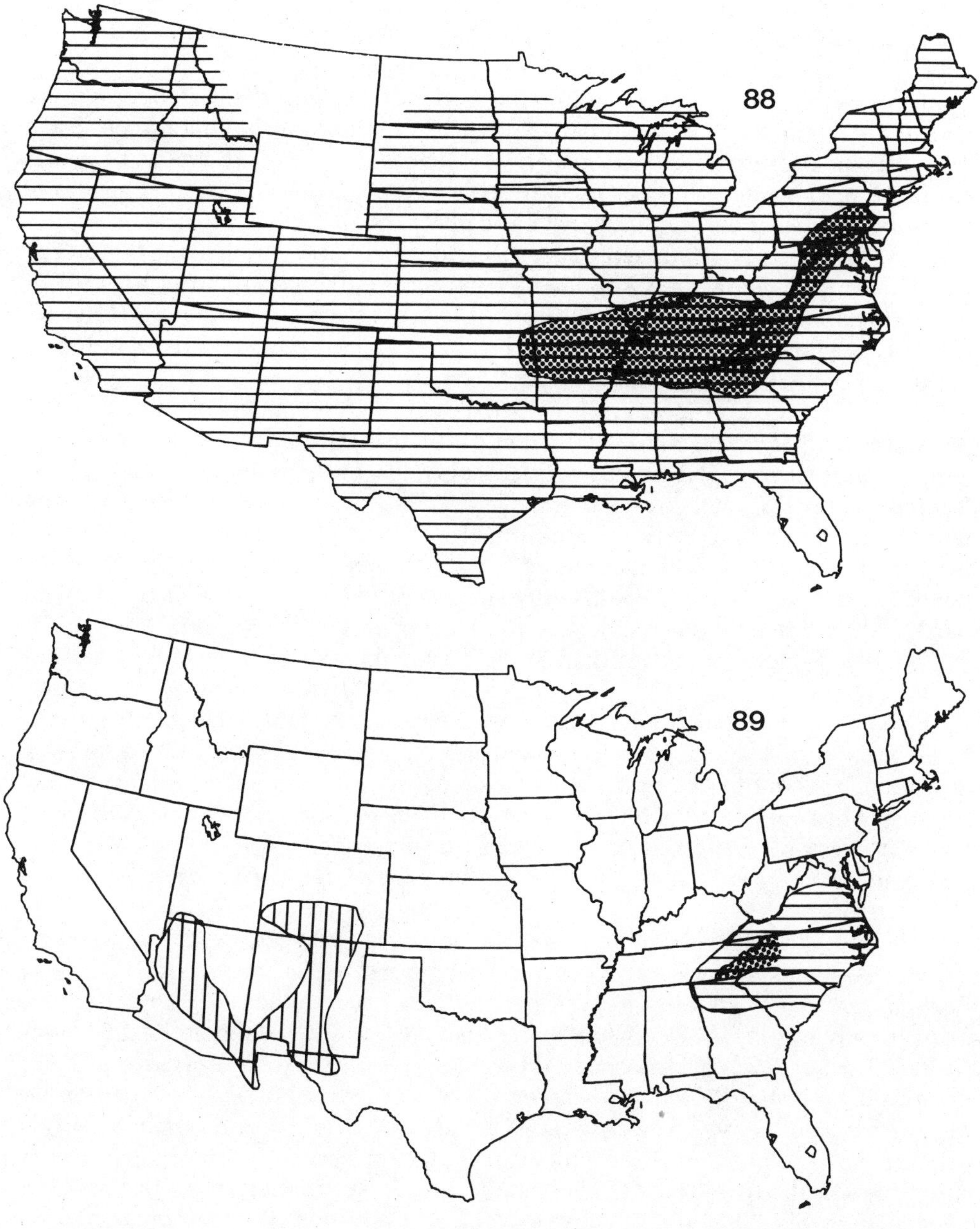

FIGS. 88–89. Distribution of the genus *Robinia*. 88. Approximate native (shaded) and introduced (horizontal striping, in the U.S.A.) distribution of *Robinia pseudoacacia*. 89. Distribution of *Robinia neomexicana* (vertical striping), *R. hispida* (horizontal striping), and *R. viscosa* (shaded area). These ranges are summarized from Hanover (1992) and the many references cited in Lavin (1992), which include especially the studies of Isely (1990), Isely and Peabody (1984), and Peabody (1984).

Inferences from the phylogenetic analysis presented in this paper suggest that *Robinia* occupies an unresolved position with respect to *Coursetia* and a monophyletic assemblage including the rest of the *Robinia*-group (i.e., *Olneya*, *Peteria*, *Genistidium*, and *Sphinctospermum*). Unlike all other genera of the *Robinia*-group, *Robinia* has retained the presumably ancestral condition of the inflorescence and base chromosome number. Flowering in the genus occurs synchronously, or nearly so, along an elongate rachis borne in the axils of coeval leaves. The other genera of the *Robinia*-group have modified this basic plan such that flowering is not synchronous, or the rachis is often highly reduced and often borne at a leafless node. The base chromosome number of *Robinia* is 10, whereas in the rest of the group it is 8 or 9 (Lavin, unpubl.). Results from chloroplast DNA restriction site analysis (Lavin & Doyle 1991; Lavin, unpubl.) also suggest that *Robinia* is the likely sister group to the rest of the *Robinia*-group, including *Coursetia*.

The cool-temperate geographical distribution of *Robinia* stands in contrast to the warm-temperate or tropical distribution of the rest of the tribe. The warm-temperate element of Robinieae includes only the more modified genera of the *Robinia*-group (i.e., *Olneya*, *Peteria*, *Genistidium*, and *Sphinctospermum*), and thus, by inference, the cool-temperate distribution of *Robinia* is derived directly from tropical ancestry. The inability of the apical meristems to withstand freezing temperatures, vegetative growth and senescence that are controlled by ambient temperatures rather than by photoperiod (Hanover 1992), and nitrate reductase activity that is higher in leaves than in the roots, like that of tropical legumes but unlike that of all other temperate ones (Sprent 1980; Johnson et al. 1991), are consistent with a recent tropical ancestry of *Robinia*. Notably, *Robinia* has never been an important source of commercial hardwood because of its crooked stems, a trait brought about by the death of its apical meristems during the winter between successive growing seasons (Hanover 1992). Even the winter-deciduous nature of *Robinia* could be the result of its shared ancestry with tropical Robinieae that predominate in the seasonally dry deciduous forests of the North American tropics (Lavin 1992).

The modern circumscription of *Robinia* includes four species, **R. pseudoacacia** L., **R. hispida** L., **R. viscosa** Ventenat, and **R. neomexicana** A. Gray (Isely & Peabody 1984). The first three species are native to the southeastern United States in Appalachia (Figs. 88–89), while the last is native to the mountains of the southwestern United States and adjacent Mexico (Fig. 89). Two groups occur within the genus: the whitish flowered *R. pseudoacacia* is a strictly diploid ($2n=20$) outcrossing species, and the predominantly pinkish flowered *R. hispida*, *R. viscosa*, and *R. neomexicana* produce, in part, triploid clonal races by apomixis or by extensive root sprouting (Peabody 1984). The taxonomic history of this latter group has been bedeviled by this clonal reproductive system; Rydberg (1924b) recognized 18 pink-flowered species and only two white-flowered species of *Robinia*. This "species diversity" recognized by Rydberg was due to the effects of the clonal mode of reproduction, which results in geographical races with slight morphological distinctions, a situation similar to that described from, among others, such genera as *Taraxacum* (e.g., see Cronquist 1988, chapter 2). Fortunately, the taxonomy of Isely and Peabody (1984) represents a more rational approach that factors in the effects that self-fertilization and cloning have on the partitioning of morphological variation within and among populations.

In contrast, the taxonomy of *R. pseudoacacia* has not been problematic and very few, if any, taxonomic synonyms and infraspecific taxa exist for this species (e.g., see Isely & Peabody 1984), undoubtedly because most individuals of this species are predominantly, if not strictly, outcrossing. Bongarten (1992) summarizes the literature that strongly suggests that, in spite of an ability to propagate clonal varieties, the genetically diverse *R. pseudoacacia* shows very little geographic patterning of morphological and genetic variation. For example, Surles et al. (1989) found that only 12% of the total allelic diversity at 18 loci was attributable to among-population variation throughout the native range in Appalachia, while 88% was attributable to within-population variation.

The following key is adapted from Isely and Peabody (1984) to illustrate the distinctions of the four species of *Robinia* that are now recognized. This treatment by Isely and Peabody should be consulted for the details pertaining to taxonomic and nomenclatural synonyms, and the identity and circumscriptions of infraspecific taxa.

Key to the Species of Robinia

1. Flowers whitish, 15–20 mm long; racemes pendent, with many flowers; legumes glabrous, the upper margin narrowly winged; peduncles, pedicels, and calyces without stipitate or sessile glands, indumentum not hispid; calyx lobes shorter than the tube. *R. pseudoacacia*.
1. Flowers pinkish, rarely whitish, mostly 20–25 mm long; racemes pendent to ascending, with few to many flowers; legumes slightly or conspicuously bristly or glandular-hispid, the upper margin not winged; peduncles, pedicels, and calyces with stipitate or sessile glands, or indumentum hispid; calyx lobes equal to or longer than the tube.
    2. Twigs and leaf rachises glabrous to sericeous or tomentose, eglandular; leaflets strigose to sericeous on both surfaces at maturity; native to the mountains of the southwestern U.S.A. and adjacent Mexico. *R. neomexicana*.
    2. Twigs and leaf rachises hispid or glandular; leaflets strigose to sericeous on only one surface at maturity, glabrous on the other; native to Appalachia, eastern U.S.A.
        3. Twigs and peduncles with sessile glands, not conspicuously hispid with trichomes 1–5 mm long; floral bracts aristate; leaflets 13–21, glabrate on adaxial surface at maturity. *R. viscosa*.
        3. Twigs and peduncles without sessile glands, generally conspicuously hispid with trichomes 1–5 mm long; floral bracts not aristate; leaflets 9–13, glabrate on abaxial surface at maturity. *R. hispida*.

The disjunct distribution of *Robinia* in Appalachia and in the mountains of the southwestern United States and the adjacent Sierra Madre Occidental in Chihuahua and Sonora, Mexico, is intriguing and suggests a once more widespread distribution. Notably, the closest relatives of *Robinia* (i.e., *Coursetia*, *Olneya*, *Peteria*, *Genistidium*, and *Sphinctospermum*) are either centered, or have an occurrence, in the desert regions of southwestern United States and northern Mexico just adjacent to the area occupied by *R. neomexicana*. It is not possible yet to estimate the infrageneric phylogeny of *Robinia*, and thus determine whether *R. neomexicana* represents an early offshoot (i.e., the sister group to the rest of the genus) in *Robinia*. One pattern that is distinctive is that *R. neomexicana* lacks the characters that uniquely mark the other individual species: *R. pseudoacacia* has whitish petals, distinctively short calyx lobes, and a winged margin on the legume; *R. hispida* has very long hispid trichomes; and *R. viscosa* has very distinctive sessile, blackish glands. *Robinia neomexicana* is the least specialized of the four species, relative to the putative ancestral condition; however, there is no evidence that *R. pseudoacacia*, *R. hispida*, and *R. viscosa* form a monophyletic group.

Because the type genus of Robinieae is *Robinia* L., and because the type species of *Robinia* is *R. pseudoacacia*, the type specimen of this species is here confirmed (Fred Barrie, pers. comm.).

Robinia pseudoacacia L., Sp. pl. 772. 1753.—TYPE: U.S.A. Virginia, 1720's to early 1730's, *Clayton 50* [Herb. Clifford 354] (lectotype, designated by F. Barrie in Jarvis et al., 1993: BM, photocopy: MONT!).

8. Coursetia DC., Ann. Sci. Nat. (Paris) 4: 92. 1825.—TYPE: *Coursetia fruticosa* (Cavanilles) Macbride.

Multistemmed trees, shrubs, and perennial, rarely annual, herbs. Roots superficial and spreading, stout and woody, occasionally fusiform-tuberous, or rarely taprooted, sprouting vegetative shoots. Stems usually spindly, erect to scandent, branches alternate; bark thick and furrowed or thin and smooth. Leaves alternate, pulvinate, imparipinnate, or paripinnate by reduction of the terminal leaflet to a seta, or unifoliolate; leaflets opposite, orbicular or narrowly to widely elliptic, rounded, retuse, or cuspidate, caducous from the more persistent rachis during the onset of the dry season or during age, adaxial surface glabrate to sericeous, with or without patchy tannin deposits, abaxial surface pilose to sericeous or lanate, with or without patchy tannin deposits; leaf rachis canaliculate, the groove without a dense indumentum, stipels mostly well developed; stipules free, small and ephemeral, or subulate and persistent, but mostly not spinescent or lignescent. Inflorescences of sessile to pedunculate, pendent to erect, axillary racemes, rarely panicles, with slow acropetal maturation, rarely having more than two flowers at anthesis simultaneously on a single rachis, the rachis with 2–100 or more nodes, sericeous to hispid, sometimes stipitate-glandular. Flowers ebracteolate, with a well-developed hypanthium; floral bracts broadly lanceolate to linear-lanceolate, readily caducous or sometimes persistent; pedicels jointed or confluent with calyx, straight and ascending during anthesis and seed set, or deflexing and twisting after anthesis in sections *Neocracca* and *Craccoides*, falling with the aborted flower, or persistent on rachis. Calyx campanulate, rounded or rarely attenuate at base, strigose to sericeous or tomentose or sometimes stipitate-glandular, with five prominent lobes equalling or longer than the tube, the two vexillary ones sometimes fused at the base, the lobes ligulate, or broadly to narrowly triangular, acute to rounded, margins not overlapping at anthesis. Standard orbicular, rarely oblong, emarginate, reflexed at 90–120°, dorsal surface glabrate, subequal to or longer than the keel petals, with two prominent callosities about the midrib in the middle of the nectar-guide, claw abruptly contracted from the blade, inflexed auricles mostly well developed; wing petals sculptured, commonly protruding beyond the keel at anthesis, and loosely attached to it by a fold in the basal auricle, the distal ends glabrate, the blades parallel to perpendicular to the plane of the keel; keel petals valvately connate distally to nearly the tip, apex acute, often rostrate. Staminal tube diadelphous to rarely pseudomonadelphous, with basal fenestrae; anthers included in the keel tip, basally attached to the filaments, the broad connective reddish brown; intrastaminal nectariferous disk absent. Pollen spheroidal, tricolporate, mesocolpium rugulate-perforate, polar regions smooth. Ovary subsessile, the stipe much shorter than the calyx tube, granuliferous to sericeous or woolly, rarely glabrous or stipitate-glandular, with (6–) 12–30 ovules; style with a latrorse to extrorse pollen brush, loosely scattered to compacted along the length of the style, or sometimes confined toward the distal tip, the style base thickened and becoming displaced to the carinal side during fruit maturation; stigma ciliate, capitate and terminating the distally tapered style. Legumes subsessile, glabrous, strigose, sericeous, villous, or woolly, rarely prickly or stipitate-glandular, inner epidermis of the endocarp of adjacent valves spongy, abutting and strongly adhering between seeds and forming uniform seed compartments, the valves thinly woody at maturity, elastically dehiscent, the dorsal and ventral margins generally sinuous, but if parallel, then the valves with distinct uniform vertical impressions marking the seed compartments. Seeds transversely oriented, lenticular to squarish, not constricted in the middle, accumulating canavanine, uniformly brownish or more commonly mottled with darker purple blotches, hilum apical, attached to a very short funiculus, testa firmly adherent. Seedlings with epigeal germination, cotyledons foliar, eophyll unifoliolate, alternate to the first metaphyll, metaphylls with intermediate leaflet number but otherwise similar to adult leaves. Base chromosome number x = 8 (Lavin 1988).

Coursetia, comprising 39 species (Lavin 1988; Sousa & Lavin 1992), is the largest genus within the *Robinia*-group and is also the only one in this group to be geographically centered in the seasonally deciduous Neotropical forests of Mexico, Central America, and the Andes in South America (Fig. 90). The rest of the *Robinia*-group has diverged from the presumably ancestral condition of inhabiting tropical forest vegetation to inhabiting warm-temperate deserts or cool-temperate montane habitats. *Coursetia* is characterized relative to the *Robinia*-group by its legume in which seeds are borne in distinct compartments, inflorescences of axillary racemes with a sequential acropetal maturation of flowers, styles bearing a pollen brush that is initiated extrorsely and often assumes a latrorse position at maturity, acute keel petals, and the loss of spinescent stipules. Morphologically, *Coursetia* is unspecialized and could be the sister group to the rest of the *Robinia*-group, although chloroplast DNA data suggests *Robinia* holds this position (see phylogeny section).

Coursetia holds the distinction of being the only genus in the tribe to have invaded South America with a large contingent of species. Essentially one-half of the genus is now restricted to South America, although the least specialized section of the genus, sect. *Madrenses*, is restricted to Central America and adjacent Mexico (Lavin 1988). Otherwise, only one species of *Gliricidia*, *G. brenningii*, and a few widespread weedy species of *Sesbania* occur as natives in South America.

Coursetia is divided into five sections described in Lavin (1988). One of these, section *Madrenses*, was not properly described and this situation is rectified here.

Coursetia sect. **Madrenses** Rydberg ex Lavin & M. Sousa, sect. nov.—Type: *Coursetia madrensis* Micheli.

Coursetia sect. *Polyphyllae* Rydberg, N. Amer. Fl. 24: 229. 1924, nom. illegit.—Type: *Coursetia polyphylla* Brandegee.

Arbores vel frutices; folia imparipinnata; foliola distaliter accrescentia, late elliptica, substantiis tanicis in maculis rubellis dispositis; stipulae plerumque incon-

FIG. 90. Distribution of *Coursetia* in North and South America (black shading).

spicuae mox caducae, vel lineares ab petiolo late divergentes; stipellae bene evolutae. Inflorescentiae racemosae, raro paniculatae, rhachidi quam foliis subtendentibus plerumque longiori; pedicelli post anthesin elonganti, recti, ascendenti. Flores petalis flavidis vel rubellis. Peniculus pollinicus latrorsus, trichomatibus secus styli dimidium distale vel per totam longitudinem laxe dispersis. Legumina glabrescentia, raro lanata vel sericea, marginibus sinuosis; cavitates seminales rotundati.

Trees and shrubs; leaves imparipinnate; leaflets distally accrescent, widely elliptic, with reddish tannin deposit occurring in patches; stipules usually inconspicuous and soon caducous, or linear and widely divergent from the petiole; stipels well developed. Inflorescences of racemes, rarely panicles, rachis usually longer than the subtending leaves; pedicels elongating after anthesis, straight and ascending. Flowers with yellowish to reddish petals. Pollen brush latrorse, the trichomes loosely scattered along the distal half to full length of the style. Legumes glabrate, rarely woolly or sericeous, margins sinuous, seed compartments rounded.

This section includes 10 species confined to southern Mexico and Central America. The nine species listed in Lavin (1988) are notably absent from Chiapas, Mexico, and most of Guatemala, but since then an additional species, *C. chiapensis*, has been discovered in west-central Chiapas, in Aguacero National

Park, as a member of the understory of dense seasonally deciduous forests dominated, in part, by the recently described *Hybosema robustum* (Sousa & Lavin 1992). This new locality partially fills the gap in a discontinuous Mesoamerican distribution. The species of sect. *Madrenses* are the least derived from the hypothetical ancestor of the genus, and the restricted distribution of this group to Mesoamerica is consistent with the view that all genera of Robinieae arose in tropical North America and that any South American representation is to be considered secondary. The constituent species are as follows.

8.1. Coursetia planipetiolata Micheli. Mexico (Guerrero).
8.2. Coursetia madrensis Micheli. Mexico (Puebla, Morelos, Guerrero).
8.3. Coursetia paniculata M. Sousa & Lavin. Mexico (Oaxaca).
8.4. Coursetia oaxacensis M. Sousa & V. Rudd. Mexico (Oaxaca, Guerrero).
8.5. Coursetia robinioides M. Sousa & Lavin. Mexico (Oaxaca).
8.6. Coursetia polyphylla Brandegee.
  8.6a. Coursetia polyphylla var. **polyphylla**. Mexico (Veracruz, Oaxaca, Guerrero).
  8.6b. Coursetia polyphylla var. **acutifolia** M. Sousa & Lavin. Nicaragua (Matagalpa).
  8.6c. Coursetia polyphylla var. **breviloba** M. Sousa & Lavin. Honduras and adjacent Guatemala and Nicaragua.
8.7. Coursetia chiapensis Lavin & M. Sousa. Mexico (Chiapas; see Sousa & Lavin 1992).
8.8. Coursetia elliptica M. Sousa & V. Rudd. Northwestern Costa Rica and adjacent Nicaragua.
8.9. Coursetia paucifoliolata M. Sousa. Nicaragua (Madriz, Estelí).
8.10. Coursetia apantensis M. Sousa. West and central Nicaragua.

Coursetia sect. **Poissonia** (Baillon) Lavin, Advances legume syst. 3: 58. 1987. *Poissonia* Baillon, Adansonia 9: 295. 1870.—TYPE: *Poissonia solanacea* Baillon [=*Coursetia orbicularis* Bentham].

Chiovendaea Spegazzini, Anales Soc. Ci. Argent. 82: 219. 1917.—TYPE: *Chiovendaea hypoleuca* Spegazzini [=*Coursetia hypoleuca* (Spegazzini) Lavin].

Coursetia series *Monophyllae* Harms, Repert. Spec. Nov. Regni Veg. 19: 16. 1923.—TYPE: *Coursetia orbicularis* Bentham.

Section *Poissonia* is the sister group to sect. *Madrenses*, but unlike this North American group, is restricted to South America along the Andean Cordillera in southern Peru, northern Argentina, and adjacent southern Bolivia. In contrast to sect. *Madrenses*, sect. *Poissonia* is greatly modified from the hypothetical ancestor of *Coursetia*, notably in its unifoliolate leaf morphology, indumentum composed of peltate trichomes, petals with a bluish pigment, and a latrorse pollen brush that is compacted at the distal end of the style. The two species of the section are:

8.11. Coursetia hypoleuca (Spegazzini) Lavin. Northern Argentina and adjacent Bolivia
8.12. Coursetia orbicularis Bentham. Southern Peru.

Coursetia sect. **Coursetia.**

Humboldtiella Harms, Repert. Spec. Nov. Regni Veg. 19: 12. 1923.—Type: *Humboldtiella ferruginea* (H. B. K.) Harms [=*Coursetia ferruginea* (H. B. K.) Lavin].

Coursetia sect. *Glandulosae* Rydberg, N. Amer. Flora 24: 229. 1924.—Type: *Coursetia glandulosa* A. Gray.

Callistylon Pittier, J. Wash. Acad. Sci. 18: 211. 1928.—Type: *Callistylon arboreum* (Grisebach) Pittier [=*Coursetia ferruginea* (H. B. K.) Lavin].

This section is diverse in both North and South America, and is distinguished by its paripinnate leaves, a distinctive trait that has evolved independently on several occasions in the tribe Robinieae: once each in *Sesbania*, *Poitea*, and *Olneya*, and twice in *Coursetia* (sect. *Coursetia* and sect. *Craccoides*). In contrast to the other sections, the species of sect. *Coursetia* produce inflorescences not in the axils of coeval leaves but at leafless nodes on woody stems. The species of the section follow.

8.13. Coursetia ferruginea (H. B. K.) Lavin. Panama, Colombia, Venezuela and adjacent Brazil, Trinidad, and Guyana.

8.14. Coursetia glandulosa A. Gray. Widespread in western Mexico and Arizona, U.S.A.

8.15. Coursetia axillaris Coulter & Rose. Northeastern Mexico and southern Texas, U.S.A.

8.16. Coursetia insomniifolia Lavin. Mexico (Coahuila).

8.17. Coursetia brachyrhachis Harms. Argentina and Bolivia.

8.18. Coursetia fruticosa (Cavanilles) MacBride. Peru.

8.19. Coursetia cajamarcana Lavin. Peru.

8.20. Coursetia maraniona Lavin. Peru.

8.21. Coursetia rostrata Bentham. Brazil (Bahia, Minas Gerais).

Coursetia sect. **Neocracca** (Kuntze) Lavin, Advances legume syst. 3: 59. 1987. *Neocracca* Kuntze, Revis. gen. pl. 3: 68. 1898.—Type: *Coursetia heterantha* (Grisebach) Lavin.

This monotypic section is distinctive in an otherwise predominantly woody and entirely perennial genus in having an annual growth habit. Lavin (1988) placed this species in a separate section with some hesitation, because, based on evidence then available, formally recognizing it would render sect. *Craccoides* paraphyletic. However, chloroplast DNA data (Lavin, unpubl.), combined with the cytological and morphological data presented earlier (Lavin 1988), strongly suggest this species is of hybrid origin from progenitors in sect. *Poissonia* and sect. *Craccoides*. Thus, its recognition as a separate section is strengthened. The evidence of hybrid origin comes, in part, from morphology, which strongly suggests a relationship of sect. *Neocracca* to sect. *Craccoides*; these traits include the very distinctive post-anthesis floral resupination process that places mature legumes in an upside-down position, legumes with square seed chambers, and inflorescences that are long-pedunculate (Fig. 91). However, chloroplast DNA (Lavin, unpubl.) and other morphological characters (bluish petal pigment, orbicular leaflets, and seedlings with two eophylls)

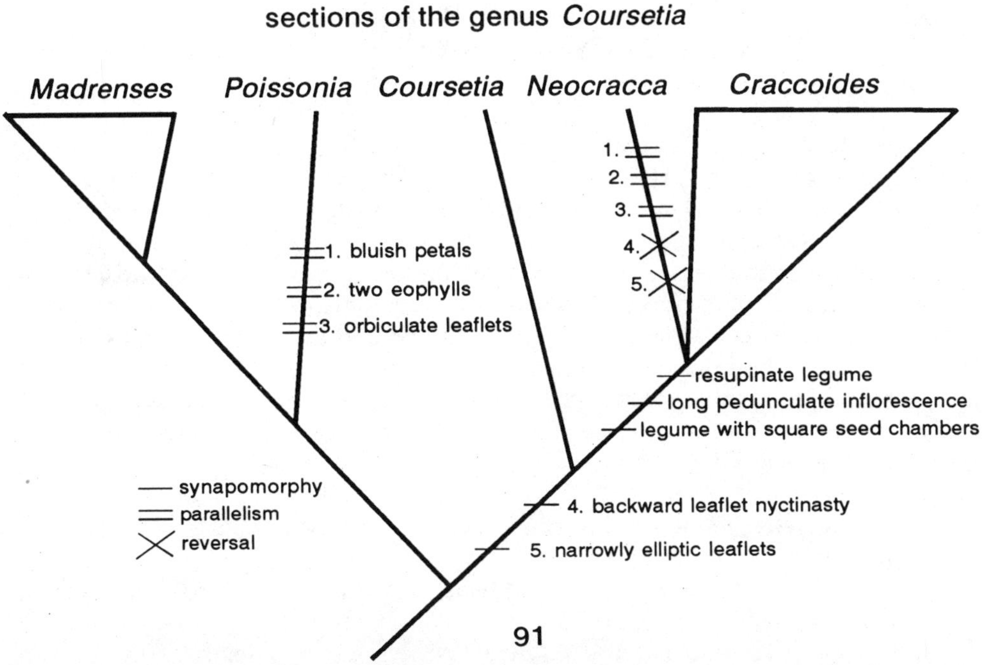

FIG. 91. Phylogeny of *Coursetia* adapted from Lavin (1988) and depicting the morphological characters relevant to the inferrence of the hybrid origin of sect. *Neocracca* from the putative parental lineages sect. *Poissonia* and sect. *Craccoides*.

strongly suggest a relationship of sect. *Neocracca* to sect. *Poissonia*. McDade (1990), by finding that hybrids in *Aphelandra* (Acanthaceae) express both maternal and paternal, and primitive and derived conditions with equal probability, has predicted that hybrids of widely divergent clades, if treated as a terminal taxon in a cladistic analysis, should often occupy a "basal" position to the clade marked by the greatest number of derived characters. This is precisely the position occupied by *Coursetia heterantha* (Fig. 91), the only species of sect. *Neocracca*, in the phylogenetic analysis of the genus (Lavin 1988; p. 37). Thus, sect. *Neocracca* is maintained as a distinct monotypic section of *Coursetia*, but of putative hybrid origin.

8.22. Coursetia heterantha (Grisebach) Lavin. Northern Argentina and adjacent Bolivia. Gómez-Sosa (1994) reports the southernmost distribution of this species occurs in the Argentine state of San Juan.

Coursetia sect. **Craccoides** (DC.) Lavin, Syst. Bot. Monogr. 21: 98. 1988. *Tephrosia* sect. *Craccoides* DC., Mém. Légum. 5: 267. 1826.—LECTOTYPE, designated by Lavin, 1988: *Coursetia caribaea* (Jacquin) Lavin.

Cracca Bentham, Vidensk. Meddel. Dansk. Naturhist. Foren. Kjøbenhavn 1853: 8. 1854, nom. conserv., non *Cracca* L., 1753. *Benthamantha* Alefeld, Bonplandia 10: 264. 1862. *Brittonamra* Kuntze, Revis. gen. pl. 1: 164. 1891.—LECTOTYPE, designated by Rydberg, 1924b: *Galega ochroleuca* Jacquin [=*Coursetia caribaea* var. *ochroleuca* (Jacquin) Lavin].

Balboa Liebmann ex Didrichsen, Vidensk. Meddel. Dansk. Naturhist. Foren. Kjøbenhavn 1853: 106. 1854.—TYPE: *Balboa diversifolia* Liebmann [=*Coursetia caribaea* var. *pacifica* (M. Sousa & Lavin) Lavin].

Like sect. *Neocracca*, sect. *Craccoides* is characterized by the distinctive resupinate legume, but furthermore has a very singular nyctinastic movement in which the leaflets fold backward with the abaxial surfaces facing outward (Lavin 1988). With 17 species, sect. *Craccoides* is the most diverse in the genus, and this diversity is distributed mainly in Mexico and in the northern Andean Cordillera of South America. Species of sect. *Craccoides* are the only ones in the genus to have invaded the pine-oak woodlands at higher altitudes and the higher latitudes in both North and South America. This latter aspect has been facilitated by the evolution in this group of a perennial herbaceous to subshrubby growth form that can die back during the cold season to the tuberous rootstocks (a similar strategy has been attained by the genus *Peteria*).

8.23. Coursetia mollis Robinson & Greenman. Pacific coastal Mexico (Figs. 92–101).

8.24. Coursetia hintonii V. Rudd. Mexico (northern Guerrero and adjacent México).

8.25. Coursetia pumila (Rose) Lavin. Widespread in Mexico. Lievens and Urbatsch (1994) discovered an earlier basionym for this species, which they propose be rejected: *Indigofera coronilloides* Martens & Galeotti, Bull. Acad. Brux. 10, part 2 (no. 7): 45. 1843.—TYPE: Mexico. Oaxaca, *Galeotti 3234* (holotype: BR, photo: MONT!).

8.26. Coursetia hidalgoana Lavin. Mexico (Hidalgo and adjacent México).

8.27. Coursetia glabella (A. Gray) Lavin. Northwestern Mexico and Arizona, U.S.A.

8.28. Coursetia barrancana Lavin. Northwestern Mexico.

8.29. Coursetia guatemalensis V. Rudd. Guatemala and adjacent Chiapas, Mexico.

8.30. Coursetia caribaea (Jacquin) Lavin.

8.30a. Coursetia caribaea var. **caribaea**. Widespread in Mexico, Central America, the Antilles, and northern South America (Figs. 102–112).

8.30b. Coursetia caribaea var. **sericea** (A. Gray) Lavin. Northwestern Mexico and Arizona, U.S.A.

8.30c. Coursetia caribaea var. **tomentosa** Lavin. Mexico (Hidalgo, San Luis Potosí).

8.30d. Coursetia caribaea var. **trifoliolata** (Rydberg) Lavin. Mexico (Puebla, Oaxaca).

8.30e. Coursetia caribaea var. **chiapensis** (Rydberg) Lavin. Mexico (Chiapas and adjacent Oaxaca).

8.30f. Coursetia caribaea var. **pacifica** (M. Sousa & Lavin) Lavin. Mexico (Oaxaca).

8.30g. Coursetia caribaea var. **astragalina** (H. B. K.) Lavin. Southern Colombia.

8.30h. Coursetia caribaea var. **ochroleuca** (Jacquin) Lavin. Ecuador and northern Peru.

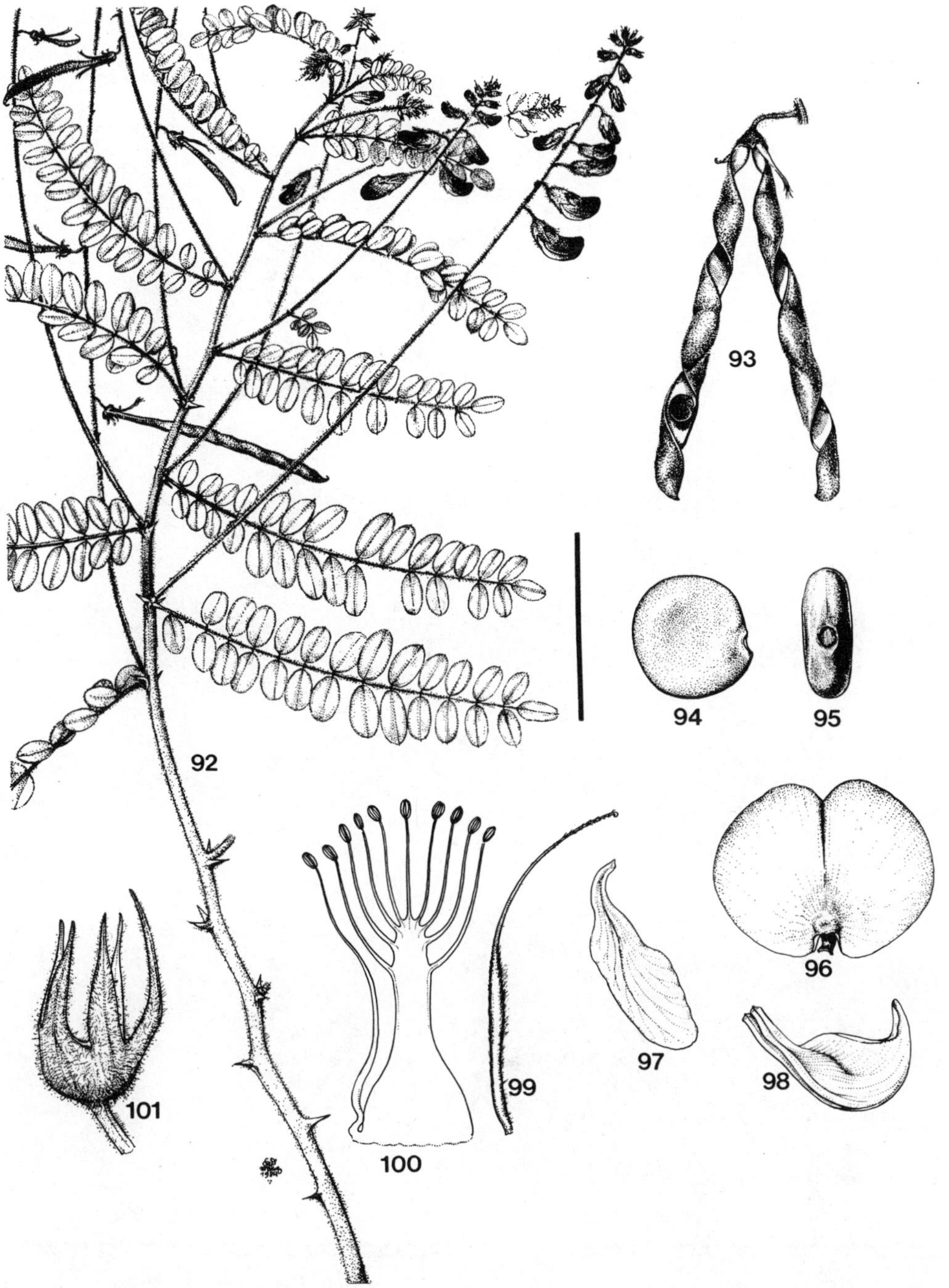

FIGS. 92–101. *Coursetia mollis*. 92. Flowering and fruiting branch. 93. Dehisced legume. 94. Side view of seed. 95. Hilum view of seed. 96. Standard petal. 97. Wing petal. 98. Keel petals. 99. Gynoecium. 100. Androecium. 101. Calyx. Reproduced from vol. 5 of *Flora Novo-Galiciana* by Rogers McVaugh. (92, bar = 5 cm; 93, bar = 3 cm; 94–95, bar = 5 mm; 96–98, bar = 20 mm; 99–101, bar = 10 mm).

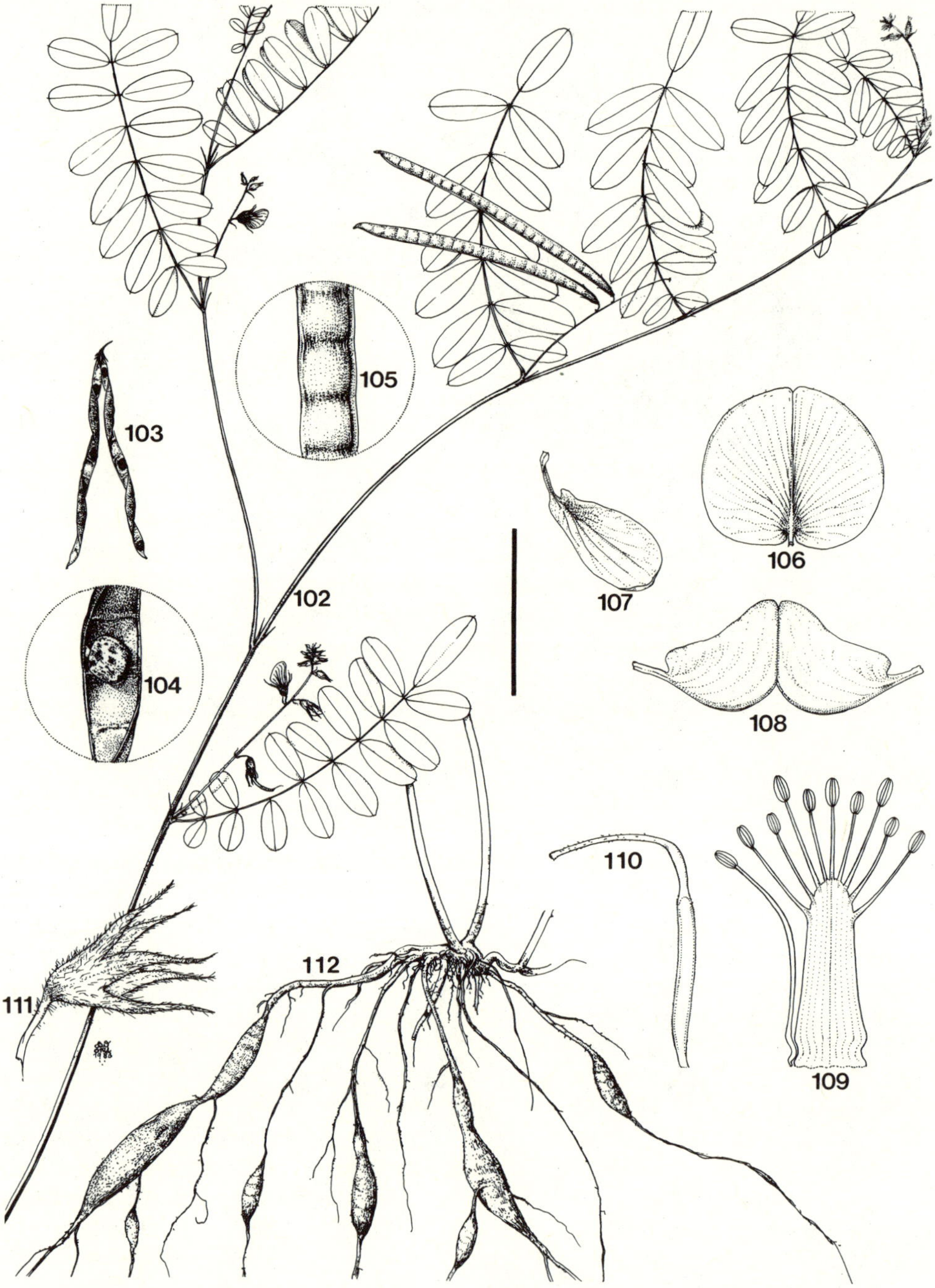

FIGS. 102–112. *Coursetia caribaea* var. *caribaea*. 102. Flowering and fruiting branch. 103. Dehisced legume. 104. Close-up of seed attached to legume valve. 105. Close external view of legume valve showing impressions of pericarp between seeds. 106. Standard petal. 107. Wing petal. 108. Keel petals. 109. Androecium. 110. Gynoecium. 111. Calyx. 112. Tuberous root system. Reproduced from vol. 5 of *Flora Novo-Galiciana* by Rogers McVaugh. (102–103, 112, bar = 5 cm; 104–105, bar = 15 mm; 106–108, bar = 10 mm; 109–111, bar = 8 mm).

8.31. Coursetia andina Lavin. Venezuela (Mérida).
8.32. Coursetia intermontana Lavin. Colombia (Boyacá).
8.33. Coursetia gracilis Lavin. Ecuador (Imbabura, Pichincha).
8.34. Coursetia vicioides (Nees & Martius) Bentham. Brazil (Bahia).
8.35. Coursetia hassleri Chodat. Northern Argentina, Bolivia, Paraguay, and adjacent Brazil.
8.36. Coursetia weberbaueri Harms. Peru (Arequipa).
8.37. Coursetia dubia (H. B. K.) DC. Central cordillera of Ecuador.
8.38. Coursetia grandiflora Bentham. Southern Ecuador and northern Peru.
8.39. Coursetia tumbezensis MacBride. Peru (Tumbes).

9. Olneya A. Gray, Mem. Amer. Acad. Arts, ser. 2, 5: 328. 1855.—TYPE: *Olneya tesota* A. Gray

Single- and multitrunked trees or sometimes large multistemmed shrubs. Roots with many superficial and spreading branches, stout and woody, sprouting vegetative stems. Stems erect to ascending, branches alternate; bark thick and furrowed, easily shredding. Leaves alternate, pulvinate, paripinnate by reduction of the terminal leaflet to a spinose seta, but most leaves with an odd number of leaflets; leaflets initially opposite but commonly alternate at maturity, narrowly obovate to elliptic, distally rounded to rarely retuse, caducous from the more persistent rachis after a drought or freezing temperatures, or during age, adaxial surface sericeous, usually without finely scattered tannin deposits, abaxial surface likewise sericeous, often with finely scattered tannin deposits, rarely in patchy deposits; leaf rachis canaliculate, the groove without a dense indumentum, stipels absent; stipules free and narrowly attached to the stem adjacent to the petiole, spinescent and persistent to caducous. Inflorescences of subsessile lax to ascending axillary racemes or panicles, congested at branch ends at leafless nodes, with a synchronous to sequential acropetal maturation, usually with two or more flowers at anthesis simultaneously on a single rachis, the rachis usually with fewer than 30 nodes, stipitate glands sparse. Flowers ebracteolate, very rarely bracteolate, with a well-developed hypanthium; floral bracts very reduced to inconspicuous generally persistent nubbins; pedicels confluent with the calyx, straight and ascending during anthesis and seed set, falling with the aborted flower. Calyx campanulate, rounded to attenuate at base, densely sericeous to tomentose, stipitate glands sparse (but dense in populations from Baja California Sur, Mexico), with five prominent lobes generally equalling or longer than the tube, the two vexillary ones often fused at the base, the lobes broadly triangular, imbricated (overlapping) in bud. Standard orbicular, emarginate, reflexed about 90° at anthesis, dorsal surface tomentose, sometimes stipitate-glandular (especially in populations from Baja California Sur, Mexico), subequal to or longer than the keel petals, with two prominent callosities about the midrib in the middle of the nectar-guide, claw abruptly contracted from the blade, inflexed auricles present; wing petals sculptured, commonly protruding beyond the keel at anthesis and loosely attached to it by a fold in the basal auricle, the distal ends tomentose (glandular in populations from Baja California Sur, Mexico), the blades almost perpendicular to the plane of the keel; keel petals valvately connate distally to nearly the tip, apex blunt. Staminal tube diadelphous, with basal fenestrae; anthers included in the keel tip, basally attached to the filaments, the broad connective reddish brown; intrastaminal nectariferous disk absent. Pollen spheroidal, tricolporate, mesocolpium rugulate-perforate, polar regions smooth. Ovary subsessile, the stipe much shorter than the calyx tube, densely sericeous, stipitate glands sparse to dense (especially so in populations from Baja California Sur, Mexico), with 10–14 (–17) ovules; style with a pollen brush uniformly surrounding most of the distal end of the style, the style base thickened and becoming displaced to the carinal side during fruit maturation; stigma ciliate, capitate and terminating the distally tapered style. Legumes subsessile, sericeous, stipitate glands sparse to dense and becoming enlarged with fruit maturation (especially so in populations from Baja California Sur, Mexico), inner epidermis of the endocarp of adjacent valves spongy and abutting but not strongly adhering between seeds and not forming uniform seed compartments, the valves thinly woody at maturity, elastically dehiscent, the dorsal and ventral margins generally parallel in fruit, except where tandem seed abortion occurs. Seeds transversely oriented, spheroidal at maturity, not constricted in the middle, accumulating canavanine, dark brownish and mottled with dark purple blotches, hilum subapical and attached to a very short funiculus, testa firmly adherent. Seedlings with epigeal germination, cotyledons thick and fleshy, eophyll 6–12-foliolate, subopposite to alternate to the first metaphyll, metaphylls similar to mature leaves. Base chromosome number $x = 9$ (Turner & Fearing 1960).

Olneya contains a single species largely confined to the Sonoran Desert of the southwestern U.S.A. and northwestern Mexico. *Olneya* is set off from its close generic relatives by a suite of characters otherwise absent or rare in the tribe Robinieae. Its calyx lobes are broadly triangular and imbricate in bud, and the standard and wing petals are tomentose and sometimes stipitate-glandular, especially on the dorsal surface of the standard and toward the distal tips of the wings. The leaf rachises are terminated by a spinose seta (i.e., a reduced terminal leaflet) but most often bear an odd number of leaflets in an alternate arrangement at maturity (a seeming "hybrid" between an impari- and paripinnate leaf). Also, the wood of *Olneya* has a massive, very dark brown heartwood, which is extremely hard and dense, with an air dry specific gravity of 1.15 (Record & Hess 1943). *Olneya* is distinctive as a dominant tree in low gravelly wash areas. It is the only taxon of Robinieae endemic to the Sonoran Desert.

The generic relationships of *Olneya* clearly lie within the *Robinia*-group, as evinced by such characters as spinescent stipules, stipitate glands, and a style bearing a pollen brush. Its geographical distribution in northern Mexico and the southern United States is also consistent with the geographical center of distribution of the *Robinia*-group. It is difficult to determine the closest relative of *Olneya*, and phylogenetic analysis suggests it can be narrowed down to *Peteria*, *Genistidium*, and *Sphinctospermum*. These four genera do not have the ancestral inflorescence condition of Robinieae, the production of elongate racemes in the axils of coeval leaves (i.e., those on hornotinous stems). Though retaining inflorescences in an axillary position, evolution in these genera has rendered the inflorescences to a functionally terminal position. In *Olneya*, the axillary racemes are congested at the vegetative branch end and on short shoots that bear very reduced nubbin-like leaves. These reduced leaves that subtend all inflorescence rachises, at least initially, reveal the true axillary nature of the inflorescence; but from a functional perspective (and a superficial glance), the inflorescence of *Olneya* appears as a panicle terminating a branch. Similar evolutionary paths presumably have been followed in *Peteria*, *Genistidium*, and *Sphinctospermum* (see discussions of these genera). Rarely do specimens of *Olneya* have single raceme rachises borne in axils of coeval leaves and on an elongating hornotinous stem (e.g., *Crosswhite 848*, MICH, NY, WIS). Though the inflorescences of *Olneya* are rarely subtended by mature leaves, this genus is nevertheless evergreen on the strictly vegetative portions of branches. According to Shreve (1964), an individual tree of *Olneya* will only shed all of its leaves during a severe drought or heavy frost. Shreve also reports that leafless living trees of *Olneya* are seldom encountered.

9. Olneya tesota A. Gray, Mem. Am. Acad. Arts, ser. 2, 5: 328. 1855.—TYPE: U.S.A. Arizona: Yuma County: on the table-lands of the Gila River, Jul 1852, *Thurber 646* (lectotype, here designated: GH!, photo: MEXU!; isolectotypes: GH! MICH! MO! NY!).

Trees and large shrubs 2–10 m tall, 2–60 cm in diameter at base. Leaves 2–10 cm long, absent from terminal branch ends during the flowering season, but then persistent in the more basal portions of a branch system; leaflets (8–) 9–21 (–24) per leaf, the larger mature ones (5–) 7–21 mm long, 2.5–7.0 (–9.0) mm wide, narrowly obovate to elliptic, both surfaces sericeous; stipules spinescent, 4–10 mm long. Inflorescences congested at branch ends at leafless nodes, an individual rachis 2–8 cm long, with usually fewer than 30 nodes, internodes mostly 0.5–6.0 mm long, generally much shorter than the subtending pedicel, tomentose to woolly, stipitate glands sparse; floral bracts reduced to very small nubbins less than 1 mm long; pedicels 3–4 mm long at anthesis. Calyx tube (3.0–) 3.5–4.5 mm long, densely sericeous to tomentose, stipitate glands sparse (except in populations from Baja California Sur, Mexico), lobes (1.5–) 2–3 mm long. Standard (9–) 10–14 mm long, (10–) 11–15 mm wide, mostly pinkish to purplish, often suffused with white, aging mostly light brown, nectar-guide greenish with whitish outline; wing petals (9–) 10–14 mm long, mostly pink to purple; keel petals (9–) 10–12 mm long, the blades mostly pink to purple. Staminal tube with filaments free for about the distal one-third. Ovary densely sericeous, stipitate glands very dense, with 10–14 (–17) ovules; style (5–) 6–7 mm long. Legumes 2–5 (–6) cm long, usually with persistent stipitate glands (these very dense in populations from Baja California Sur, Mexico), mostly with 1–2 seeds, sometimes with 3, rarely with 4–5. Seeds (5–) 6–10 mm in diameter, initially light brown with dark purplish mottling, but soon becoming uniformly dark brown. Figs. 113–117.

Phenology. Flowering from March through June, but any one individual flowering for no more than about one week. Most specimens from the southern part of the range were collected in flower during March and April, whereas most of those from the northern part of the range were collected in flower during May and June; mature fruits persist for several months, mostly July through September, rarely later, throughout the geographical range. A major pollen vector of *Olneya* is the bee *Centris pallida*, which is also the principal pollen vector for *Cercidium microphyllum* and *C. floridum* (Neff et al. 1977; Simpson 1977). These bees swarm the dense canopy of flowers during the heat of mid-day. Populations of *Olneya* experience massive flowering only about two years in every five (Shreve 1964). Individuals of *Olneya* may be long-lived, up to 200 years old (Johnson 1992).

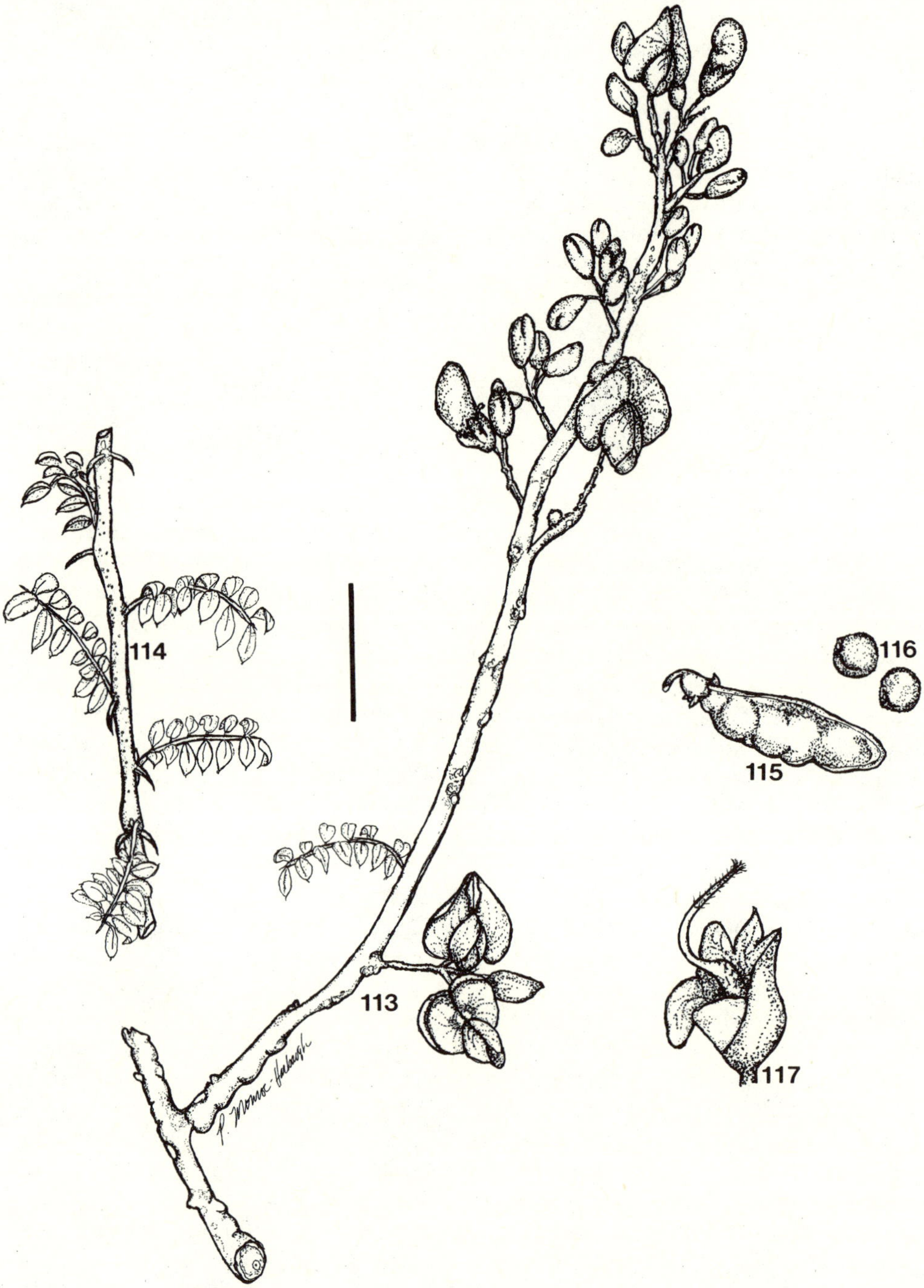

FIGS. 113–117. *Olneya tesota*. 113. Flowering stem. 114. Leafy stem. 115. Mature fruit. 116. Seeds. 117. Calyx and gynoecium. (113–114, bar = 15 mm; 115–116, bar = 3 cm; 117, bar = 7 mm).

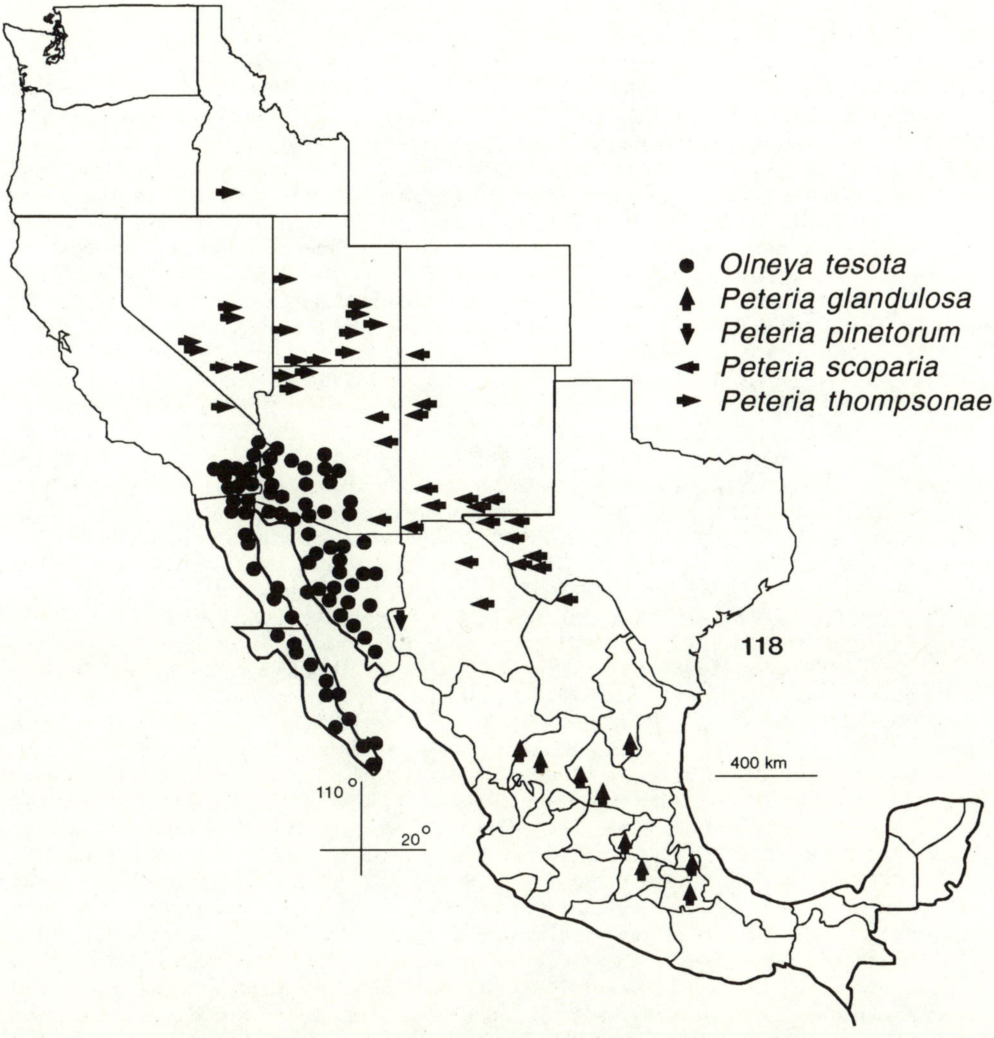

FIG. 118. Distribution of *Olneya* and *Peteria* in the U.S.A. and Mexico.

Distribution (Fig. 118). Mexico (Sonora and adjacent Sinaloa, Baja California, and Baja California Sur), and U.S.A. (southern Arizona and adjacent California), rarely found outside the Sonoran Desert; gravelly washes and open lower bajadas; 0–1300 m. Mason and Mason (1987), Shreve (1964), and others suggest that *Olneya* is an indicator of frost-free areas, a factor that must set the northern limits of its geographical distribution. Factors setting the southern limits of *Olneya*, at the very northern edge of the Pacific coastal deciduous forest, may include increasing moisture; Shreve notes that the most common cause of death of large trees of *Olneya* is the prolonged saturation of soil by rains and floods.

REPRESENTATIVE SPECIMENS. **U.S.A.** ARIZONA: La Paz Co.: 24 mi W of Salome on rd to Ehrenberg, *Shreve 7386* (ARIZ). Maricopa Co.: near Biltmore Hotel, Phoenix, *Abrams 12685* (CAS); N of Gila

Bend, *Clark 11692* (GH); Black Canyon at New River, 32 mi N of Phoenix, *Crosswhite 848* (MICH, NY, WIS); South Mts, Lone Butte Quad, *Daniel 4293* (CAS); Hwy 69 (Black Canyon), 25 mi N of Phoenix, *Deaver 5959* (ARIZ, CAS); Phoenix, *Eastwood 6166* (CAS); near Phoenix, *Eastwood 16668* (CAS); on rd to Phoenix, *Eastwood 17719* (CAS); Harquahala Mts, *Fischer 5847* (ARIZ); Salt River Reservation, 5 mi NE of McDowell Rd, *Francis & Enos 73* (ARIZ, NY); Phoenix, *Graham* (CAS); Gila, *Hayes 203* (F); 100 mi W of Tucson, E of Gila Bend, *Lavin & Lavin 4642* (NY, TEX); 7 mi W of Tonopah, *Lavin & Lavin 4657* (TEX); Scottsdale Rd at Currie's Corner, *Lehto 3836* (NY); mesa S of Apache Trail, *McKelvey 340* (GH); table-lands, Phoenix, *Nelson 10294* (CAS, MO, NY); plains of Maricopa, *Pringle 15138* (CAS, F, GH, MICH, MO, NY), *Pringle 15138a* (CAS, F, GH, MICH, MO, NY, WIS); Gila River Indian Reservation, *Rea 1190* (ARIZ); about 5 mi S of Cave Creek, *Taylor & Taylor 30304* (NY); 5 mi E of Gila Bend, *Wolf 2295* (CAS, GH). Pima Co.: Pisinimo, *Agatha 526* (WIS); Covered Wells, *Marie 323* (WIS); Anklam rd, Tucson Mts, *Caldwell 164* (ARIZ); Alamo Canyon, Ajo Mts, *Cooper 577* (ARIZ); Tucson, *Demaree 8039* (MO, NY); desert near Tucson Mts, SW of Tucson, *Graham 27.6.1912* (CAS); near Sawtooth Mtn, W of Tucson, *Graham 2.7.1904* (CAS); rd to Picture Rocks in Tucson Mts, *McKelvey 476* (GH); 1.1 mi W of Tucson, *Mears & Mears 16786* (TEX); Quitobaquito, Organ Pipe Cactus National Monument, *Nichol 28.4.1939* (ARIZ); SE of Cat Mtn, 5 mi SW of Tucson, *Sherff 9.8.1908* (F); Tucson Mts, *Thornber 25.1.1905* (ARIZ), *Thornber 24.3.1908* (ARIZ); 1 mi E of Gates Pass, Tucson Mts, *Turner 61–2* (ARIZ); 1.7 mi E of Gates Pass, Tucson Mts, *Van Devender 88–209* (ARIZ). Pinal Co.: 2 mi N of Florence, *Atwood 2221* (NY); near Florence, *Collom 60* (GH, MICH, MO, NY, WIS); Rivers, *Clear 14.7.1945* (CAS); mesas, Sacaton, *Hastings & Thornber 9020* (ARIZ); Casa Grande, *Jones 98913* (CAS, UC); near Sacaton, *Loomis 1953* (ARIZ); between Florence and Florence Junction, *McKelvey 900C* (GH); Goldmine Mtn, *Mittleman 4* (ARIZ); near Sacaton, *Peebles 8.7.1943* (ARIZ); San Tan Mts, *Peebles 7909* (F, GH, NY, WIS); Sacaton, *Peebles 9311* (ARIZ, CAS, MICH, UC), *Peebles & Harrison 4226* (US); 3 mi E of jct. of Hwy 80 and Old Mesa Rd, *Solomon 346* (MO); Sacaton, *Thornber 23.3.1902* (ARIZ); on mesa near Sacaton, *Tidestrom 12620* (RENO). Yavapai Co.: Bumblebee, 5 mi S of town, *Benson 9672* (ARIZ); Rock Springs, Black Rock, *Demaree 44292* (UC). Yuma Co.: 10 mi E from the Cal.-Ariz. line, E of Blythe, *Barkley & Blondeau 4218* (ARIZ, CAS, F, GH, NY, UC); W of Quartzite, *Billings 1784* (RENO); 4 mi W of mts, E of Yuma, *Booth* (ARIZ); Crystal Hill Campground, 6 mi E of Hwy 95 in foothills of Kofa Mts, *Daniel 172* (MICH); Ehrenberg, *Jepson 20-X–1912* (CAS, GH, UC); 2 mi SE of Imperial Dam, Gila Canal, *McMurry 1398* (ARIZ, CAS); Yuma, along bluffs, *Thornber 26.9.1912* (ARIZ); 18 mi E of Yuma, *Wolf 2291* (CAS, GH).—California: Imperial Co.: Signal Mtn, *Abrams 1907* (CAS, GH); near Tecate, *Dudley 112* (CAS); 20 mi N of Ogilby, *Hitchcock & Muhlick 23211* (CAS, F, UC); Palo Verde Mts, *McLaughlin 3472* (ARIZ); NE base of Signal Mtn, *McMinn 1447* (CAS, UC); Indian Wash, along rd connecting Hwy 8 and S78, *Tilforth et al. 1318* (NY). Riverside Co.: 5 mi E of Mecca, *Alexander & Kellogg 2194* (GH, UC); E of Mecca, *Alexander & Kellogg 2231* (CAS, GH, MO, NY, UC); Chuckwalla Desert, 45 mi W of Blythe, *Carter 1431* (CAS, GH, MICH, NY, MO, UC, US); Painted Canyon, Mecca, *Clary 313* (MO); Desert Center, *Garth C645* (ARIZ); Mecca to Blythe, entrance to Box Canyon, *Everett & Balls 23116* (NY, UC); Cottonwood Springs, *Everman 1924* (CAS); Chocolate Mts, at Shaver's Well, *Ferris 986* (CAS); W of Desert Center, *Jones 1926* (CAS); Box Canyon Rd exit from Interstate 10, *Keil 14066* (TEX); 75 mi W of Blythe, *Lavin & Lavin 4653* (TEX); 50 mi W of Blythe, *Lavin & Lavin 4654* (NY, TEX); Colorado Desert, Mecca, *McKibben 1904* (UC); mouth of Painted Canyon, *Munz & Keck 9345* (GH); Hidden Springs canyon, *Nelson & Nelson 3311* (CAS, UC); Mecca, Salton Basin, *Parish 8123* (CAS, UC); Cottonwood Springs, *Parish 10833* (GH); Colorado Desert, *Parish 10857* (ARIZ, CAS, GH, MO, NY, UC); near Mecca, *Rixford 1911* (CAS); Hidden Springs Rd, Coachella Valley, *Rowntree 1937* (CAS); 6 mi E of Mecca, *Sponslu 6968* (MO); 12 mi W of Blythe, near rd, *Stark 1540* (CAS); Chuckawalla Bench, *Schellenger 12* (UC); Fort Yuma, *Schott 1854* (F, NY); Chuckawalla Mts, 1 mi W of Astec Well, *Shultz & Shultz 8334* (NY); Painted Canyon, Mecca, *Spencer 165* (GH), *Spencer 1920* (MO); Mecca, *Spalding 11.10.1906* (ARIZ); 10 mi E of Desert Center, *Wolf 3656* (CAS, UC). San Bernadino Co.: Earp, *Darrow 1942* (ARIZ); 10 mi N of Vidal, *Ferris & Bacigalupi 8241* (CAS); Whipple Mts, *Jepson 5223* (TEX). San Diego Co.: near the narrows, *Gander 2532* (GH); Yaqui Well, *Jepson 8790* (UC); Borrego Mtn, *Jepson 17097* (UC); Ocotillo Wells, *Moldenke & Moldenke 25335* (TEX); San Felipe Wash, Valecito Quadrangle, *Wieslander 168* (UC); Hedges Mine, *Thackery 454* (F, GH, MO, NY, TEX, UC, WIS); San Diego to the Colorado, *Schott s.n.* (NY); San Felipe Canyon, *Spencer 1925* (GH). County unknown: Colorado Desert, *Parish & Parish 800* (CAS, NY); California, *Orcutt 2014* (US); Colorado Desert, *Parry 1881* (F); Picacho, *Gee 1902* (UC); Colorado Desert, *Pringle 1881* (F); on mesa, *Webster 1919* (US). **Mexico.** Baja California Sur: Comondú,

Bryant 1889 (UC); La Paz, *Brandegee 1892* (UC); San José, *Bryant 1892* (CAS, GH); W of cemetery, "Las Lomas de Anita," Loreto, *Carter & Reese 4583* (CAS, MICH, MO, NY, TEX), *Carter & Sousa 5148* (CAS, GH, MEXU, MICH, MO, NY, TEX); vicinity of Loreto, SW of cemetery, *Carter 5846* (MO, NY, TEX); SW end of Cerralvo Island, Gulf of California, *Chambers 948* (CAS, UC); 5 mi from San Ignacio on the Calmalli Rd, *Ferris 8623* (CAS, MICH, NY); Isla Espíritu Santo, *Gaviño 16* (MEXU); Las Tinajas and vicinity, E of Los Picachos de Santa Clara, *Gentry 7589* (CAS); Loreto, *Johnston 3780* (CAS, GH, NY, UC); Loreto, arroyo de Gua, *Johnston 3786* (CAS); Espíritu Santo Island, *Johnston 3969* (CAS); Ceralbo [Cerralvo] Island, *Johnston 4041* (CAS, GH, NY, UC); Pte. Balandras, cerca de Pichilingue, La Paz, *Magallanes 1714* (MEXU, MICH); S end of Cerralvo Island, *Moran 3609* (CAS); San José Island, S end of Amortajada Bay, *Moran 3766* (CAS, UC); La Paz, Pichilingue Island, *Rose 16521* (GH, NY); Cerralvo Island, Ruffo's Ranch, *Soule 5* (CAS), *Soule 20* (CAS); Isla Santa Catalina, *Sousa Peña 230* (MEXU); Rancho La Mezclera, Mpio. La Paz, *Tenorio & Romero 12846* (MEXU); S of Loreto, *Warnock 2086* (TEX); between Bahía de la Concepción and Comondú, *Wiggins 5463* (CAS, GH, MICH, NY, UC); Puerto Balandra, Carmen Island, *Wiggins 17524* (CAS); SW of Isla Ceralbo [Cerralvo], *Wiggins 17724* (CAS); Marques Bay, Isla Carmen, *Wiggins 19049* (CAS).—BAJA CALIFORNIA NORTE: San Felipe, *Brandegee 1894* (UC); south end of Bahía de Los Angeles, *Cowan 2288* (MEXU, US); Banded Agate Mtn, *Fosberg 8310* (CAS, MO); 15 mi S of Mexicali, *Henrickson & Rutherford 1985* (CAS, GH, MICH); Los Angeles Bay, *Johnston 3444* (CAS, GH, MO, UC); near Mexicali, *Jones 1928* (G, GH); Isla Angel de la Guarda, Pond Island Bay, *Moran 12971* (CAS); 37 mi W of Mexicali, *Munz 9583* (UC); ±5 km N of Bahía de Los Angeles, *Schuh & Massie 37* (NY); San Francisquito, Mpio. Ensenada, *Tenorio & Romero 13058* (MEXU); SE part of Bahía de Los Angeles, *Wiggins & Wiggins 14890-A* (CAS).—SONORA: Las Duraznilla, *Brandegee 17.5.1892* (UC); 0.3 mi by rd S of airstrip at Tepache, *Carter et al. 71–50* (ARIZ, UC); SE side of Sierra del Rosario, Gran Desierto, *Coss et al. 20369* (ARIZ); 48 mi SE of Santa Ana, *Dreyer 217* (ARIZ); Ensenada La Cruz, S end of Isla Tiburón, *Felger & Lowe 2598* (ARIZ); Sierra del Rosario, Gran Desierto, *Felger 75–40* (ARIZ); 6.7 mi N of La Unión, Sierra Viejo, *Felger 7891* (ARIZ); Sierra Viejo, *Felger 7903* (ARIZ); vicinity of Tecomate, Bahía Agua Dulce, Isla Tiburón, *Felger & Russell 8905* (ARIZ); Isla Tiburón, Punta Tortuga, *Felger et al. 10329* (ARIZ); Isla Tiburón, Ensenada Blanca, *Felger et al. 10365* (ARIZ); Ensenada Grande, Bahía San Pedro, *Felger & Cooper 10380* (ARIZ); Isla Tiburón, vicinity of Tecomate, *Felger et al. 11118* (ARIZ), *Felger 12512* (ARIZ); 4 mi by rd NW of Rancho Noche Buena, Sierra Seri, *Felger & Bezy 14031* (ARIZ); Las Cuevitas, 10 mi N of Desemboque San Ignacio, *Felger 14260* (ARIZ); Sierra del Rosario, *Felger et al. 20742* (ARIZ); 40 mi N of Hermosillo, *Ferris 8763* (ARIZ); 15 mi S of Pitiquito, Dtto. Altar, *Gentry 4475* (ARIZ, MO); Picu Mts, Dtto. Altar, *Gentry 4481* (ARIZ); 7–9 mi W of Navojoa, *Gentry 7968* (ARIZ, MICH, UC); Bahía Algodones, San Carlos, *Goldberg & Cornell 77–109* (ARIZ); Batamotal, *Goldman 395* (MO); vicinity of Libertad, *Graham 3798* (CAS); Libertad, *Graham 3878* (CAS); 7 mi S of Altar, *Graham 3907* (CAS); 25 mi N of Hermosillo, *Hansen et al. 1371* (MICH, WIS); 30 mi N of Hermosillo, *Hansen et al. 1376* (MICH, TEX, US, WIS); La Tuiaja, *Hartman 241* (GH); 12 km S of Puerto Libertad, *Henrickson 1546* (MICH); San Esteban Island, *Johnston 3205* (CAS); Tiburón Island, *Johnston 3277* (CAS); E rim of MacDougal Crater, Sierra Pinacate, *Kamb 1998* (ARIZ, CAS, TEX, UC); El Alamo, near Magdalena, *Kennedy 7012* (CAS, F, MO, UC); Isla San Esteban, *Lott & Atkinson 2446* (CAS, NY); 10 mi SE of Navojoa, *Mahler & Thieret 6006* (NY); Paso de Luis, *Mearns 2700* (CAS); Quitovac, between Sonoyta and Caborca, *Nabhan & Rea 286* (ARIZ); 57.6 km ESE of San Luis R. C., between Gran Desierto dunes and Gila Mts, *Newberry 353* (NY); 50 mi S of Nogales, *Niering & Whitaker 8.5.1963* (ARIZ); Onavas, *Pennington 261* (TEX); 2 mi S of La Rivera, Dtto. Buena Vista, *Peters 115A* (UC); foothills and mesas, *Pringle 11.8.1884* (F, GH, MICH, NY, WIS); vicinity of Navojoa, *Rose et al. 13104* (NY); vicinity of Guaymas, Los Ranchos, *Rose et al. 15048* (NY); Mpio. Puerto Peñasco, Río Sonoyta, on rd to Sierra El Pinacate, *Sanders & Neisess 5707* (ARIZ, TEX); ravines on the Gila, *Schott s.n.* (NY); 41 mi N of Hermosillo, *Shreve 6560* (ARIZ, MICH, NY); Torim, *Studhalter 1520* (US); Tiburón Island, vicinity of Tecomate, *Whiting & Gentry 9011, 9012* (ARIZ); 9 mi W of Pozo Cerna, *Wiggins 6041* (CAS); 9.6 mi S of Llano, *Wiggins 11632* (CAS, TEX, UC).

Felger and Moser (1985) present an extensive discussion of the ethnobotanical uses of *Olneya* by the Seri Indians of Sonora, Mexico. It is clear that *Olneya* is an economically important species to these people of the Sonoran Desert; it has many designated uses, including nutritional (the seeds and unripe pods) and medical, and in the production of firewood, tools, weapons, musical instruments, and artistic sculptures from the wood. Also, tinctures made from the green wood of *Olneya* were used by the Seris in vision quests for power. Such important and varied ethnobotanical uses of *Olneya* are consistent with such uses of many other species of Robinieae in the New World, especially of *Robinia*, *Gliricidia*, and *Sesbania*.

In spite of the great range of variation in flower and seed sizes and numbers of leaflets per leaf, geographical patterning of morphological variation exists in *Olneya tesota* only with regard to stipitate-glandular indumentum and perhaps ovule number per ovary. In populations from Baja California Sur, this glandular indumentum is very dense on the calyx and ovary, and continues to grow on the ovary as it matures into fruit. Also, the highest ovule-per-ovary-ratio so far counted in *Olneya* (15–17) is found on these same specimens, as compared to 10–14 in other parts of the range. Adequate fruiting specimens which were amenable to counting ovules were not numerous, however, and this difference could be due to sampling error. Notably, in spite of the many ovules found in the ovaries of *Olneya*, only 1–2, rarely 3–5, seeds are set, and these vary in a seemingly random fashion as to the position along the placenta (i.e., they are not always the distal- or proximal-most along the placenta).

The type locality is stated as "table lands of the Gila" on the label of the holotype. The Gila River drains predominantly Maricopa and Gila counties in Arizona, but the type locality probably lies in Yuma County; in the protologue it is cited as the "lower Gila," and on the isotype at GH as "Head of Gulf of California." in Gray's handwriting.

10. Peteria A. Gray., Pl. wright. 1: 50. 1852, non *Peteria* Rafinesque, 1820, nom. rej.—TYPE: *Peteria scoparia* A. Gray.

Perennial herbs. Stems erect to ascending, occasionally sprawling, lignescent at base; bark none. Roots tuberous, thick and spreading, sprouting leafy shoots. Leaves alternate, pulvinate, imparipinnate; leaflets mostly opposite, occasionally alternate at maturity, very narrowly to widely elliptic, retuse to acuminate, caducous from the more persistent rachis during the onset of the dry season or winter, adaxial surface glabrate to sericeous, with or without finely scattered tannin deposits, abaxial surface strigose to sericeous, with or without finely scattered tannins; leaf rachis canaliculate, the groove with a dense indumentum, stipels absent; stipules free and narrowly attached to the stem adjacent to the petiole, spinescent and persistent. Inflorescences of erect, sessile terminal racemes, with sequential acropetal maturation, but often with more than one flower at anthesis simultaneously on a single rachis, the rachis with 5–40 or more nodes, strigose to sericeous, sometimes stipitate-glandular. Flowers ebracteolate, with a well-developed hypanthium; floral bracts linear-lanceolate, usually persistent, sometimes spine-tipped; pedicels confluent with the calyx, straight and ascending during anthesis and seed set, falling with the aborted flower. Calyx campanulate, rounded at base, strigose to sericeous, usually with stipitate glands, with five prominent lobes equalling or shorter than the tube, the two vexillary ones often fused at the base, the lobes narrowly triangular, margins not overlapping at anthesis. Standard orbicular to oblong, emarginate, reflexed mostly 90° at anthesis, dorsal surface glabrate, subequal to longer than the keel petals, with two prominent callosities about the midrib in the middle of the nectar-guide, claw abruptly contracted from the blade, inflexed auricles present; wing petals sculptured, commonly protruding beyond the keel at anthesis and loosely attached to it by a fold in the basal auricle, the distal ends glabrate, the blades mostly parallel to the plane of the keel; keel petals valvately connate distally nearly to the tip, apex blunt. Staminal tube diadelphous, with basal fenestrae; anthers included in the keel tip, basally attached to the filaments, the broad connective reddish brown; intrastaminal nectariferous disk absent. Pollen spheroidal, tricolporate, mesocolpium rugulate-perforate, polar regions smooth. Ovary subsessile to short-stipitate, the stipe about as long as the calyx tube, glabrate, without stipitate glands, with 8–12 ovules; style with a pollen brush densely tufted toward the distal tip of the style, the style base thickened and becoming displaced to the carinal side during fruit maturation; stigma ciliate, capitate and terminating the distally tapered style. Legumes subsessile to stipitate, the stipe rarely longer than the calyx tube, glabrous, stipitate glands absent, inner epidermis of the endocarp of adjacent valves smooth to slightly spongy, mostly not abutting and not forming uniform seed compartments, the valves thinly woody at maturity, elastically dehiscent, the dorsal and ventral margins generally parallel in fruit, except where tandem seed abortion occurs. Seeds transversely oriented, generally cylindrical at maturity, not constricted in the middle, accumulating canavanine, uniformly brownish but mottled with dark purple blotches, hilum apical and attached to a very short funiculus, testa firmly adherent. Seedling with epigeal germination, cotyledons thick and fleshy, eophyll unifoliolate, orbicular, alternate to the first metaphyll, metaphylls with intermediate leaflet number relative to the mature leaves. Base chromosome number x = 8 (Lavin, unpubl.) and 9 (Goldblatt 1981).

Peteria is at once distinguished from the rest of the tribe Robinieae by its inflorescence comprising terminal racemes. Also, its perennial herbaceous habit, well-developed spinescent stipules, and the relatively dense arrangement of trichomes along the canal of the leaf rachis are a unique combination not otherwise found in the tribe. A unique character, but not readily observed, is the presence of large rhizomatous tubers from which the above-ground stems arise. Label data indicate that these tubers were eaten by Native Americans, who in Mexico gave it the name "camote del monte" or wild sweet potato. The specific relationships of *Peteria* clearly lie in the *Robinia*-group, as evinced by its stipitate-glandular indumentum, spinescent stipules, style with a pollen brush, and chloroplast DNA evidence (Lavin & Doyle 1991). Among these genera, *Peteria* is superficially similar to the genus *Robinia* in its long tubular calyx with relatively short calyx lobes, which is borne on relatively short pedicels. In contrast, *Peteria* also approaches *Olneya*, *Genistidium*, and *Sphinctospermum* in its leaves that lack stipels.

A relationship of *Peteria* to *Genistidium* and *Sphinctospermum* is postulated based on inflorescence morphology. Often the lower-most flowers in the terminal raceme of *Peteria* are subtended by leaves. Above this, the flower bracts are often spinescent and in some cases each bract is subtended by two stipules. This arrangement suggests that the terminal raceme of *Peteria* represents the consolidation of many solitary axillary flowers that are borne in the axils of very reduced leaves, very similar to the arrangement found in *Sphinctospermum* and *Genistidium*. Axillary vegetative growth in *Peteria* is seldom extensive, and sympodial stem growth is unknown in this genus. In this regard, the terminal inflorescence of *Peteria* is not the same as that found in other more distantly related legume groups noted for their terminal inflorescences (e.g., *Tephrosia*, *Dalea* L., etc.). No inflorescence rachis of the species of *Peteria* has been observed to be "opposite the leaves," as described by Porter (1956), which would imply sympodial growth.

Peteria is distinctive in the *Robinia*-group not only in its terminal racemes, but also in its glabrous ovary, which is stipitate to a greater degree than in the other genera of this group. The ovarian stipe in *Peteria* is often nearly as long as the calyx tube, whereas in the rest of the *Robinia*-group, the ovarian stipe is essentially absent or much shorter than the calyx tube, and the ovary always bears some sort of indumentum. With regard to a glabrous, stipitate ovary, *Peteria* distinctly converges on the genera of the *Gliricidia*-group. Porter (1956) distinguishes *P. pinetorum*, in part, by its legume that is borne on a stipe up to 5 mm long. This, however, is not distinctive of this species but occurs sporadically in the other species of the genus.

Within the tribe Robinieae, *Peteria* is second only to *Robinia* in having invaded cool-temperate habitats. Although the *Robinia*-group stands out among the rest of Robinieae in being generally centered in warm-temperate North America, two species of *Peteria*, *P. thompsonae* and *P. scoparia*, occur in cool-temperate regions of the United States. *Peteria thompsonae* is restricted to this climate, and its range extends as far north as southwestern Idaho, where it is found in low-elevational Great Basin desert habitats. Perhaps the evolution of the completely herbaceous growth form arising from underground tubers has facilitated this cool-temperate invasion from presumably warm-temperate or even tropical ancestry. With *Peteria*, *Sphinctospermum*, and a few species of *Coursetia* exhibiting a strictly herbaceous growth habit, it is possible that the tribe Robinieae could have been a major New World source of temperate herbaceous groups. Why this lineage of Robinieae did not radiate into temperate regions of the New World, and why the Millettioid-Galegeoid lineage did is perhaps an unanswerable question. *Peteria* is one of the few representatives of a strictly New World Robinioid temperate herbaceous lineage. That *Peteria* and other herbaceous species of Robinieae (e.g., *Coursetia caribaea* var. *sericea*) are commonly misidentified as "*Astragalus* sp." attests to the morphological convergence of temperate herbaceous Robinieae and temperate herbaceous Galegeae.

Porter (1956) provides a good review of the taxonomic history of the genus *Peteria*. The scope of his analysis was to clarify the taxonomic identity of the species that occurred in Nevada, *P. thompsonae* and *P. nevadensis*. Though he concluded that the two were not distinct, he also took the opportunity to describe a collection of H. S. Gentry from the Río Mayo region in southwestern Chihuahua as a new species, *P. pinetorum*, which is still recognized in this treatment. Beyond this, Porter mentions little if anything regarding the overall circumscription of the genus, and he did not further develop the idea of Johnston (1941) that *Peteria* might be most closely related to *Genistidium*. As such, the treatment presented here is largely in agreement with that of Porter in that the same four species are recognized. Many more recent collections, however, have expanded the known ranges of all species, except *P. pinetorum*, and furthermore have refined the list of those characters that are most useful for distinguishing the species and suggesting relationships among them.

The monophyly of the genus *Peteria* is clearly marked by such derived characters (polarized relative to the genera of the *Robinia*-group) as terminal racemes, glabrous ovaries, perennial herbaceous stems arising from tuberous root stocks, standard petals that are not strongly reflexed (i.e., more than about 60–70°) at anthesis and that do not bear a nectar-guide, spinescent floral bracts (even if occasionally weakly so), and a canaliculate leaf rachis in which the groove bears distinctly more trichomes than the other parts of the rachis. Species of *Peteria* are morphologically marked by only a few characters each, and there are very few, if any, other characters that can be polarized to infer monophyletic subgroups. Nevertheless, *P. thompsonae* is tentatively considered the sister species to the rest of the genus, because the other three species possess very narrow-elliptic leaflets. The broadly elliptic to orbicular leaflets of *P. thompsonae* match more closely the condition in the outgroup genera (i.e., the others in the *Robinia*-group). Other such potentially phylogenetically informative characters are unknown at present. In the following key, the species are arranged in a presumed phylogenetic order with *P. thompsonae* listed first. The order of the remainder is based on overall similarity. *Peteria scoparia* is near *P. thompsonae* in having flowers with whitish petals and, at least in some northern populations, slightly less narrowly elliptic leaflets (up to 4 mm wide). *Peteria pinetorum* is most similar to *P. glandulosa* in having often more than 13 narrowly elliptic leaflets per leaf and in being more uniformly stipitate-glandular on the calyx and pedicels. *Peteria glandulosa*, in contrast, is more similar to *P. scoparia* in having strigose inflorescence rachises that are stipitate-glandular toward the distal tip, as opposed to *P. pinetorum*, which has a villous rachis that is stipitate-glandular from the base. Thus, the phenetic order of overall similarity is *P. thompsonae*, *P. scoparia*, *P. glandulosa*, and *P. pinetorum*.

Key to the Species of Peteria

1. Leaflets broadly elliptic to orbicular, if narrowly elliptic, then leaflets usually well over 12 mm long; flowers at anthesis usually congested along the rachis, flowering internodes 0.1–5.0 (–12) mm long, usually more than 2 consecutive flowers simultaneously at anthesis; standard petals 16–22 mm long; U.S.A. (northern Arizona, Nevada and adjacent California and Idaho, and Utah). — 10.1. *P. thompsonae.*
1. Leaflets narrowly elliptic, not over 10 mm long; flowers at anthesis not congested along the rachis, flowering internodes (3–) 5–35 mm long, rarely with more than 1 flower reaching anthesis at a time; standard petals 14–16 mm long (petals unknown in *P. pinetorum*); U.S.A. (southern Arizona, New Mexico and adjacent southwestern Colorado) and Mexico.
    2. Leaflets 7–17 (–19) per leaf; inflorescence rachis glabrate, or strigose and stipitate-glandular toward the distal end; U.S.A. and Mexico (central and eastern Chihuahua, Coahuila, México, Nuevo León, Puebla, Querétaro, San Luis Potosí, and Zacatecas), desert scrub and grasslands.
        3. Petals mostly whitish, sometimes with a purplish tinge; leaves with the terminal leaflet often reduced to a spinose seta; leaflets commonly, but not always, deciduous by anthesis or fruiting stage; U.S.A. and Mexico (central and eastern Chihuahua and Coahuila). — 10.2. *P. scoparia.*
        3. Petals pure purplish to mostly whitish; leaves with terminal leaflet always well developed; leaflets mostly persistent through anthesis and fruiting stage; Mexico (Distrito Federal, Durango, México, Nuevo León, Puebla, Querétaro, San Luis Potosí, and Zacatecas). — 10.3. *P. glandulosa.*
    2. Leaflets 15–25 (–29) per leaf; inflorescence rachis villous and distinctly stipitate-glandular from the base; Mexico (southwestern Chihuahua), pine forests. — 10.4. *P. pinetorum.*

10.1. Peteria thompsonae S. Watson, Amer. Naturalist 7: 300. 1873.—Type: U.S.A. Utah: Kane Co., Kanab, 1872, *Thompson s.n.* (holotype: GH!; isotypes: NY! US!).

Peteria nevadensis Tidestrom, Proc. Biol. Soc. Wash. 36: 183. 1923.—TYPE: U.S.A. Nevada: Clark Co., on mesa 16 mi NE of Las Vegas, 840 m, 23 Apr 1919, *Tidestrom 9083* (holotype: US!; isotypes: GH! NY!).

Perennial herbs 2–10 dm tall. Larger leaves (5.5–) 7.0–15.5 cm long; leaflets (9–) 15–31 per leaf (terminal leaflet well-developed), persistent on rachis through fruiting stage, the larger mature ones 12–21 mm long, 7–13 mm wide, broadly elliptic to orbicular, rarely narrowly elliptic in eastern Utah, both surfaces glabrate (but then not along the margin) to strigose or sericeous, with or without tannin deposits not confined to the midrib but finely scattered on both surfaces of older leaflets; leaf rachis more densely sericeous above than below; stipules spinescent, (3–) 4–6 mm long. Inflorescence generally a congested terminal raceme, the lower flowers sometimes subtended by leaves, rachises (5–) 10–30 (–40) cm long, with (10–) 20–100+ nodes, the internodes at anthesis mostly 0.1–5.0 (–12) mm long, sparsely strigose, stipitate glands usually common towards the distal end; floral bracts herbaceous to subspinescent, 5–11 mm long; pedicels sericeous, with stipitate glands, 2–3 mm long at anthesis, 3–4 mm long in fruit. Calyx tube (6–) 7–11 mm long, with stipitate glands, lobes 7–11 mm long. Standard 16–22 mm long, whitish, aging mostly yellowish to light brown; wing petals 16–22 mm long, whitish; keel petals 14–19 mm long, the blades whitish, aging purplish at tip. Staminal tube with filaments free for about the distal one-fourth. Ovary glabrous, with 8–12 ovules; style 4.5–6.0 mm long. Legumes 4.0–6.5 cm long, with 2–5 seeds. Seeds 4–6 mm in diameter at maturity, light brown, often with dark purplish mottling at maturity. Figs. 119–120.

Phenology. Flowering from May through June in California, Nevada, Idaho, and Utah, and from the end of April through June and sporadically into July or August in Arizona; fruiting by the end of May through June, occasionally into July and August, especially in Arizona.

Distribution (Fig.118). Centered but only locally common in the Great Basin of the western U.S.A. in northern Arizona, California, Idaho, Nevada, and Utah; sporadic in gravelly washes and sandy soils, especially along roadsides; 830–1800 m. In all states but Arizona, *P. thompsonae* co-occurs with *Abronia*, *Artemisia*, *Atriplex*, *Bromus*, *Chrysothamnus*, *Eriogonum*, *Glyptopleura*, *Nama*, *Nemacladus*, *Poa*, *Penstemon*, *Salvia*, *Sarcobatus*, *Townsendia*; in Arizona, it co-occurs commonly with a distinctively different set of species, including members of *Ambrosia*, *Artemisia*, *Coleogyne*, *Ephedra*, *Eurotia*, *Juniperus*, *Larrea*, *Pinus*, Poaceae, *Prosopis*, *Thamnosma*, and *Yucca*.

ADDITIONAL SPECIMENS EXAMINED. **U.S.A.** ARIZONA: Mohave Co.: 0.5 mi NE of White Saddle Spring, *Armstrong 1039* (NY); Parashont Trail Canyon, 11 mi S of Mt. Trumbull Village, *Atwood 6031* (MO, NY); Pakoon Wash, 1 mi N of Pakoon Ranch, *Atwood 10957* (NY); 10 mi W of Fredonia, Hwy 389, *Deaver 6253* (ARIZ, CAS); White Saddle Spring, *Gierisch 4668* (ARIZ, NY); 4.5 mi W of Fredonia on Hwy 389–59, *Higgins 10978* (NY); Antelope Valley, Hwy 389 10.5 mi W of Fredonia, *Holmgren 3384* (CAS, NY); Kaibab Indian Reservation, Moccasin, *Kelly s.n.* (CAS); just S of Hwy 389 on rd to Mt. Trumbull, *Lavin 6179a* (RM); Kaibab Payute Indian Reservation, Mt. Trumbull Rd 0.4 mi S of Hwy 389, *Lehto L23825* (ARIZ, NY); 9.8 mi W of Fredonia, *Moore et al. 6794* (ARIZ, NY); Parashont Canyon, 3.6 mi by Parashont Canyon Rd S of Grand Canyon National Park boundary, *Reichenbacher & Wilson 1730* (ARIZ); Burro study area, Gyp Hill, *Smith 1032* (ARIZ); Cottonwood Wash rd, 19 mi SE of Mesquite, *Thorne & Atwood 5248* (NY).—CALIFORNIA: Inyo Co.: Kingston Range, border with San Bernadino Co., Mesquite Valley Rd, 8 mi NW of Smith Talc Rd, *Castagnoli et*

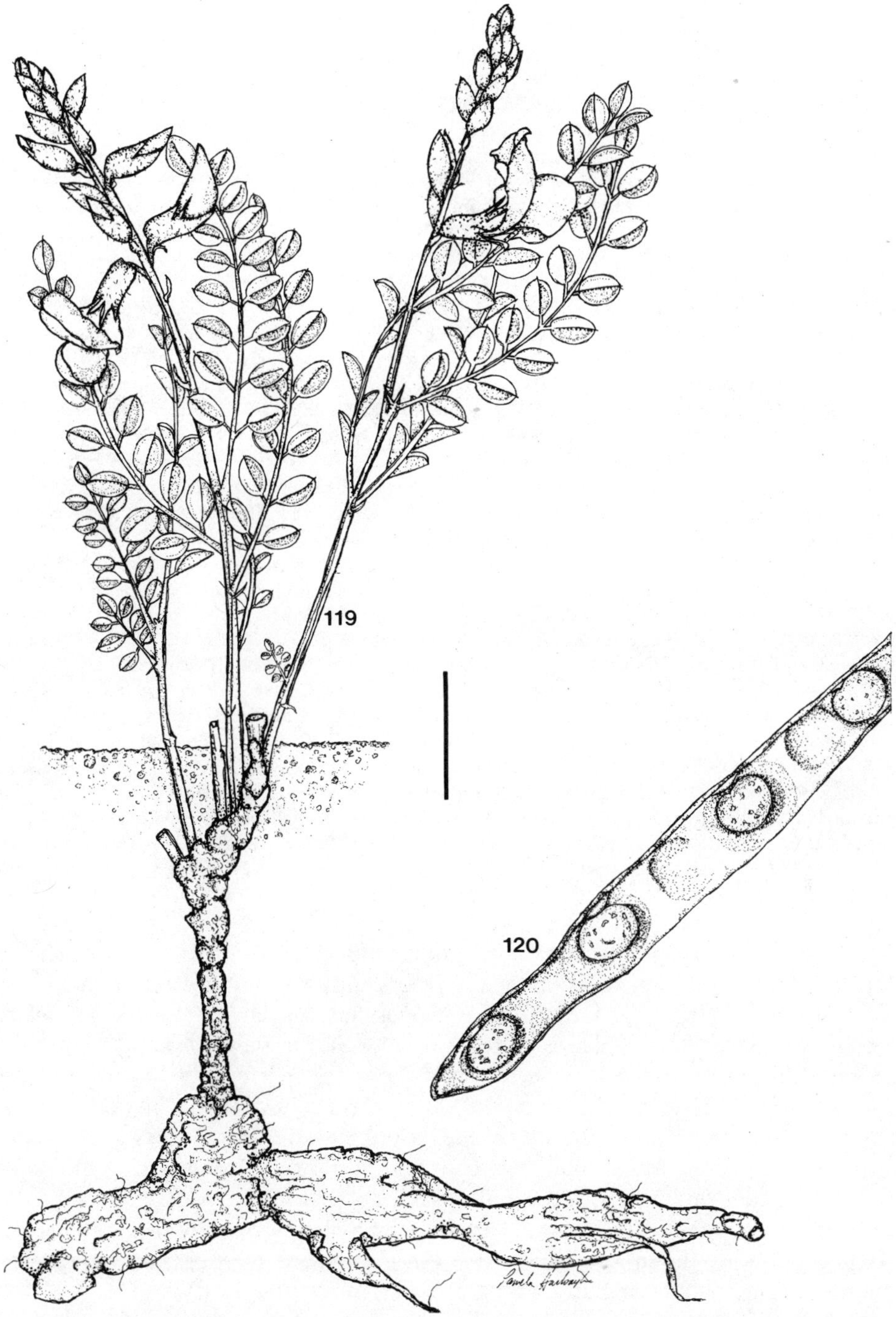

FIGS. 119–120. *Peteria thompsonae*. 119. Flowering individual attached to underground tuber system. 120. Close-up of the internal legume valve showing seeds attached. (119, bar = 2 cm; 120, bar = 10 mm).

al. 150 (CAS).—IDAHO: Owyhee Co.: 13 mi S of Bruneau, along Hwy 51, *Baker 14023* (NY); approx. 5 mi SW of Bruneau Hot Springs, *Brunsfeld 2461* (NY); 13 mi SE of Bruneau, at Hot Creek, *Christ & Christ 16715* (NY); about 12 mi SE of Bruneau by rd by Indian Bathtub, *Halse 1934* (ARIZ, CAS, NY); in and along the dirt rd going down to the Bruneau River, *Mooers 756* (NY); SW of Hot Springs, *Ripley & Barneby 6511* (CAS, NY); Indian Bathtub, S of Bruneau, *Rosentreter 150* (NY); near Indian Bathtub, S of Bruneau, *Rosentreter 1328* (NY); near Bruneau, *Wuerthner 83–27* (NY).—NEVADA: Esmeralda Co.: mining area 2 mi W of Hwy 95, 10 mi S of Tonopah, *Neese & White 9049* (NY); Goldfield Sands, 2 mi N of Goldfield, *Pavlik 153* (NY); Goldfield, *Shockey s.n.* (CAS); 0.8 mi N of Alkali Springs, 10.8 mi NW of Goldfield, *Tiehm 6351* (CAS, MO, NY, RENO); 10.1 mi N of Silver Peak, *Tiehm 6379* (CAS, MO, NY, RENO). Lincoln Co.: T9S R61E S28, Lower Pahranagat Lake on Old Corn Creek Rd, 11.7 mi W toward Desert Game Range, *Thorne & Welsh 924A* (NY). Nye Co.: South Frenchman Flat, *Beatley 3256* (CAS); South Frenchman Flat, *Beatley 8248* (CAS, NY, RENO, US), *Beatley 8626* (CAS, NY, RENO); North Stonewall Drainage, S of Wildhorse Springs, E of Goldfield Hills, *Beatley 12883* (CAS, NY, RENO, US); 20 mi SW of Currant, *Eastwood & Howell 9422* (CAS, F, GH); 3/4 mi N of Locke's Ranch, *Maguire & Holmgren 25127* (ARIZ, CAS, MO, NY); foothills of Pancake Range near Silverton, *Ripley & Barneby 3647* (CAS, NY); Railroad Valley, 22 mi SW of Currant, *Welsh et al. 414* (RENO); 3 mi SW of Locke's, Railroad Valley, *Williams & Williams 78753* (NY, RENO); 0.6 mi NE of Locke's, *Williams 781433* (NY, RENO); Railroad Valley, 3.4 mi SW of Locke's, NW of Hwy 6, *Williams & Tiehm 8081* (NY); S of Locke's, Railroad Valley, 1.6 mi S of Hwy 6, *Williams & Tiehm 81196* (CAS, MO, NY, RENO), *81206* (RENO); Frenchman Playa, Nevada Test Site, *Williams 149* (RENO).—UTAH: Emery Co.: San Rafael Desert, 15.6 mi S of Green River, *Holmgren et al. 8785* (NY, RENO); Green River, *Jones s.n.* (MO, NY, UC), *Jones s.n.* (CAS, NY, UC), *Jones s.n.* (MICH, NY); San Rafael Swell, *Jones s.n.* (CAS, GH, NY, UC). Iron Co.: 5 mi SE of Uvada, *Barneby 17816* (NY). Juab Co.: Partoun, 5 mi S of Trout Creek, *Lavin & Hedrick 6157* (RM); 11 mi S of Troutcreek, *Maguire & Becraft 2664* (GH, MO, UC); 8 mi S of Troutcreek, *Maguire & Becraft 2665* (MO, UC); 2.5 mi SW of Partoun, *Welsh et al. 16771* (NY). Kane Co.: Kanab, *Jones 5386* (RENO); W side of Kanab, N side of Hwy 89, *Neese & Welsh 16880* (NY); left hand fork of Last Chance Canyon, along Escalante rd, 10 mi N of Four Mile Bench, *Welsh & Welsh 12811* (MO); near the cliffs, Kanab, *Tidestrom 2409* (NY, WIS). San Juan Co.: 1 mi E of Upheaval Dome, Island in the Sky, Canyonlands National Park, *Welsh 5382* (MO, NY, WIS). Washington Co.: NW of Smithsonian Butte, *Higgins 15574* (NY); Springdale, *Jones 5229 or 5230* (MO, UC, WIS); Little Zion, *Jones s.n.* (MICH); just N of Springdale toward Zion National Park, *Lavin 6182a* (RM); 2 mi S of Rockville on dirt rd to Hwy 59, *Lavin 6183a* (RM); Springdale, *Ripley & Barneby 4932* (CAS, NY); Zion Canyon, Hwy 9, Virgin River Bridge, *Thorne & Chandler 5612* (RENO); Wire Mesa, 3 mi S of Rockville, *Welsh et al. 22908* (NY). Wayne Co.: Blue benches SW of Hanksville, N of Henry Mts, Bert Avery Seep, *Neese 5942* (NY).

Peteria thompsonae is remarkable among the species in this genus for having the greatest range in size variation for the petals and calyx, inflorescence rachis, and leaflet size and number per leaf. This species has the largest flowers of all the species (standard petals measure up to 22 mm long), but the lower end of the petal size variation, 16 mm, overlaps with that of the other species. The inflorescence of *P. thompsonae* differs from that of the other species in that the flowers are congested at anthesis, because the internodes are generally short, usually less than 5 mm long, and often more than two consecutive flowers reach anthesis simultaneously. In the other species, the inflorescences have more widely spaced nodes and bear one to two flowers at anthesis simultaneously. Furthermore, if the season remains favorable, the inflorescence rachis of *P. thompsonae* can continue to grow, sometimes up to 40 cm long and with a hundred or more nodes.

Subtle differences in leaf morphologies and occasionally other characters are correlated with geographical distribution. In northern Arizona (Mohave County), populations are distinctive in having a longer flowering season, the end of April through August; the leaflets are commonly more numerous (25–31) and consistently more glabrate, and bear more scattered tannin deposits than in populations from other regions. Furthermore, the flowers in these populations have a generally smaller calyx tube (6–7 mm long) and and a greater number of ovules per ovary (10–12). Populations from southern Utah (Kane and Washington counties) have leaves similar to those of the Arizona populations in that they tend to be more distinctly marked with finely scattered tannin deposits and have glabrate to sparsely strigose surfaces, but leaflet number per leaf is not as consistently high as in the Arizona populations. Also, the flowering season is shorter, and the calyx dimensions are usually larger.

Populations from eastern Utah (Emery, Wayne, and San Juan counties) are distinct in having leaves with narrowly elliptic leaflets that are glabrate at least on the upper surface. This difference in leaflet dimension is clearly discontinuous with the remainder of the populations of the species, which bear broadly elliptic to orbicular leaflets. No formal taxonomic distinction is made, however, until a quantitative morphometric and genetic analysis of *P. thompsonae* indicates that the differences in these geographical races deserve taxonomic recognition.

10.2. Peteria scoparia A. Gray, Pl. wright. 1: 50. 1852.—Type: U.S.A. Texas: Brewster Co., mountain valleys beyond the pass of the Limpia [probably near Alpine, fide Porter, 1956], 1849, *Wright 138* (lectotype, designated by Porter, 1956: GH!; isolectotype: GH!).

Perennial herbs 2–8 dm tall. Larger leaves 4–5 cm long; leaflets 11–13 per leaf (terminal leaflet often reduced to a spinose seta), readily caducous from or persistent on the rachis by flowering or fruiting stage, the larger mature ones 3–6 (–9) mm long, 1–2 (–4) mm wide, linear to very narrowly elliptic, adaxial surface glabrate, abaxial surface glabrate to strigose, especially along the tanniniferous midrib and margins; leaf rachis generally much more densely sericeous above than below; stipules spinescent, 3–5 mm long. Inflorescence an open terminal raceme, the lower flowers sometimes subtended by leaves, rachises 12–25 cm long, with 8–20 nodes, the internodes at anthesis mostly (3–) 6–35 mm long, glabrate, or strigose and stipitate-glandular toward distal end; floral bracts weakly spinescent, 3–5 mm long; pedicels glabrate to strigose or sericeous, with or without stipitate glands, 3–8 mm long at anthesis, 5–8 mm in fruit. Calyx tube 5.5–7.0 mm long, with or without stipitate glands, lobes 2.0–3.5 (–4.0) mm long. Standard 15–16 mm long, mostly whitish with a purplish tinge, aging light brown; wing petals 14.0–15.5 mm long, mostly whitish; keel petals 12–14 mm long, the blades mostly whitish with a purplish tip. Staminal tube with filaments free for about the distal one-fourth. Ovary glabrous, with 10–11 ovules; style 4.5–6.0 mm long. Legumes 3.5–5.0 (–6.2) cm long, with 2–5 seeds. Seeds 3.5–6.0 mm in diameter at maturity, light brown, often with dark purplish mottling at maturity.

Phenology. Flowering occurs from June through early September; fruits persist during August and September.

Distribution (Fig. 118). Primarily centered in the Chihuahuan Desert and adjacent regions in the U.S.A. (in Texas in Brewster, Culberson, Hudspeth, Jeff Davis, and Presidio counties; in New Mexico in Colfax, Doña Ana, Grant, Hidalgo, Luna, and McKinley counties; in Arizona in Apache and Cochise counties; in Colorado in Montezuma County) and Mexico (Chihuahua, Coahuila); on limestone substrates, often on steep slopes and atop rimrocks, in desert landscapes dominated by Poaceae, *Agave*, *Brickellia*, *Hechtia*, *Juniperus*, *Yucca*, *Leucaena*, *Pinus*, *Prosopis*, *Quercus*, and *Vauquelinia;* 1000–2000 m.

ADDITIONAL SPECIMENS EXAMINED. **U.S.A.** ARIZONA: Apache Co.: draws in sandy badlands, 9 mi S of St. Johns, *Barneby 12970* (CAS); Petrified Forest, *Toole & Goodding s.n.* (US), *Goodding 4183* (ARIZ). Cochise Co.: on hill S of Stockton Hill, *García s.n.* (ARIZ); on limestone hills near Tombstone, *García 124* (ARIZ).—COLORADO. Montezuma Co.: Mesa Verde, *Brandegee 1103* (UC).—NEW MEXICO: Colefax Co.: on flat valley floor between Mesita Leon and Mesita de la Venta, *Losure 176* (ARIZ). Dona Ana Co.: chiefly in the Valley of the Rio Grande, below Doñana, Camp Bache, near Rock Creek, *Bigelow et al. 251* (NY); Tortugas Mtn E of Mesilla Park, *Standley s.n.* (MO); Tortugas Mts, on mesa W of Organ Mts, *Wooten s.n.* (ARIZ, CAS, RENO, UC). Grant Co.: Telegraph Mts, *Wooton s.n.* (US). Hidalgo Co.: Big Hatchet Mts, Peterson Ranch, *Turner & Powell 6096* (TEX). Luna Co.: Victorio Mts, 3 mi S of Gage exit on I–10 at old mining town along limestone ridge, *Worthington 12401* (NY, TEX); between Frontera and Mimbres, *Wright 962* (GH, MO, NY, UC). McKinley Co.: 5 mi NE of Jones Ranch Day School, 8 mi W of Hwy 32 and roadhouse, *Peabody & Sears 1992* (BRY); Ramah Navaho, *Vestal & Vestal 151* (GH), *Vestal & Vestal 480* (GH). Otero Co.: Otero Mesa, 15 mi E of Orogrande, El Paso Canyon, *Meents & Moir s.n.* (NMC).—TEXAS: Brewster Co.: rare on Sul Ross College Hill, Alpine, *Warnock T429* (ARIZ, TEX). Culberson Co.: Guadalupe Mts, *Havard 51* (GH); lower Pine Canyon, Gaudalupe Mts, *Warnock 20837* (SRSC); along Pine Springs Canyon, Guadalupe Mts, *Waterfall 5240* (GH, MO, NY). El Paso Co.: Franklin Mts, *Cory 45066* (TEX). Hudspeth Co.: Hueco Tanks State Park, East Mtn, *Van Devender s.n.* (ARIZ); low hills approaching Sierra Diablo Plateau, 11.5 mi N of Allamore, *Waterfall 5350* (GH, MO, NY); cliffs in the Eagle Mts, SE of the old Love Ranch, 35 mi SE of Sierra Blanca, *Waterfall 6717* (GH, MO, NY). Jeff Davis Co.: entrance to Goat Canyon, Mt. Livermore, *Hinckley 423* (F, NY, TEX); Goat Canyon, rocky slope near mouth of Goat Canyon, 6 mi SW of Mt. Livermore, *Hinckley s.n.* (NY); Goat Canyon, George Jones Ranch, Davis Mts, *Hinckley s.n.* (ARIZ, GH); ca 22 mi W of Fort Davis, *Powell 3379* (MO, SRSC, TEX); ca. 9 mi ESE of Ft. Davis along NW edge of Henderson Mesa, on the Pollard Ranch, 3 mi NE of the hwy, 6 mi E of the CDRI Visitor's Center, *Zimmerman 2440* (SRSC). Presidio Co.: on limestone hills just N of Chinati Peak, between Dead Horse and Pinto Canyon, *Butterwick & Lott 3829* (TEX); at signal block on railroad, 5 mi E of Marfa, *Hinckley 2823* (NY, SRSC); SE of old Jim Shannon ranch, at the southern end of Sierra Tierra Vieja, *Hinckley 3025* (SRSC); 7.1 mi E of Alamito Creek bridge at Marfa city limits, 17.7 mi W of Alpine on Hwy 90, *Lott 190* (MO, SRSC); along hwy about 3 mi S of Marfa toward Shafter, *Warnock 10578* (SRSC, TEX). County unknown: *Hayes 159* (GH); *Hinckley 18163* (TEX). **Mexico.** CHIHUAHUA: 3 mi E of Hacienda San Diego, near hot spring, *Moore 425* (BRY); 4 km from Pte. Estaciön Sueco, Mpio. M. Ahumada, *Valdés VR–546* (MEXU, NY); near lake Encinillas, N of Chihuahua, *Wislizenus 126* (MO).—COAHUILA: SW flank of Sierra del Carmen, *Chiang et al. 9273* (CAS, MO, NY); between Rancho de Figueroa and Hacienda del Barrendo, *Mittleman & Engard M14* (NY).

Peteria scoparia occasionally has paripinnate leaves, formed by the reduction of only the terminal leaflet to a spinose seta; the subtending leaf rachis is unmodified. In other members of Robinieae with paripinnate leaves, such as *Sesbania*, *Olneya*, and certain species of *Coursetia* and *Poitea*, the terminal leaflet as well the subtending rachis are reduced to a seta.

Peteria scoparia has been described as having early-deciduous leaflets (Porter 1956); while this is true for many populations, it is definitely not a consistent character for the species. Specimens of *P. scoparia* from southwestern Colorado and the northernmost populations in New Mexico and Arizona display a more luxuriant foliage that approaches *P. thompsonae*; that is, the leaflets are persistent through anthesis and measure up to 4 mm wide. Although these northernmost specimens of *P. scoparia* are vegetatively very similar to *P. thompsonae*, they have inflorescences with widely spaced flowers, standard petals up to 16 mm long, and leaflets up to 4 mm wide.

Stipitate glands especially on the calyx, pedicel, and distal end of the raceme rachis vary from absent to abundant. Such variability correlates very little with geography. The calyces from Mexican specimens are somewhat more densely glandular than those of the northern populations, but very likely this an artifact caused by the small Mexican sample.

10.3. Peteria glandulosa (A. Gray ex S. Watson) Rydberg, N. Amer. Fl. 24: 183. 1924. *Peteria scoparia* var. *glandulosa* A. Gray ex S. Watson, Proc. Amer. Acad. Arts 17: 342. 1882.—Type: Mexico. San Luis Potosí: in arenosis circa urbem, 1876, *Schaffner 834* (lectotype, designated by Porter, 1956: GH!).

Perennial herbs about 1–4 dm tall. Larger leaves 2–6 cm long; leaflets 7–17 (–19) per leaf (terminal leaflet well developed), mostly persistent on the rachis through fruiting, the larger mature ones 3–7 mm long, 1.5–3.5 mm wide, narrowly to broadly elliptic, adaxial surface glabrate, abaxial surface glabrate or strigose to sericeous, especially along the tanniniferous midrib and margins; leaf rachis generally much more densely sericeous above than below; stipules spinescent, 3–7 mm long. Inflorescence an open terminal raceme, the lower flowers often subtended by leaves, rachises 3–20 cm long, with 15–32 nodes, the internodes at anthesis mostly 5–13 mm long, strigose, stipitate glands sparse but more common towards the distal end; floral bracts spinescent, 2.5–5.0 mm long; pedicels sericeous, with stipitate glands, 3–4 mm long at anthesis, (2–) 4–7 mm in fruit. Calyx tube 6–8 mm long, with stipitate glands, lobes (2.5–) 4–6 mm long. Standard 14–16 mm long, 13–14 mm wide, predominantly purplish but sometimes described as whitish, aging mostly light brown to purplish or yellowish; wing petals 14–16 mm long, mostly purplish; keel petals 12–13 mm long, the blades mostly purplish. Staminal tube with filaments free for about the distal one-fourth. Ovary glabrous, with 10–12 ovules; style 5–6 mm long. Legumes 3.0–5.5 cm long, with 2–6 seeds. Seeds 3–6 mm in diameter at maturity, light brown, often with dark purplish mottling at maturity.

Phenology. Flowering during March and April in México, and from May through August in more northern states; fruiting specimens are recorded from June through August.

Distribution (Fig. 118). Endemic to Mexico (Durango, México, Nuevo León, Puebla, Querétaro, San Luis Potosí, Zacatecas); low thorn scrub and dry grassland habitats dominated by *Muhlenbergia*, *Acacia*, *Opuntia*, *Mimosa*, and *Adolphia*; 1300–2400 m. Populations are generally described as very localized and consisting of very few individuals.

Additional Specimens Examined. **Mexico.** Distrito Federal: Zacatenco, *Bopp O. 133* (MEXU).—Durango: Sierra del Registro, 7 mi NW of Nombre de Díos, *Ripley & Barneby 14158* (NY).—México: 1 km al W de San Cristóbal Ecatepec, *Cárdenas s.n.* (MICH); 1.5 km al W de San Cristóbal Ecatepec, *Cisneros 1680* (MICH).—Nuevo León: Zaragosa, *Hinton 23109* (TEX).—Puebla: Mpio. Alchichica, alrededores de la Laguna de Alchichica, *Calzada et al. 2392* (NY); El Riego, *Purpus 1188* (F, GH, MO, NY, UC); Esperanza, *Purpus 2474* (UC); La Cañada, *Pringle 7482* (F, GH, MO, UC, US).—Querétaro: San Juan del Río, *Pringle 7285* (GH).—San Luis Potosí: Mpio. Salinas, Km 120 de la carretera San Luis Potosí–Zacatecas, después 8 km al N sobre el camino a Vivorillas, sitio La Huerta de Nopales, *García M. s.n.* (MEXU); chiefly in the region of San Luis Potosí, *Parry & Palmer 172* (GH, NY); San Luis Potosí,

Schaffner 622 (CAS, NY, US).—ZACATECAS: 48 mi NW of Fresnillo along Hwy 45, *Lavin & Sundberg 5037* (TEX).

Peteria glandulosa is distinguished from *P. scoparia* by its flowers that are usually purplish (though occasionally whitish) at anthesis, calyx lobes that measure 4–6 mm long (at least in the northern part of the range where populations come in close geographical proximity to those of *P. scoparia*), leaves with sometimes up to 19 leaflets that are consistently persistent through anthesis, and calyces and pedicels that are consistently stipitate-glandular. *Peteria scoparia* differs in having consistently whitish petals at anthesis, calyx lobes that measure 2.5–4.0 mm long, leaves with no more than 13 leaflets that are often deciduous by anthesis, and calyces and pedicels that are inconsistently stipitate-glandular.

The southernmost collections of *P. glandulosa* from the Distrito Federal and México (*Bopp O. 133*, *Cárdenas s.n.*, and *Cisneros 1680*) are distinctive, and converge to a greater degree on *P. scoparia*, but only with regard to the length of the calyx lobes. Populations represented by these specimens apparently flower earlier, during March and April, than do other more northern ones, and they produce flowers with relatively short calyx lobes, 2.5–4.0 mm long, which reach anthesis well before all of the stem leaves fully expand. Notably, from the base of the stem to the tip of the inflorescence, the paired stipules and flower bracts (mostly subtended by stipules) are fully developed and strongly spinescent. In all of the vegetative nodes, strong axillary branch growth is underway while the leaf subtending that node is just beginning to expand. These specimens suggest that axillary growth actively suppresses that of the subtending leaf, lending weight to the idea that the individual flowers of *Peteria* are not part of a terminal raceme but are more properly solitary axillary flowers, much like those found in *Genistidium* and *Sphinctospermum*; growth of the axillary flower suppresses the subtending leaf to just a bract, occasionally with paired stipules.

The type collection consists of a single herbarium sheet at GH, *Schaffner 834*, contrary to Porter's report (1956). Additional specimens collected by Schaffner at CAS, NY, and US are numbered "622" and from the year 1879. The holotype (#834) is in flower and young fruit, whereas the other specimens (#622) are mostly past flower and in fruit. Clearly, these latter specimens are not isotypes.

10.4. Peteria pinetorum C. L. Porter, Rhodora 58: 351. 1956.—TYPE: MEXICO. Chihuahua: Los Cascarones, Río Mayo, transition, pine slopes, 11 Sep 1936, *Gentry 2658* (holotype: GH!; isotypes: ARIZ! F! MO! UC! US!).

Perennial herbs about 1 m tall. Larger leaves 5–9 cm long; leaflets 15–25 (–29) per leaf (terminal leaflet well developed), persistent on rachis through fruiting, the larger mature ones 6–10 mm long, 2.0–2.5 mm wide, very narrowly elliptic, both surfaces glabrate, but strigose to sericeous along the tanniniferous midrib of the abaxial surface; leaf rachis slightly more densely sericeous above than below; stipules spinescent, 3–6 mm long. Inflorescence an open terminal raceme, the lower flowers subtended by leaves, rachises 11–24 cm long, with 15–25 nodes, the internodes at anthesis mostly 5–11 mm long, villous, stipitate glands common from near the base to the distal tip; floral bracts spinescent, 3–5 mm long; pedicels glabrate, with stipitate glands, probably less than 9 mm long at anthesis, 9–10 mm in fruit. Calyx tube 6.0–6.5 mm long, with stipitate glands, lobes 3.5–4.0 mm long. Petals and staminal tube unknown. Ovary glabrous, with 9–10 ovules; style length unknown. Legumes 5–7 cm long, with 2–7 seeds. Seeds about 3–5 mm in diameter at maturity, light brown but probably attaining dark purplish mottling at maturity.

Peteria pinetorum is known only from the type locality in the western foothills of the Sierra Madre Occidental in southwestern Chihuahua, probably about 2000 m. Label data suggest the habitat is at the lower transition zone of pine forests with probably thornscrub. This species may have a similar phenology to *P. glandulosa* and *P. scoparia*, to which it is morphologically most similar.

Peteria pinetorum differs from other species of the genus by its inflorescence rachises, which are villous and bear stipitate glands commonly from the base to the distal tip. Also, its leaves are distinctive with 15–25 (–29) narrowly elliptic leaflets. They are like those of *P. scoparia* and *P. glandulosa*, but these two species have usually 15 or fewer leaflets per leaf. Flowers of *P. pinetorum* (inferred from fruiting calyx measurements, 3.5–4.0 mm long) seem to measure on the smaller end of the variation scale for *Peteria*. The groove of the canaliculate rachis, though more densely hairy than the rest of the rachis, is not as densely hairy as in the other species of *Peteria*. A putative significant difference in the size of the stems (Porter 1956), about 1 m tall, is probably not distinctive of this species, because this height is attained by *P. thompsonae* (at least in Arizona populations), and nearly attained by *P. scoparia*, which has stem lengths recorded up to 0.8 m tall. *Peteria pinetorum* is morphologically most similar to *P. glandulosa*. The leaves of *P. glandulosa* sometimes have 19 narrowly elliptic leaflets, a number converging on that of *P. pinetorum*. Also, the calyx and pedicels of *P. pinetorum* are more uniformly and densely glandular, like those of *P. glandulosa*, and unlike those of *P. scoparia*, which are with or without sparse stipitate glands. Finally, the floral bracts of *P. pinetorum* are distinctly spinescent, like those of *P. glandulosa*, but unlike the more weakly spinescent ones of *P. scoparia*. Petal color, once known, should prove useful, and perhaps diagnostic of this species. That of *P. glandulosa* is often purple at anthesis, whereas that of *P. scoparia* is whitish at anthesis, fading light purplish upon wilting. Likewise, calyx lobe length, measured from old calyces of *P. pinetorum* as 3.5–4.0 mm long, is similar to *P. scoparia*, although the southernmost populations of *P. glandulosa* have calyx lobes of this size. Measurements of fresh calyx lobes would be useful for evaluating the relationships of this species.

11. Genistidium I. M. Johnston, J. Arnold Arb. 22: 113. 1941.—TYPE: *Genistidium dumosum* I. M. Johnston.

Small bushes to straggling shrubs. Stems erect to ascending or straggling, branches alternate; bark smooth or superficially furrowed. Root system tap-like, root sprouting may give rise to clones (Poole 1992). Leaves alternate, pulvinate, unifoliolate to mostly palmately trifoliolate; leaflets opposite, caducous with the equally persistent rachis during the onset of the dry season or during age, adaxial surface sparsely sericeous, usually without finely scattered tannin deposits, abaxial surface sericeous, often without finely scattered tannin deposits; leaf rachis canaliculate, the groove without a dense indumentum, stipels absent; stipules free and narrowly attached to the stem adjacent to the petiole, weakly spinescent and persistent. Inflorescence of solitary flowers in axils of trifoliolate leaves at the base of a flowering branch and in axils of unifoliolate leaves towards the distal end of the same branch, the rachis very reduced, with one node, eglandular. Flowers ebracteolate, with a well-developed hypanthium; floral bracts early caducous; pedicels confluent with the calyx, straight and ascending during anthesis and seed set, falling with the aborted flower. Calyx campanulate, rounded to attenuate at base, sericeous, eglandular, with five prominent lobes about as long as the tube, more or less evenly spaced, sometimes the vexillary lobes slightly fused, narrowly triangular, valvate in bud. Standard orbicular, emarginate, reflexed about 90° at anthesis, dorsal surface glabrate, subequal to longer than the keel petals, with two callosities weakly developed about the midrib in the middle of the nectar-guide, claw abruptly contracted from the blade, inflexed auricles present; wing petals sculptured, commonly protruding beyond the keel at anthesis and loosely attached to it by a fold in the basal auricle, the distal ends glabrate, the blades mostly parallel to the plane of the keel; keel petals valvately connate distally nearly to the tip, apex blunt. Staminal tube diadelphous, with basal fenestrae; anthers included in the keel tip, basally attached to the filaments, the broad connective reddish brown; intrastaminal nectariferous disk absent. Pollen spheroidal, tricolporate, mesocolpium rugulate-perforate, polar regions smooth. Ovary subsessile, the stipe much shorter than the calyx tube, sericeous, eglandular, with 4–6 ovules; style with a pollen brush uniformly surrounding most of the distal end of the style, the base thickened and becoming displaced to the carinal side during fruit maturation; stigma ciliate, capitate and terminating the distally tapered style. Legumes subsessile, finely sericeous, eglandular, inner epidermis of the endocarp of adjacent valves spongy and abutting but not strongly adhering between seeds and not forming uniform seed compartments, the valves thinly woody at maturity, elastically and explosively dehiscent, the dorsal and ventral margins generally parallel in fruit, except where tandem seed abortion occurs. Seeds transversely oriented, lenticular, not constricted in the middle, accumulating canavanine, uniformly light brown with purplish mottling, hilum apical and attached to a very short funiculus, testa firmly adherent. Seedlings with epigeal germination, cotyledons foliar, eophyll unifoliolate, metaphylls unifoliolate to trifoliolate and alternate to the eophyll. Base chromosome number $x = 8$.

Genistidium comprises one species, *G. dumosum*, confined to very local areas in the Chihuahuan Desert of central Coahuila and southern Nuevo León, Mexico, and in the Big Bend National Park region of Brewster County, Texas, U.S.A. That *Genistidium* has no reported economic uses may be the result of its very localized distribution. *Genistidium* may have forage value, as suggested by herbarium specimens with cropped branch tips and by the fact that all of its relatives commonly serve as fodder for domesticated and wild animals. Populations of *Genistidium* from western Texas are disjunct from those in central Coahuila, and from descriptions on herbarium label data, appear to have a subtly distinct habit compared to Mexican populations. Though individuals of *G. dumosum* are described (and pictured) as having a very brushy growth habit, those from Texas populations have very spindly and straggling branches. Greenhouse plants raised from seeds from Texan populations have a distinct spindly or straggling growth form, but additional field work is needed to evaluate this difference.

Genistidium is easily distinguished from the other genera of Robinieae by its predominantly trifoliolate leaves. Also, the flowering branch is distinctive in that it bears solitary axillary flowers in the axils of trifoliolate leaves towards the base and in the axils of unifoliolate leaves toward the distal end. The brushy habit and green bark is also distinctive (a Mexican common name for *Genistidium* is *retama*). Two other species of Robinieae come close to this morphology. *Sphinctospermum constrictum* has inflorescences of solitary axillary flowers and stipules that are persistent and lignescent (but not the spinescent type found in most of the *Robinia*-group). *Sphinctospermum* is, however, an annual herb with simple (not unifoliolate) leaves. *Coursetia insomniifolia*, which occurs in the Chihuahuan Desert within the range of *Genistidium*, has a somewhat brushy habit, green bark, small yellow flowers, and legumes with few seeds (Lavin 1988); however, *C. insomniifolia* has paripinnate leaves (the terminal seta is distinctive) and glabrous legumes. *Genistidium* also stands out among the genera of the *Robinia*-group in having an ovary with only 4–6 ovules. A range this low is very uncommon in the tribe and otherwise occurs only in the genus *Hebestigma*, two species of *Lennea*, and one of *Sesbania*, all of which are very distant relatives of *Genistidium*.

The assignment of *Genistidium* to the *Robinia*-group is unequivocal based on its floral morphology, which includes a subsessile ovary, a style with a pollen brush, a calyx with well-developed lobes, and chloroplast DNA data evidence (Lavin, unpubl.). *Sphinctospermum* is considered the closest relative of *Genistidium*, because both genera share a very reduced inflorescence morphology composed of solitary axillary flowers and a base chromosome number of x=8. The tetraploid number (2n=32; Lavin, unpubl.) of *Genistidium* was counted from seeds collected from one population by Mike Powell in the Big Bend region. It would certainly be worthwhile to obtain counts from plants to the south in central Coahuila to determine variation in ploidy level.

11. Genistidium dumosum I. M. Johnston, J. Arnold Arb. 22: 113. 1941.—TYPE: MEXICO. Coahuila: Sierra de San Antonio, canyon at San Antonio de los Alamos, on volcanic tuff at summit of major cliffs, 2–3 Sep 1940, *Johnston & Muller 944* (holotype: GH!; isotypes: GH! MEXU!).

Shrubs 0.2–1.0 m tall. Stems from a thick tuberous caudex; branches straight and erect or spindly and straggling, many from the base, glabrous, greenish distally, young growth strigose to sericeous. Leaves unifoliolate to mostly palmately trifoliolate, the larger 0.7–2.0 (–3.0) cm long; leaflets 1 or 3 per leaf, the larger 5–20 (–25) mm long, 2–6 (–8) mm wide, linear to narrowly oblanceolate, adaxial surface sparsely sericeous, abaxial surface sericeous, apex rounded, midrib extending into an apiculate tip, margins not much recurved, abaxial surface of dried leaflets rarely with tannin deposits; stipules 1–2 mm long, linear-subulate, weakly spinescent, persistent; stipels absent. Inflorescences of solitary flowers; rachises 0.2–2 mm long, strigose, with 1 (–2) nodes; floral bracts 1.0–1.5 mm long, linear-lanceolate, strigose, early caducous; pedicels 1.5–2.5 mm long at anthesis, strigose. Calyx campanulate, rounded at base, sericeous, tube 2.5–3.5 mm long, the lobes narrowly triangular, 3–4 mm long. Standard yellowish, aging yellow, nectar-guides greenish, the blade 6–8 mm long, 6–8 mm wide; wing petals 6–7 mm long, yellowish; keel 5.5–6.5 mm long, fused to the tip along the carinal margin, yellowish. Staminal tube straight, diadelphous, filaments subequal. Style 4–5 mm long at anthesis, diverging from ovary at a 90° angle, with a pollen brush uniformly surrounding the style along the distal one-half to full length; ovary densely sericeous, subsessile. Legume 1.5–3.0 cm long, 4–6 mm wide, finely sericeous; seeds 1–3, 1.5–3.0 mm in diameter, light brown with purplish mottling. Figs. 121–122.

Phenology. Collected in flower and fruit from June through September in Texas, and from March through October in Coahuila and Nuevo León. Poole (1992) reports that flowering is dependent on precipitation and has seen populations in western Texas flower as early as April and as late as November. Poole further reports that although populations can be seen to set abundant fruit, age-class distribution is narrow; only mature adult individuals but no juveniles or seedlings have yet been observed in the wild. Powell (in Poole 1992) notes that in experimental gardens, the seedlings of *Genistidium* are easily overgrown by other species. Poole doubts this would be a problem in its native habitat where sparse vegetation cover predominates.

Distribution (Fig. 123). U.S.A. (Texas, Brewster Co.) and Mexico (central Coahuila and central Nuevo León); mostly in rocky, sometimes clayey, soils usually derived from limestone in the Chihuahuan Desert scrub vegetation; 900–1600 m. The three western Texas and two central Coahuilan populations of *Genistidium* comprise few scattered or very locally clustered individuals most often associated with *Agave*, *Euphorbia*, *Hechtia*, *Selaginella*, *Krameria*, *Bouteloua*, *Opuntia*, and *Porophyllum* (Poole 1992).

Additional Specimens Examined. **U.S.A.** Texas: Brewster Co.: between Terlingua and Lajitas, *Lundell 14765* (LL, TEX); north end of Reed Mesa, 2 mi W of Terlingua, *Powell 3186* (SRSC); NW end of Reed Plateau, between Terlingua and Lajitas, *Powell 3339* (SRSC, TEX); N end of Reed Plateau, in highway gap, *Powell 3454* (SRSC); 2 mi W of Terlingua, near turn-off to Villa de la Mina, *Powell & Powell 3639* (SRSC); limestone hills W of Terlingua, *Powell & Powell 3751* (SRSC, TEX); 3 mi W of Terlingua, Reed Plateau, *Warnock et al. 1161* (GH, LL, SRSC, TEX); along Terlingua–Lajitas rd, *Warnock 7585* (SRSC); 3 mi W of Terlingua, *Warnock 20938* (SRSC). **Mexico.** Coahuila: Picacho de San José, steep north ridge below summit, *Johnston 8203* (GH); Picacho de San José, below summit, vicinity of Santa Elena Mines, Sierra de las Cruces, *Stewart 1116* (GH, TEX); SE end of Sierra Mojada, 6 km SE of Esmeralda, S of Atravesada, *Stewart 2217* (GH, TEX); Rancho San Antonio de los Alamos, N slope of Cerro del Sauz, *Wendt et al. 796* (TEX); Puerto Colorado, between Sierra de la Fragua and Sierra Colorado, *Wendt et al. 1951* (TEX).—Nuevo León: Sandia–Soledad, Aramberri, *Hinton 23005* (TEX).

Johnston (1941) placed *Genistidium* in subtribe Craccinae of the tribe Galegeae (sensu Rydberg), a group characterized by flowers with an intrastaminal nectariferous disk and arranged in terminal racemes (or pseudoracemes). Within this subtribe, he allied *Genistidium* with *Tephrosia* (then *Cracca* L.) and *Peteria*; a relationship to the former has since been shown to be artificial (Polhill & Sousa 1981; Lavin 1987; Lavin & Doyle 1991). *Peteria*, however, may be a close relative of *Genistidium*, and its terminal raceme may attest to this. Terminal racemes are known in Robinieae only in *Peteria*, but the inflorescence of *Genistidium* superficially approaches one and may represent the intermediate step by which *Peteria* attained a terminal inflorescence from an originally axillary condition. The solitary flowers of *Genistidium* are arranged on a flowering branch, which has successively and distally more diminutive leaves, trifoliolate at the base and progressively smaller and unifoliolate towards the tip. The further reductions of these leaves to bracts would result in a terminal raceme much like that of *Peteria*. In fact, species of *Peteria* often have a flower or two at the base of the terminal raceme borne in the axil of a leaf, reduced or not, and the lowest bracts are often more robust (and sometimes stipulate) relative to more distal ones (see discussion under *Peteria*).

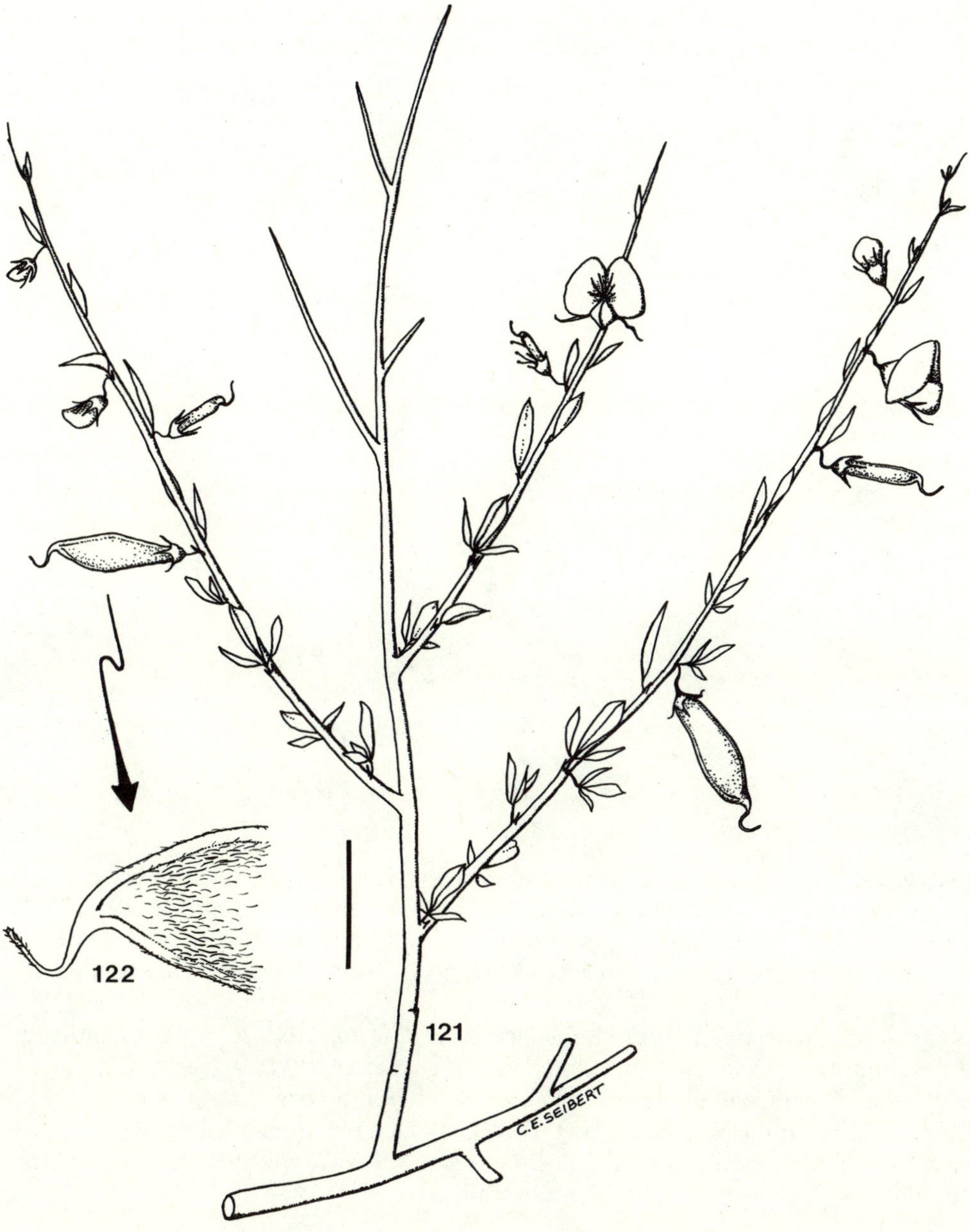

FIGS. 121–122. *Genistidium dumosum*. 121. Flowering and fruiting branch. 122. Close-up of the distal end of the mature legume showing a sericeous external surface. (121, bar = 3 cm; 122, bar = 3 mm).

In spite of the disjunct distribution of *G. dumosum*, the species was discovered within a short time-span by I. M. Johnston and C. H. Muller in 1940 and R. M. Stewart in 1941 in central Coahuila, and by B. Warnock in 1940 (or 1941) in western Texas (Poole 1992). This is rather remarkable since populations of this species comprise few individuals (48 total individuals from western Texas, Poole 1992).

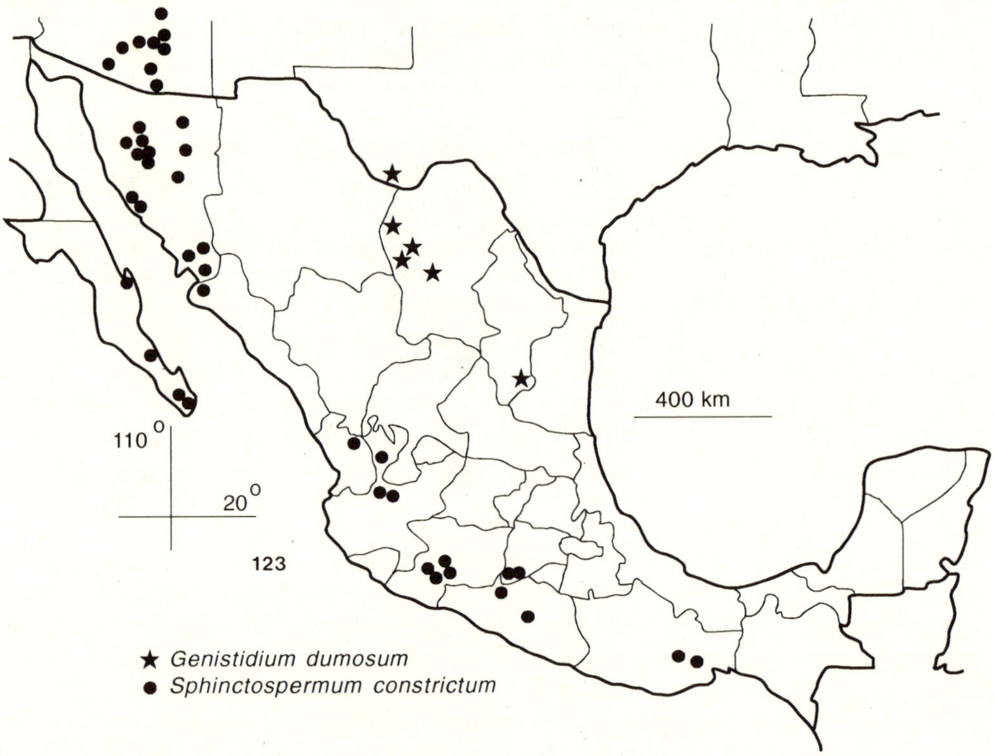

FIG. 123. Distribution of *Genistidium* and *Sphinctospermum* in the U.S.A. and Mexico.

Because the populations and number of individuals in each population are both so few, *G. dumosum* is a federally listed rare plant in the United States (Poole 1992). Poole has investigated many potential habitats in western Texas near to those known but to no avail. Warnock (1970) reports that *Genistidium* is browsed by deer.

The yellowish petal color characteristic of *G. dumosum* is probably uniform throughout the species, but one collection, *Wendt et al. 1951*, suggests at least one population from Coahuila has purplish flowers. Wendt (pers. comm.) suggest that this should be confirmed. Notably, this collection comes from what appears to be a basaltic substrate rather than a limestone one, which is typical for the rest of the populations of *Genistidium* (pers. observation).

12. Sphinctospermum Rose, Contr. U.S. Natl. Herb. 10: 107. 1906.—TYPE: *Sphinctospermum constrictum* (S. Watson) Rose.

Annual herbs. Stems erect to ascending, branching alternate and well developed from the base. Tap roots often becoming lignescent. Leaves alternate, pulvinate, simple, linear-lanceolate, both surfaces glabrate to sparsely strigose, without tannin deposits on abaxial surface; leaf rachis and stipels absent; stipules subulate, persistent. Inflorescences of solitary flowers borne on a very short glabrate rachis with one node. Flowers ebracteolate, with a conspicuous hypanthium; floral bracts linear-lanceolate, usually caducous before anthesis; pedicels confluent with calyx, straight and ascending during anthesis and seed set; falling with the aborted flower. Calyx narrowly attenuate at base, glabrous to sparsely strigose, stipitate glands absent, with five prominent lobes generally equalling or longer than the tube, the two vexillary ones often fused at the base, the lobes narrowly triangular, margins not overlapping at anthesis. Standard orbicular, emarginate, reflexed about 90° at anthesis, dorsal surface glabrate, subequal to longer than the keel petals, with two prominent callosites about the midrib in the middle of the nectar-guide, claw abruptly contracted from the blade, inflexed auricles barely developed; wing petals sculptured, commonly protruding beyond the keel at anthesis, loosely attached to it by a fold in the basal auricle, the distal ends glabrate, the blades essentially parallel to the plane of the keel; keel petals valvately connate distally nearly to the tip, apex blunt. Staminal tube diadelphous, with basal fenestrae; anthers included in the keel tip, basally attached to the filaments, the broad connective brownish red; intrastaminal nectariferous disk absent. Pollen spheroidal, tricolporate, mesocolpium rugulate-perforate, polar regions smooth. Ovary subsessile, the stipe much shorter than the calyx tube, glabrous, with 10–12 ovules; style with a pollen brush uniformly surrounding the distal end of the style, the style base thickened and not becoming displaced to the carinal side during fruit maturation; stigma ciliate, capitate and terminating the distally tapered style. Legumes subsessile, glabrous, inner epidermis of the endocarp of adjacent valves papery, abutting and firmly adhering between seeds and forming uniform seed compartments, the valves thinly woody at maturity, elastically dehiscent, the dorsal and ventral margins parallel in fruit. Seeds transversely oriented, squarish, with a central constriction, accumulating canavanine, uniformly brownish to grayish, hilum subapical but recessed, attached to a very short funiculus, testa firmly adherent. Seedling with epigeal germination, cotyledons foliar, falcate, eophyll simple (unifoliolate), linear-lanceolate, alternate to the first metaphyll, metaphylls similar to mature leaves. Base chromosome number x = 8 (Wood 1949; Lavin, unpubl.).

The identity of the monotypic *Sphinctospermum* has never been problematic; it is the only genus of Neotropical descent with linear unifoliolate leaves, squarish seeds with a central constriction, an annual growth habit, and an inflorescences of solitary flowers (Figs. 124–132; Lavin 1990; Lavin & Doyle 1991). In the Old World, this morphology is approached by *Tephrosia* subgen. *Macronyx* (Dalz.) Hooker; indeed, this similarity prompted at one time the exclusion of *Sphinctospermum* from the tribe Robinieae and the hypothesis of its close relationship to *Tephrosia* of the tribe Millettieae (Lavin 1987). However, a comprehensive reevaluation of the morphological data revealed that the tribal placement of *Sphinctospermum* was equivocal with respect to its placement in Millettieae or Robinieae (Lavin 1990). Chloroplast DNA evidence, though, was indisputable in suggesting the relationships of *Sphinctospermum* are with the *Robinia*-group of Robinieae (i.e., the barbistyled Robinieae), and when combined with traditional data, the total body of evidence supported a close relationship of *Sphinctospermum* with this group of Robinieae (Lavin & Doyle 1991). In particular, the evidence, aside from that of chloroplast DNA, that strongly supports a relationship with the *Robinia*-group includes a style with a pollen brush and a calyx with five well-developed lobes.

The exact relationships of *Sphinctospermum* within the *Robinia*-group are as yet less certain. For example, the *Robinia*-group is characterized by an indumentum composed of stipitate glands, a bulbous style base that is displaced to the carinal side during maturation, and a standard petal with inflexed auricles at the juncture of the claw and the blade; the absence of such characters in *Sphinctospermum* strongly suggests that it may be the earliest offshoot (i.e., the sister group) of the rest of the *Robinia*-group (Lavin 1990). However, any one genus of the *Robinia*-group may lack any one of these traits; in particular, *Genistidium*, which is also characterized by inflorescences of solitary axillary flowers and also lacks an indumentum composed of stipitate glands. It is such evidence that is used to suggest a sister-group relationship of *Sphinctospermum* with *Genistidium*, an endemic of the northern Chihuahuan Desert (see phylogeny).

Sphinctospermum is one of three monotypic genera in the *Robinia*-group (see also *Olneya* and *Genistidium*). The reason that these species are ranked as separate genera, rather than combined into a single more widely circumscribed genus, is that there is little evidence that can be used for grouping them, compared to the evidence that shows how distinct each is; that is, synapomorphies are few among *Olneya*, *Genistidium*, and *Sphinctospermum*, whereas each is marked by numerous autapomorphies. In contrast, the opposite condition exists for the genera *Coursetia* (Lavin 1988) and *Poitea* (Lavin 1993), where synapomorphies for all the constituent species in each genus were many, relative to the few synapomorphies for species subgroups within each genus.

12. Sphinctospermum constrictum (S. Watson) Rose, Contr. U.S. Natl. Herb. 10: 107. 1906. *Tephrosia constricta* S. Watson, Proc. Amer. Acad. Arts. 24: 46. 1889. *Cracca constricta* (S. Watson) Tidestrom, Proc. Biol. Soc. Wash. 48: 40. 1935.—TYPE: MEXICO. Sonora: in ravines in the mountains about Guaymas, Oct 1887, *Palmer 261* (lectotype, designated by Lavin, 1990: GH!, photo: CAS!; isolectotype UC!).

Stems 1.5–7.5 dm tall. Leaf blade (1–) 2–8 (–9) cm long, 1.5–2.5 mm wide; stipules 1–2 mm long. Inflorescences rachis 0.1–1.0 mm long; pedicels glabrate to sparsely strigose, 1–2 mm long; floral bracts linear-lanceolate, caducous, 1–2 mm long. Calyx campanulate, glabrous to sparsely strigose, 1.5–2.0 mm long; lobes 1.5–2.0 mm long. Standard 5–6 mm long, 5–6 mm wide, pinkish to mauve, nectar-guide yellowish; wing petals 5–6 mm long, pinkish to mauve; keel petals 5–6 mm long, pinkish to mauve. Legumes 20–30 mm long; seeds 3–12, 1.0–1.5 mm long. Figs. 124–132.

Phenology. Flowering occurs during the wet season, August through November; mature fruits persist throughout this time.

Distribution (Fig. 123). Pacific coastal deciduous forests of Mexico from Oaxaca northward and extending into the Sonoran Desert of Sonora and southeastern Arizona, U.S.A. The habitat of *Sphinctospermum* varies widely, and this species can be found in tropical deciduous forests, grasslands, thorn scrub, and occasionally oak woodlands. It especially grows in occasionally disturbed vegetation, such as along sides of roads and arroyos.

Taxonomic discussions and representative collections of the monotypic *Sphinctospermum* have been enumerated previously (Lavin 1990).

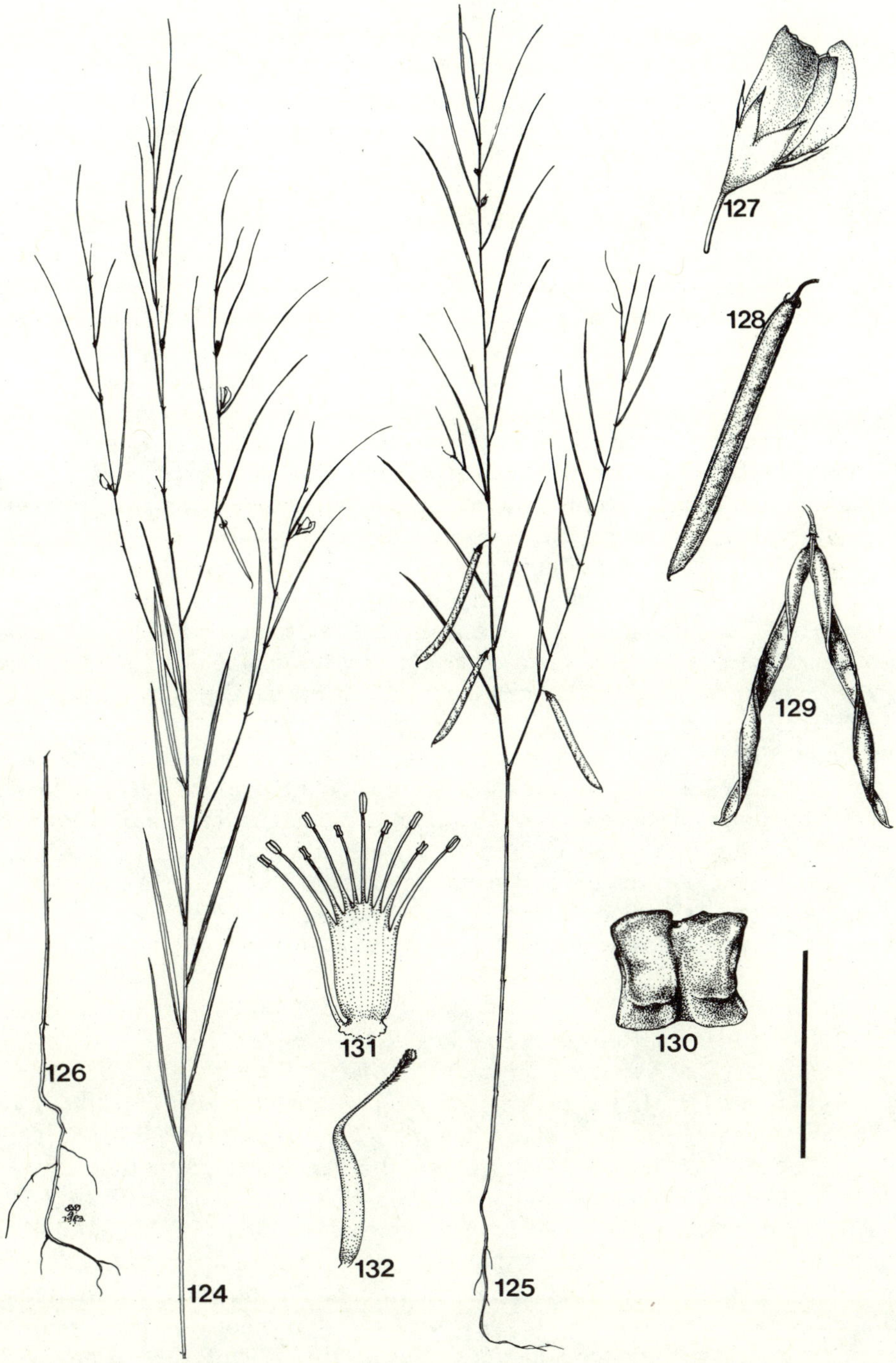

FIGS. 124–132. *Sphinctospermum constrictum*. 124. Flowering stem. 125. Fruiting stem. 126. Annual root system. 127. Flower. 128. Mature legume. 129. Dehisced legume. 130. Seeds. 131. Androecium. 132. Gynoecium. Reproduced from vol. 5 of *Flora Novo-Galiciana* by Rogers McVaugh. (124–126, bar = 10 cm; 127, bar = 7 mm; 128–129, bar = 15 mm; 130, bar = 2 mm; 131–132, bar = 5 mm).

Excluded Names

Robinia spilophylla Standley, nomen nudum.

Gliricidia spilophylla (Standley) Rydberg, nomen nudum.

Taxa Excluded from the Robinieae

Gliricidia lutea J. R. Johnston., Proc. Amer. Acad. Arts 40: 687. 1905. = *Margaritolobium luteum* (J. R. Johnston) Harms, Repert. Spec. Nov. Regni Veg. 19: 67. 1923.—Type: Venezuela. Isla Margarita: on dry hillside, El Valle, 15 Aug 1903, *Johnston 34* (lectotype, here designated: GH!; isolectotypes: F! GH! NY! UC!).

This species is clearly not most closely related to the species placed now in *Gliricidia*, nor to any other placed in Robinieae. The inflorescence is a pseudoraceme, the pod is indehiscent, and the ovate stipules, which completely envelope the newly expanding leaves, are nearly identical to those known from *Austrosteenisia* Geesink, *Kunstleria* Prain, *Muellera* L., and *Piscidia* L., which all belong to the tribe Millettieae; no such stipule morphology is found among the genera of Robinieae. Geesink (1984), in his treatment of the genera of Millettieae, maintained this species in the monotypic genus *Margaritolobium* Harms, largely because of its unusual combination of *Lonchocarpus*- and *Millettia*-like characteristics. Like *Lonchocarpus*, the mature legumes, apparently only known from the lectotype, are indehiscent. As in *Millettia* (especially sect. *Podocarpae*), the buds are scaly, a hypanthium is well developed, the calyx is ebracteolate, and the vexillary filament is free. The geographic position of this monotypic genus, endemic to Margarita Island off the northern coast of Venezuela, is somewhat unusual for genera of Millettieae, which have centers of diversity in Mexico, the Greater Antilles, Africa, and southeastern Asia. The generic alliance of *Margaritolobium* is yet equivocal, but certainly it is not closely related to any genus in Robinieae.

Gliricidia michelii Rusby, Mem. Torr. Bot. Club 6: 22. 1896. *Apurimacia michelii* (Rusby) Harm, Repert. Spec. Nov. Regni Veg. 19: 10. 1923.—Type: Bolivia. La Paz: Vic. Sorata, Nov 1892, *Bang 1621* (holotype: NY!; isotypes: GH! MICH! MO! NY! US!). = *Apurimacia boliviana* (Britton) Lavin, Adv. legume syst. 3: 64. 1987. *Coursetia boliviana* Britton, Bull. Torr. Bot. Club 16: 260. 1889.—Type: Bolivia. La Paz: Sorata, 8000 ft, *Rusby 1043* (lectotype, designated by Lavin, 1987: NY!; isolectotypes: F! NY! US!).

The geographical range of *A. boliviana* includes southern Peru and adjacent Bolivia. It has bracteolate flowers that lack the hypanthium, but have a densely sericeous abaxial standard surface. Furthermore, the flowers are arranged in pseudoracemose inflorescences, in particular with the biflorous arrangement described for *Lonchocarpus* (see Sousa & de Sousa, 1981). Thus, there is little doubt that *Apurimacia* Harms is allied with the primarily North American *Lonchocarpus*, and its geographic distribution along the drier parts of the Andes is common to other originally North American groups (e.g., *Coursetia*, Lavin, 1988). *Apurimacia* contains just two species, *A. boliviana* and *A. dolichocarpa* (Griseb.) Burkart, the latter from the Sierra Chica in Córdoba, Argentina (Burkart 1952). There is no evidence to suggest that these two species are congeneric; perhaps they have different origins from within the *Lonchocarpus* line.

Poecilanthe Bentham, J. Proc. Linn. Soc. Bot. 4, suppl.: 80. 1860.—TYPE: *Poecilanthe grandiflora* Bentham.

Amphiodon Huber, Bol. Museu Goeldi 5: 398. 1909.—TYPE: *Amphiodon effusus* Huber.

The evidence for excluding *Poecilanthe* from a close relationship with the tribe Robinieae was presented earlier (Lavin 1987), and much of that discussion is still considered valid. *Poecilanthe* does not accumulate canavanine in seed, its woody legume is 2–3-ovulate and indehiscent or tardily dehiscent, the vexillary stamen is fused basally and distally free and thus forms no basal fenestrae, the keel petals are distally free in most species, the anther connective is inconspicuous, the leaf rachis is not distinctly canaliculate (though subtly so in unifoliolate species), the leaflets are exstipellate and obligately alternate, and the stipules are absent or very inconspicuous. The last three vegetative characters stand in complete contrast to Robinieae, which have stipulate leaves (highly reduced in *Hebestigma*), a canaliculate leaf rachis, leaflets that are stipellate (often reduced to nubbins, as in *Gliricidia* and *Hebestigma*) and opposite (secondarily alternate at maturity in occasional species, and then not uniformly throughout a given species). Such vegetative morphology of Robinieae is found commonly in Millettieae, whereas that of *Poecilanthe* is found commonly in Dalbergieae.

Cladistic analysis of such characters (see phylogeny section) strongly suggests that *Poecilanthe* is very distantly related to Robinieae. Evidence from chloroplast DNA sequences confirms this finding based on morphology (Lavin, unpubl.). Furthermore, Greinwald et al. (in press) have isolated the alkaloids from ten species of *Poecilanthe*. The co-occurrence of the quinolizidine alkaloids epilupinine and cytisine is distinctive of *Poecilanthe* but has otherwise been reported only for *Harpalyce* DC. of the tribe Brongniartieae; no such types of alkaloids have ever been isolated from members of Robinieae.

Galegeae subtribe Craccinae Rydberg, pro parte, N. Amer. Fl. 24: 156. 1923.—TYPE: *Cracca* L., non *Cracca* Bentham, 1854, nom. conserv. [=*Tephrosia* Persoon].

Cracca L. is a nomenclatural synonym of *Tephrosia* Persoon (tribe Millettieae). Thus Craccinae is not actually part of the tribe Robinieae as herein circumscribed. Rydberg (1923a) included the genus *Peteria* in Craccinae, but see the discussion of that genus for reasons of its postulated relationship with genera of Robinieae rather than with *Tephrosia*. Also, *Cracca* L. is often confused with *Cracca* Bentham, the latter a taxonomic synonym of *Coursetia*. The two Craccas represent distantly related legume taxa whose complex nomenclatural history has been outlined by Lavin (1988, and references cited therein).

ACKNOWLEDGMENTS

This study was supported by the National Science Foundation (Grant No. BSR–9118785 and an Epscor grant) to Montana State University. Michael Powell contributed seed of the rare *Genistidium dumosum,* and Allen Zimmerman collected anatomical material of *Olneya tesota*. Jim Grimes, Denis

Kearns, Walter Kittredge, Melissa Luckow, and Beryl Simpson assisted with critical library work. Tony Simons and Colin Hughes furnished their valuable insights into the taxonomy of *Gliricidia*. Fernando Chiang kindly translated the Latin diagnoses. Roger Polhill provided much of the foundation upon which this work was built. Christiane Anderson, Mike Crisp, and Tom Wendt made valuable comments that greatly improved the manuscript. Karin Douhtit, Pam Harbaugh, María Rickards, and Cathy Seibert drew the illustrations; those by K. Douhtit first appeared in *Flora Novo-Galiciana*, volume 5, by Rogers McVaugh and were provided on loan by the curators of MICH. Fred Barrie kindly sent a photo of the holotype of *Robinia pseudoacacia*. The curators of the following herbaria provided loans of specimens and, in some cases, verified the absence of particular type specimens: ARIZ, B, BM, BR, BRY, C, CAS, CM, F, FHO, FI, FLAS, G, GH, GOET, H, HAC, HAJB, HAL, JBSD, K, L, LINN, M, MA, MEXU, MICH, MO, MONT, NY, OXF, RENO, RM, S, SRSC, TEX, UC, UPRRP, US, W, WAG, and WIS. The support of Alfonso Delgado Salinas, Maru García Peña, Alex Delgado García, Andrea Delgado García, Kristi Lavin, Amanda Lavin, and Gabriel Lavin provided the first author the opportunity to conduct this study–I thank them most of all.

LITERATURE CITED

Bell, E. A. 1981. Non-protein amino acids in the Leguminosae. In *Advances in legume systematics*, Part 2, ed. R. M. Polhill and P. H. Raven, 677–705. Kew: Royal Botanic Gardens.

Bongarten, B. C. 1992. Genetic variation in Black Locust within its native range. In *Black locust: biology, culture, & utilization*, ed. J. W. Hanover, K. Miller, and S. Plesko, 78–97. East Lansing: Department of Forestry, Michigan State University Press.

Boyd, A. 1992. *Musopsis* n. gen.: a banana-like leaf genus from the Early Tertiary of eastern North Greenland. Amer. J. Bot. 79: 1359–1367.

Bremer, K. 1988. The limits of amino acid sequence data in angiosperm phylogenetic reconstruction. Evolution 42: 795–803.

———. 1992. Ancestral areas: a cladistic reinterpretation of the center of origin concept. Syst. Biol. 41: 436–445.

Burnham, R. J. 1994. A preliminary report of Miocene plant fossils from central and southern Ecuador. Amer. J. Bot. 81 (suppl.): abstract 250.

Burkart, A. 1952. *Las Leguminosas Argentinas*. 2d ed. Buenos Aires: ACME Agency.

———. 1969. Un nuevo género de Papilionoideae-Galegeae del Ecuador. Darwiniana 15: 523–530.

Buskirk, R. E. 1985. Zoogeographic patterns and tectonic history of Jamaica and the northern Caribbean. J. Biogeography 12: 445–461.

Buzek, C. 1992. Fruits of *Pterocarpus tertiarius* Weyland from the North-Bohemian Basin, Czechoslovakia. In *Advances in legume systematics: Part 4, the fossil record*, ed. P. S. Herendeen and D. L. Dilcher, 19–31. Kew: Royal Botanic Gardens.

Chaney, R. 1940. Tertiary forests and continental history. Bull. Geol. Soc. Amer. 51: 469–488.

———. 1947. Tertiary centers and migration routes. Ecological Monographs 17: 140–148.

Chaney, R., and D. I. Axelrod. 1959. Miocene floras of the Columbia Plateau. Carnegie Institution of Washington, Contributions to Paleontology 617: 1–237.

Chappill, J. A. 1995. Cladistic analysis of the Leguminosae: the development of an explicit phylogenetic hypothesis for the legumes. In *Advances in legume systematics: Part 7, phylogeny*, ed. M. D. Crisp and J. J. Doyle. Kew: Royal Botanic Gardens. In press.

Cracraft, J. 1988. Deep-history biogeography: retrieving the historical pattern of evolving continental biotas. Syst. Zool. 37: 221–236.

———. 1994. Tropical biogeography: a zoological perspective. Trends in Ecology and Evolution 9: 347–348.

Crisp, M. D., and P. H. Weston. 1987. Cladistics and legume systematics, with an analysis of the Bossiaeeae, Brongniartieae, and Mirbelieae. In *Advances in legume systematics*, Part 3, ed. C. H. Stirton, 65–130. Kew: Royal Botanic Gardens.

Croizat, L., G. Nelson, and D. E. Rosen. 1974. Centers of origin and related concepts. Syst. Zool. 23: 265–287.

Cronquist, A. 1988. *The evolution and classification of flowering plants*. 2d ed. Bronx: The New York Botanical Garden.

Dunn, S. T. 1912. A revision of the genus *Millettia* Wight et Arn. J. Linn. Soc., Bot. 41: 123–243.

Evans, S. V., L. E. Fellows, and E. A. Bell. 1985. Distribution and systematic significance of basic non-protein amino acids and amines in the Tephrosieae. Biochem. Syst. Ecol. 13: 271–302.

Farris, J. S. 1988. Hennig86, version 1.5, an interactive program for phylogenetic analysis. Distributed by the author. Stoney Brook: State University of New York.

———. 1994. RNA version 940821, copyright by James S. Farris, Molekylärsystematiska laboratoriet Naturhistoriska riksmuseet, S 104 05 Stockholm.

Farris, J. S., M. Källersjö, A. G. Kluge, and C. Bult. 1994. Permutations. Cladistics 10: 65–76.

Felger, R. S., and M. B. Moser. 1985. *People of the desert and sea—ethnobotany of the Seri Indians.* Tucson: The University of Arizona Press.

Felsenstein, J. 1993. Phylip: Phylogeny Inference Package, version 3.52c. Distributed by the author. Seattle: Department of Genetics, University of Washington.

Ferguson, I. K., and J. J Skvarla. 1981. The pollen morphology of the subfamily Papilionoideae (Leguminosae). In *Advances in legume systematics*, Part 2, ed. R. M. Polhill and P. H. Raven, 859–896. Kew: Royal Botanic Gardens.

Forero, E., and A. H. Gentry. 1988. Neotropical plant distribution patterns with emphasis on northwestern South America: a preliminary overview. In *Proceedings of a workshop on Neotropical distribution patterns*, ed. P. E. Vanzolini and W. R. Heyer, 21–37. Rio de Janeiro: Academia Brasileiro de Ciéncias.

Geesink, R. 1981. Tephrosieae. In *Advances in legume systematics*, Part 1, ed. R. M. Polhill and P. H. Raven, 245–260. Kew: Royal Botanic Gardens.

Geesink, R. 1984. *Scala Millettiearum. A survey of the genera of the tribe Millettieae (Legum.-Pap.) with methodological considerations*. Leiden: E. J. Brill/Leiden University Press.

Gentry, A. H. 1982. Neotropical floristic diversity: phytogeographical connections between Central and South America, Pleistocene climatic fluctuations, or an accident of the Andean Orogeny? Ann. Missouri Bot. Gard. 69: 557–593.

Gillett, J. B. 1963. *Sesbania* in Africa (excluding Madagascar) and southern Arabia. Kew Bull. 17: 91–159.

Goldblatt, P. 1981. Cytology and the phylogeny of Leguminosae. In *Advances in legume systematics*, Part 2, ed. R. M. Polhill and P. H. Raven, 427–463. Kew: Royal Botanic Gardens.

Gomes, C. M. R., O. R. Gottlieb, R. G. Gottlieb, and A. Salatino. 1981. Phytochemistry in perspective: chemosystematics of the Papilionoideae. In *Advances in legume systematics*, Part 2, ed. R. M. Polhill and P. H. Raven, 859–896. Kew: Royal Botanic Gardens.

Gómez-Sosa. E. 1994. *Coursetia*. In *Flora de San Juan*, ed. R. Kiesling, 1: 315–316. Buenos Aires: Vazquez Mazzini Editores.

Grande, L. 1985. The use of paleontology in systematics and biogeography, and time control refinement of historical biogeography. Paleobiology 11: 234–243.

Greinwald, R., P. Bachmann, G. P. Lewis, L. Witte, F. Czygan. In press. Alkaloids of the genus *Poecilanthe* (Fabaceae–Dalbergieae). Biochem. Syst. Ecol.

Grimaldi, D. A. 1988. Relicts in the Drosophilidae. In *Zoogeography of Caribbean insects*, ed. J. K. Liebherr, 183–213. Ithaca, New York: Cornell University Press.

Hanover, J. W. 1992. Black Locust: an historical and future perspective. In *Black locust: biology, culture, & utilization*, ed. J. W. Hanover, K. Miller, and S. Plesko, 7–18. East Lansing: Department of Forestry, Michigan State University.

Harms, H. 1923. Leguminosae americanae novae. V. Repert. Spec. Nov. Regni Veg. 19: 68–69.

Herendeen, P. S. 1992. The fossil history of the Leguminosae from the Eocene of southeastern North America. In *Advances in legume systematics: Part 4, the fossil record*, ed. P. S. Herendeen and D. L. Dilcher, 303–316. Kew: Royal Botanic Gardens.

———. 1995. Phylogenetic relationships of the tribe Swartzieae. In *Advances in legume systematics: Part 7, phylogeny*, ed. M. D. Crisp and J. J. Doyle. Kew: Royal Botanic Gardens. In press.

Herendeen, P. S., W. L. Crepet, and D. L. Dilcher. 1992. The fossil history of the Leguminosae: phylogenetic and biogeographic implications. In *Advances in legume systematics: Part 4, the fossil record*, ed. P. S. Herendeen and D. L. Dilcher, 303–316. Kew: Royal Botanic Gardens.

Howard, R. A. 1988. *Charles Wright in Cuba 1856–1867*. Alexandria, Virginia: Chadwyck-Healey.

Hughes, C. 1987. Biological considerations in designing a seed collection strategy for *Gliricidia sepium* (Jacquin) Steudel (Leguminosae). Commonwealth Forestry Review 66: 31–45.

Hughes, C., and B. T. Styles. 1989. The benefits and risks of woody legume introdutions. In *Advances in legume biology*, ed. C. H. Stirton and J. L. Zarucchi. Monogr. Syst. Bot. Missouri Bot. Gard. 29: 505–531.

Isely, D. 1990. *Vascular Flora of the Southeastern United States*, Vol. 3, Part 2. Chapel Hill: University of North Carolina Press.

Isely, D., and F. J. Peabody. 1984. *Robinia* (Leguminosae: Papilionoideae). Castanea 49: 187–202.

Jarvis, C. E., F. R. Barrie, D. M. Allan, and J. L. Reveal. 1993. A list of Linnean generic names and their types. Regnum Veg. 127: 1–100.

Johnston, I. M. 1941. New phanerogams from Mexico, IV. J. Arnold Arb. 22: 113–114.

Johnson, K. H., B. C. Bongarten, and L. R. Boring. 1991. Effects of nitrate on in vivo nitrate reductase activity of seedlings from three open-pollinated families of *Robinia pseudoacacia*. Tree Physiology 8: 381–389.

Johnson, M. B. 1992. *Olneya tesota*—a potential food crop for hot arid zones. Nitrogen Tree Fixing Association Highlights 92–08.

Källersjö, M., J. S. Farris, A. G. Kluge, and C. Bult. 1992. Skewness and permutation. Cladistics 8: 275–287.

Lavin, M. 1986. The occurrence of canavanine in seeds of the tribe Robinieae. Biochem. Syst. Ecol. 14: 71–73.

———. 1987. A cladistic analysis of the tribe Robinieae (Papilionoideae, Leguminosae). In *Advances in legume systematics*, Part 3, ed. C. H. Stirton, 31–64. Kew: Royal Botanic Gardens.

———. 1988. Systematics of *Coursetia* (Leguminosae-Papilionoideae). Syst. Bot. Monographs 21: 1–167.

———. 1990. The genus *Sphinctospermum* (Leguminosae): taxonomy and tribal relationships as inferred from a cladistic analysis of traditional data. Syst. Bot. 15: 544–559.

———. 1992. Evolutionary relationships of the genus *Robinia*. In *Black locust: biology, culture, & utilization*, ed. J. W. Hanover, K. Miller, and S. Plesko, 61–77. East Lansing: Department of Forestry, Michigan State University.

———. 1993. Biogeography and systematics of *Poitea* (Leguminosae). Syst. Bot. Monographs 37: 1–87.

———. 1995. The tribe Robinieae and allies: model groups for assessing early Tertiary northern latitude diversification of the Leguminosae. In *Advances in legume systematics: Part 7, phylogeny*, ed. M. D. Crisp and J. J. Doyle. Kew: Royal Botanic Gardens. In press.

Lavin M., and A. Delgado. 1990. Pollen brush of Papilionoideae (Leguminosae): morphological variation and systematic utility. Amer. J. Bot. 77: 1294–1312.

Lavin, M., and J. J. Doyle. 1991. Tribal relationships of *Sphinctospermum* (Leguminosae): integration of traditional and chloroplast DNA data. Syst. Bot. 16: 162–172.

Lavin, M., and M. Luckow. 1993. Origins and relationships of tropical North America in the context of the boreotropics hypothesis. Amer. J. Bot. 80: 1–14.

Lavin, M., J. J Doyle, and J. D. Palmer. 1990. Evolutionary significance of the loss of the chloroplast DNA inverted repeat in the Leguminosae subfamily Papilionoideae. Evolution 44: 390–402.

Lavin, M., S. Mathews, and C. Hughes. 1991. Intraspecific chloroplast DNA variation in *Gliricidia sepium* (Leguminosae): intraspecific phylogeny and tokogeny. Amer. J. Bot. 78: 1576–1585.

Lewis, G. P. 1988. *Sesbania* Adans. in the Flora Zambesiaca region. Kirkia 13: 11–51.

Liebherr, J. K. 1986. *Barylous*, a new genus (Coleoptera: Carabidae) endemic to the West Indies with Old World affinities. J. New York Entomol. Soc. 94: 83–97.

Lievens, A. W., and L. E. Urbatsch. 1994. Proposal to reject the name *Indigofera coronilloides* (Fabaceae: Faboideae). Taxon 43: 483–484.

Liston, A. 1995. Use of the polymerase chain reaction to survey for the loss of the inverted repeat in the legume chloroplast genome. In *Advances in legume systematics: Part 7, phylogeny*, ed. M. D. Crisp and J. J. Doyle. Kew: Royal Botanic Gardens. In press.

Luckow, M. 1995. A phylogenetic analysis of the *Dichrostachys* group (Mimosoideae: Mimoseae). In *Advances in legume systematics: Part 7, phylogeny*, ed. M. D. Crisp and J. J. Doyle. Kew: Royal Botanic Gardens. In press.

MacFadden, B. J. 1990. Chronology of Cenozoic primate localities in South America. J. Human Evol. 19: 7–22.

MacFadden, B. J. 1992. *Fossil horses: systematics, paleobiology, and evolution of the family Equidae*. Cambridge: Cambridge University Press.

Macklin, B., and D. O. Evans. 1990. Perennial *Sesbania* species in agroforestry systems. Proceedings of a Workshop held in Nairobi, Kenya. Waimanalo: Nitrogen Fixing Tree Association.

Madison, M. 1977. A revision of *Monstera* (Araceae). Contr. Gray Herb. 207: 3–100.

Maddison, W. P., and D. R. Maddison. 1992. *MacClade: Analysis of phylogeny and character evolution. Version 3.01.* Sinauer Associates, Sunderland, Massachusetts.

Marshall, L. G., and T. Sempere. 1993. Chapter 12: Evolution of the Neotropical Cenozoic land mammal fauna in its geochronologic, stratigraphic, and tectonic context. In *Biological relationships between Africa and South America*, ed. P. Goldblatt, 329–392. New Haven: Yale University Press.

Mason, C. T., and P. B. Mason. 1987. *Mexican roadside Flora*. Tucson, Arizona: The University of Arizona Press.

Mathews, S., M. Lavin, and R. A. Sharrock. 1995. Evolution of the phytochrome gene family and its utility for phylogenetic analyses of angiosperms. Ann. Missouri Bot. Gard. 82(2): 296–321.

Matthew, W. D. 1915. Climate and evolution. Ann. New York Acad. Sci. 24: 171–318.

McDade, L. 1990. Hybrids and phylogenetic systematics I. Patterns of character expression in hybrids and their implication for cladistic analysis. Evolution 44: 1685–1700.

McKenna, M. C. 1972. Was Europe connected directly to North America prior to the middle Eocene? In *Evolutionary biology*, Vol. 6, ed. T. Dobzhansky, M. Hecht, and W. Steere, 179–188. New York: Appleton-Century-Crofts.

McKenna, M. C. 1975. Fossil mammals and early Eocene North Atlantic continuity. Ann. Missouri Bot. Gard. 62: 335–353.

———. 1983a. Cenozoic paleogeography of North Atlantic land bridges. In *Structure and developement of the Greenland-Scotland Ridge*, ed. M. H. P. Bott, S. Saxov, M. Talwani, and T. Thiede, 351–399. New York: Plenum Press.

———. 1983b. Holocene land mass rearrangement, cosmic events, and Cenozoic terrestrial organisms. Ann. Missouri Bot. Gard. 70: 459–489.

Micheli, M. 1894. *Gliricidia* in Légumineuses nouvelles. Bull. Herb. Boiss. 2: 442–443.

Neff, J. L., B. B. Simpson, and A. R. Moldenke. 1977. Action of natural selection on the breeding system and floral characters. In *Convergent evolution in warm deserts, an examination of strategies and patterns in deserts of Argentina and the United States*, ed. G. H. Orians and O. T. Solbrig, 204–224. Stroudsberg, Pennsylvania: Dowden, Hutchinson and Ross, Inc.

Nelson, G., and P. Ladiges. 1991. Standard assumptions for biogeographical analysis. Aust. Syst. Bot. 4: 41–58.

Nelson, G., and N. Platnick. 1981. *Systematics and biogeography: cladistics and vicariance*. New York: Columbia University Press.

Olmstead, R. 1989. Phylogeny, phenotypic evolution, and biogeography of the *Scutellaria angustifolia* complex (Lamiaceae): inference from morphological and molecular data. Syst. Bot. 14: 320–338.

Page, R. D. M. 1990. Component analysis: a valiant failure? Cladistics 6: 119–136.

———. 1993. COMPONENT, version 2.0. Tree comparison software for use with Microsoft® Windows™. London: Natural History Museum.

Page, R. D. M., and C. Lydeard. 1994. Towards a cladistic biogeography of the Caribbean. Cladistics 10: 21–41.

Palmer, J. D. 1987. Chloroplast DNA evolution and biosystematic uses of chloroplast DNA variation. Amer. Nat. 130: S6-S29.

Parrish, J. T. 1987. Global palaeogeography and palaeoclimate of the Late Cretaceous and Early Tertiary. In *The origin of angiosperms and their biological consequences*, ed. E. M. Friis, W. G. Chaloner, and P. R. Crane, 51–73. Cambridge: Cambridge University Press.

———. 1993a. The Palaeogeography of the opening South Atlantic. In *The Africa-South America connection*, ed. W. George and R. Lavocat, 8–27. New York: Oxford University Press.

———. 1993b. Palaeoclimatic history of the opening South Atlantic. In *The Africa-South America connection*, ed. W. George and R. Lavocat, 28–43. New York: Oxford University Press.

Peabody, F. 1984. A revision of the genus *Robinia* (Leguminosae: Papilionoideae). Ph.D. dissertation. Ames: Iowa State University.

Pianka, E. R. 1994. *Evolutionary ecology*. 5th ed. New York: Harper Collens College Publishers.

Pitman, W. C., S. Cande, J. LaBrecque, and J. Pindell. 1993. Chapter 2: Fragmentation of Gondwana: the separation of Africa from South America. In *Biological relationships between Africa and South America*, ed. P. Goldblatt, 15–34. New Haven: Yale University Press.

Pittier, H. 1922. On the species of *Dalbergia* of Mexico and Central America. J. Washington Acad. Sci. 12: 54–64.

Polhill, R. M. 1971. Some observations on generic limits in Dalbergieae-Lonchocarpinae. Kew Bull. 25: 259–273.

———. 1981a. Dalbergieae. In *Advances in legume systematics*, Part 1, ed. R. M. Polhill and P. H. Raven, 233–242. Kew: Royal Botanic Gardens.

———. 1981b. Galegeae. In *Advances in legume systematics*, Part 1, ed. R. M. Polhill and P. H. Raven, 357–363. Kew: Royal Botanic Gardens.

———. 1994. Classification of the Leguminosae. In *Phytochemical dictionary of the Leguminosae*, ed. I. Southern et al. London: Chapman & Hall.

Polhill, R. M., and M. Sousa. 1981. Robinieae. In *Advances in legume systematics*, Part 1, ed. R. M. Polhill and P. H. Raven, 283–288. Kew: Royal Botanic Gardens.

Poole, J. M. 1992. Status report on *Genistidium dumosum*. Texas Natural Heritage Program, Texas Parks and Wildlife Department, 4200 Smith School Road, Austin, Texas.

Porter, C. L. 1956. The genus *Peteria* (Leguminosae). Rhodora 58: 344–355.

———. 1957. Range extensions in the genus *Peteria* (Leguminosae). Rhodora 59: 216.

Prain, D. 1904. The species of *Dalbergia* of South-eastern Asia. Ann. Roy. Bot. Gard., Calcutta 10: 1–114, 91 pl.

Prothero, D. R. 1994. *The Eocene-Oligocene transition: paradise lost*. New York: Columbia University Press.

Raven, P. H., and D. I. Axelrod. 1974. Angiosperm biogeography and past continental movements. Ann. Missouri Bot. Gard. 61: 539–673.

Raven, P. H., and R. M. Polhill. 1981. Biogeography of the Leguminosae. In *Advances in legume systematics*, Part 1, ed. R. M. Polhill and P. H. Raven, 27–34. Kew: Royal Botanic Gardens.

Record, S. J., and R. W. Hess. 1943. *Timbers of the New World*. New Haven: Yale University Press.

Reid, E. M., and M. E. J. Chandler. 1933. *The London clay flora*. London: London British Museum (Natural History).

Rohlf, F. J. 1993. NTSYS-pc, Numerical Taxonomy and Multivariate Analysis System, version 1.80. A computer program distributed by Exeter Software, 100 North Country Rd., Building B, Setauket, New York 11733.

Ronquist, F. 1994. Ancestral areas and parsimony. Syst. Biol. 43: 267–274.

Rosen, D. E. 1976. A vicariance model of Caribbean biogeography. Syst. Zool. 24: 431–464.

———. 1978. Vicariant patterns and historical explanation in biogeography. Syst. Zool. 27: 159–188.

———. 1985. Geological hierarchies and biogeographic congruence in the Caribbean. Ann. Missouri Bot. Gard. 72: 636–659.

Rudd, V. E. 1981. Aeschynomeneae. In *Advances in legume systematics*, Part 1, ed. R. M. Polhill and P. H. Raven, 347–354. Kew: Royal Botanic Gardens.

Rydberg, P. A. 1923a. Galegeae subtribe Craccinae. N. Amer. Fl. 24(3): 156–184.

———. 1923b. Genera of North American Fabaceae I. Tribe Galegeae. Amer. J. Bot. 10: 485–498.

———. 1924a. Genera of North American Fabaceae II. Amer. J. Bot. 11: 470–482.

———. 1924b. Galegeae subtribes Sesbaniinae, Corynellinae, and Robiniinae. N. Amer. Fl. 24(4): 202–249.

Rzedowski, J. 1993. Diversity and origins of the Phanerogamic flora of Mexico. In *Biological diversity of Mexico: origins and diversity*, ed. T. P. Ramamoorthy, R. Bye, A. Lot, and J. Fa, 129–144. Oxford: Oxford University Press.

Sachet, M.-H. 1987. The littoral species of *Sesbania* (Leguminosae) in the South Pacific islands and its relatives. Bull. Mus. Hist. Nat. (Paris), sér. 9: 3–87.

Schot, A. M. 1994. A revision of *Callerya* Endl. (incl. *Padbruggea* Miq., and *Whitfordiodendron* Elm.) (Papilionaceae: Millettieae). Blumea 39: 1–40.

Schrire, B. D. 1989. A multidisciplinary approach to pollination biology in the Leguminosae. In *Advances in legume biology*, ed. C. H. Stirtion and J. L. Zarucchi. Monogr. Syst. Bot. Missouri Bot. Gard. 29: 183–242.

Sclater, P. L. 1858. On the general geographic distribution of the members of the class Aves. J. Linn. Soc. London, Zool. 2: 130–145.

Shreve, F. 1964. Vegetation of the Sonoran Desert. In *Vegetation and flora of the Sonoran Desert*, ed. F. Shreve and I. L. Wiggins, 3–186. Stanford: Stanford University Press.

Simberloff, D. 1987. Calculating probabilities that cladograms match: a method of biogeographical inference. Syst. Zool. 36: 175–195.

Simberloff, D., K. L. Heck, E. D. McCoy, and E. F. Conner. 1981. There have been no statistical tests of cladistic biogeographic hypotheses. In *Vicariance biogeography: a critique*, ed. G. Nelson and D. E. Rosen, 40–63. New York: Columbia University Press.

Simpson, B. B. 1977. Breeding systems of dominant perennial plants of two disjunct warm desert ecosystems. Oecologia 27: 203–226.

Simpson, G. G. 1953. *Evolution and geography*. Eugene: Oregon State System of Higher Education.

———. 1965. *The geography of evolution: collected essays*. Philadelphia: Chilton.

———. 1980. *Splendid isolation: a curious history of South American mammals*. New Haven: Yale University Press.

Soto N., J. C., and M. Sousa S. In press. Plantas medicinales de la cuenca del Río Balsas. Cuadernos, Instituto de Biología, U. N. A. M., Mexico.

Sousa, M., and A. Delgado. 1993. Mexican Leguminosae: phytogeography, endemism, and origins. In *Biological diversity of Mexico: origins and diversity*, ed. T. P. Ramamoorthy, R. Bye, A. Lot, and J. Fa, 459–511. Oxford: Oxford University Press.

Sousa, M., and M. Lavin. 1992. Nuevas Robinieae (Leguminosae) de Chiapas, Mexico. Anales Inst. Biol. Univ. Nac.. México, Ser. Bot. 63: 139–145.

Sousa, M., and M. P. de Sousa. 1981. New World Lonchocarpinae. In *Advances in legume systematics*, Part 1, ed. R. M. Polhill and P. H. Raven, 261–281. Kew: Royal Botanic Gardens.

Sprent, J. I. 1980. Root nodule anatomy type of export product and evolutionary origin in some Leguminosae. Plant, Cell Environ. 3: 35–43.

Surles, S. E., J. L. Hamrick, and B. C. Bongarten. 1989. Allozyme variation in black locust (*Robinia pseudoacacia*). Can. J. For. Res. 19: 471–479.

Stritch, L. R. 1984. Nomenclatural contributions to a revision of the genus *Wisteria*. Phytologia 56: 183–184.

———. 1985. A revision of the leguminous genus *Wisteria* Nuttall. Ph.D. Dissertation. Carbondale: Southern Illinois University.

Swofford, D. L. 1993. PAUP: Phylogenetic analysis using parsimony, version 3.1. Computer program distributed by the Illinois Natural History Survey, Champaign, Ilinois.

Thorne, R. F. 1973. Floristic relationships between tropical Africa and tropical America. In *Tropical forest ecosystems in Africa and South America: a comparative review*, ed. B. J. Meggers, E. S. Ayensu, and W. D. Duckworth, 27–48. Washington, D.C.: Smithsonian Institution Press.

Tiffney, B. 1985a. Perspectives on the origin of the floristic similarity between eastern Asia and eastern North America. J. Arnold Arb. 66: 73–94.

———. 1985b. The Eocene North Atlantic land bridge: its importance in Tertiary and modern phytogeography of the northern hemisphere. J. Arnold Arb. 66: 243–273.

Tucker, S. C. 1987a. Floral initiation and development in legumes. In *Advances in legume systematics*, Part 3, ed. C. H. Stirton, 183–239. Kew: Royal Botanic Gardens.

———. 1987b. Pseudoracemes in papilionoid legumes: their nature, development, and variation. Bot. J. Linn. Soc. 95: 181–206.

Turner, B. L. 1955. Chromosome numbers in the genus *Sesbania*: evidence for a conservative treatment. Rhodora 57: 213–218.

Turner, B. L., and O. S. Fearing. 1960. Chromosome numbers in the Leguminosae III. Species of the southwestern United States and Mexico. Amer. J. Bot. 47: 603–608.

Upchurch, G. R., and J. A. Wolfe. 1987. Mid-Cretaceous to Early Tertiary vegetation and climate: evidence from fossil leaves and woods. In *The origin of angiosperms and their biological consequences*, ed. E. M. Friis, W. G. Chaloner, and P. R. Crane, 75–105. Cambridge: Cambridge University Press.

Urban, I. 1900. *Gliricidia* and *Hebestigma* in VI. Leguminosae novae vel minus cognitae. I. Symb. antill. 2: 287–290.

Van Der Hammen, T., and A. Cleef. 1983. *Trigonobalus* and the tropical amphi-pacific element in the North Andean forest. J. Biogeography 10: 437–440.

Vrba, E. 1993. Chapter 13: Mammal evolution in the African Neogene and a new look at the Great American Interchange. In *Biological relationships between Africa and South America*, ed. P. Goldblatt, 393–432. New Haven: Yale University Press.

Warnock, B. 1970. *Wildflowers of the Davis Mountains and the Marathon Basin, Texas*. Sul Ross State University, Alpine, Texas.

Wendt, T. 1988. *Chiangiodendron* (Flacourtiaceae: Pangieae), a new genus from southeastern Mexico representing a new tribe for the New World flora. Syst. Bot. 58: 29–54.

———. 1989. Las selvas de Uxpanapa, Veracruz–Oaxaca, Mexico: evidencia de refugios florísticos cenozoicos. Anales Inst. Biol. Univ. Nac. México, Ser. Bot. 58: 29–54.

———. 1993. Composition, floristic affinities, and origins of the canopy tree flora of the Mexican Atlantic slope rain forests, In *Biological diversity of Mexico: origins and diversity*, ed. T. P. Ramamoorthy, R. Bye, A. Lot, and J. Fa, 595–680. Oxford: Oxford University Press.

West, R. M., and M. R. Dawson. 1978. Vertebrate paleontology and the Cenozoic history of the North Atlantic region. Polarforschung 48: 103–119.

Wheeler, E., and P. Baas. 1992. Fossil wood of the Leguminosae: a case study in xylem evolution and ecological anatomy. In *Advances in legume systematics: Part 4, the fossil record*, ed. P. S. Herendeen and D. L. Dilcher, 281–301. Kew: Royal Botanic Gardens.

Withington, D., N. Glover, J. L. Brewbaker. 1987. *Glircidia sepium*: management and improvement. Proceedings of a workshop held in Turrialba, Costa Rica. Waimanalo: Nitrogen Fixing Tree Association.

Wolfe, J. A. 1972. An interpretation of Alaskan Tertiary floras. In *Floristics and paleofloristics of Asia and eastern North America*, ed. A. Graham, 201–233. Amsterdam: Elsevier Publ. Co.

———. 1975. Some aspects of plant geography of the northern hemisphere during the late Cretaceous and Tertiary. Ann. Missouri Bot. Gard. 74: 785–803.

———. 1977. Paleogene floras from the Gulf of Alaska region. U. S. G. S. Professional Paper 997: 1–108, 30 pl.

Wolfe, J. A., and W. C. Wehr. 1991. Significance of the Eocene fossil plants at Republic, Washington. Washington Geology 19: 18–24.

Zona, S. 1990. A monograph of *Sabal* (Arecaceae: Coryphoideae). Aliso 12: 583–666.

NUMERICAL LIST OF SPECIES

1. Sesbania—not listed
2. Hebestigma cubense
3.1a. Lennea viridiflora var. viridiflora
3.1b. Lennea viridiflora var. novogaliciensis
3.2. Lenna melanocarpa
3.3. Lennea modesta
4.1. Hybosema robustum
4.2. Hybosema ehrenbergii
5. Poitea (see Lavin 1993)
6.1. Gliricidia brenningii
6.2. Gliricidia maculata
6.3. Gliricidia sepium
7. Robinia (see Isely & Peabody 1984)
8. Coursetia (see Lavin 1988)
9. Olneya tesota
10.1. Peteria thompsonae
10.2. Peteria scoparia
10.3. Peteria glandulosa
10.4. Peteria pinetorum
11. Genistidium dumosum
12. Sphinctospermum constrictum (see Lavin 1990)

INDEX TO NUMBERED COLLECTIONS EXAMINED

The numbers in parentheses refer to the corresponding species in the text and in the Numerical List of Species presented above.

Abarea, B. 4321 (6.3).
Abrams, L. 1907 (9).
Abrams, L. R. 12685 (9).
Acevedo, R., & F. Vázquez 415 (3.1a).
Agatha, M. 526 (9).
Aguilar, J. I. 762 (4.2).
Aguilar, J. A. et al. 193 (6.2).
Ahshapanek, D. 186 (6.3); 262 (4.2); 304 (6.3); 660 (6.3).
Alain, Bro. 1948 (2); 2773 (2).
Alain, Bro., & Bro. Clemente 1426 (2).
Alava, R., & S. A. Cook 1519 (6.3); 1694 (6.3).
Alcorn, J. B. 2898 (6.3).
Alexander, A. M., & L. Kellogg 2194 (9); 2231 (9).
Allard, H. A. 18936 (6.3).
Allen, P. H. 1077 (6.3); 4261 (6.3); 566 (6.3); 871 (6.3).
Anderson, W. R., & C. W. Laskowski 4470 (6.3).
Andrews, L. M. 580 (6.3).
Araquistain, M., & D. Castro 1861 (6.3).
Araquistain, M., & P. Moreno 1280 (6.3); 1328 (6.3); 1499 (6.3); 1602 (6.3).
Areces, A. 25031 (2).
Areces, A. et al. 31705 (2).

Arias, L. M., & A. Vara 232 (6.2).
Armstrong, J. 1039 (10.1).
Arnason, T., & J. Lambert 17162 (6.2).
Arnoldo-Broeders, M. 3879 (6.3).
Arvigo, R. et al. 377 (6.2).
Atwood, J. T. 1421 (6.3); 5403 (6.3).
Atwood, N. D. 2221 (9); 6031 (10.1); 10957 (10.1).
Bailey, L. H. 283 (9).
Baker, C. F. 190 (6.3).
Baker, R., & W. Burger 166 (6.3).
Baker, W. H. 14023 (10.1).
Balsinhas, A. 3248 (6.3); 3258 (6.3).
Barkley, F. A., & R. Blondeau 4218 (9).
Barkley, F. A. 14141 (6.3); 17M061 (4.2); 17M133 (6.3).
Barlow, F. D. 2/3 (6.3); 7/5C (6.2).
Barneby, R. C. 12970 (10.2); 17816 (10.1).
Barrera, A. et al. 450 (6.2).
Bartlett, H. H. 11305 (6.2); 12224 (3.2); 12374A (6.2); 12855 (3.2); 12955 (6.3).
Basurto, P., & G. Durán 10 (3.2); 48 (3.2); 174 (3.2); 225 (3.2); 249 (3.2); 308 (3.2); 366 (3.2); 465 (3.2); 529 (3.2); 558 (3.2); 562 (6.3); 595 (3.2); 630 (3.2); 710 (3.2).
Basurto, P., & R. Patrón 53 (6.3); 204 (3.2).
Beatley, J. 3256 (10.1); 8248 (10.1); 8626 (10.1); 12883 (10.1).
Beetle, A. A. 26255 (6.3).
Benson, L. 9672 (9).
Benz, B., & L. Jn. Pierre 1157 (6.3).
Bequaert, J. 85 (6.2).
Bermejos, J. 39771 (6.3).
Berry, P. E. 3511 (6.3).
Bigelow, J. M. et al. 251 (10.2).
Billings, W. D. 1784 (9).
Bisse, J. et al. 36381 (2).
Blum, K. E. 1349 (6.3).
Bopp O., M. 133 (10.3).
Brandegee, T. S. 1103 (10.2); 1889 (9); 1892 (9); 1894 (9); 17.5.1892 (9).
Breedlove, D. E. 7664 (3.3); 9660 (4.2); 10226 (4.2); 12033 (4.2); 12877 (3.3); 22899 (4.2); 23892 (6.3); 23925 (6.3); 24703 (4.2); 26203 (4.2); 26375 (4.2); 29005 (6.2); 29312 (3.3); 33292 (6.2); 34358 (4.2); 34358 (4.2); 34764 (4.2); 34764 (4.2); 37343 (4.2); 39746 (4.2); 39954 (4.1); 41303 (4.2); 50139 (4.2); 50275 (4.2); 50902 (3.3).
Breedlove, D. E., & E. McClintock 23453 (6.3); 23576 (6.3).
Breedlove, D. E., & P. H. Raven 8435 (6.3).
Breedlove, D. E., & A. R. Smith 32729 (4.2).
Breedlove, D. E., & R. F. Thorne 21426 (4.2); 30414 (6.2 & 6.3).
Brenes, A. M. 6 (6.3).
Breteler, F. J. 3158 (6.3).
Britton, N. L. 2477 (6.3).
Britton, N. L., & E. G. Britton 7961 (6.3).
Britton, N. L., & T. E. Hazen 165 (6.3).
Britton, N. L., & J. A. Shafer 3132 (6.3).
Britton, N. L., & P. Wilson 4528 (6.3); 5475 (2); 5517 (2); 5561 (2); 5633 (2).
Britton, N. L. et al. 33 (2); 5845 (2); 6699 (2).
Broadway, W. E. 2433 (6.3); 5552 (6.3); 6294 (6.3).
Brunsfeld, S. J. 2461 (10.1).
Bryant, W. E. 1889 (9); 1892 (9).
Bunch, P. et al. 0487 (6.3).
Bunting, G. S., & L. Licht 634 (6.3); 694 (6.3); 932 (6.3); 1122 (6.3).
Burley, J. S. 208 (6.3).
Busey, P. 697 (6.3).
Butterwick, M., & E. Lott 3829 (10.2).
Cabrera, E. 8381 (3.2); 1252 (6.2).
Cabrera, E., & G. Ibarra 1152 (6.2).
Cabrera, E., & H. de Cabrera 2151 (6.2); 2869 (4.2); 3769 (4.2); 12992 (6.2).
Cabrera, E. et al. 4484 (3.2); 11171 (6.2).
Calderón, S. 109 (6.3); 2518 (3.1a).
Calderón, G. M. 113 (3.2); 678 (3.2); 841 (3.2).
Caldwell, M. E. 164 (9).
Calzada, J. I 283 (3.2); 1619 (3.1a); 1725 (3.3); 3249 (3.2).
Calzada, J. I., & J. Arellano 2225 (6.2).
Calzada, J. I., & J. Elizondo 8486 (6.3).
Calzada, J. I. et al. 2392 (10.3); 2577 (6.3); 2581 (3.2); 4394 (6.3).
Cano P., F. A. 47 (4.2).
Carlson, M. C. 110 (6.3); 758 (4.2); 1936 (6.3).
Carpenter, J. B. 478 (6.3).
Carrasquilla, L. 173 (6.3).
Carter, A. 1431 (9); 5846 (9).
Carter, A., & J. Reese 4583 (9).
Carter, A., & M. Sousa 5148 (9).
Carter, A. M. et al. 71–50 (9).
Casa, F., & M. Valverde 10673 (2).
Castagnoli, S. et al. 150 (10.1).
Castillo C., G., & L. Tapia 796 (3.1a).
Castillo C., G. et al. 1923 (6.3).
Cedillo T., R. 545 (6.3); 3047 (3.1a); 3852 (3.2).
Chambers, K. L. 948 (9); 2705 (6.3).
Chanek, M. 97 (3.2); 113 (6.3).
Chavelas P., J. ES195 (6.2); ES844 (6.2).
Chavelas, J., & L. A. Pérez 389 (3.2).
Chiang, F. et al. 9273 (10.2).
Christ, J. H., & C. B. Christ 16715 (10.1).
Cisneros, R. C. 1680 (10.3).
Clark, O. M. 11692 (9).
Clarke, H. D. et al. 2029–1 (6.3).
Clarkson, L. L., & W. J. E. Kress 75–269 (6.3).
Clary, Mrs. B. L. 313 (9).
Clear, B. 14.7.1945 (9).
Clint, Mrs. M. M–373 (3.2).

Collom, R. E. 60 (9).
Comanor, P. L. 838 (6.3).
Combs, R. 600 (2); 736 (6.3).
Cominote, J. 66 (6.3).
Contreras, E. 486 (6.2); 3580 (3.2); 7051 (3.2); 7053 (3.2); 7671 (6.2); 7869 (3.3); 7873 (3.3); 7979 (3.3); 9639 (3.2); 10037 (3.3); 10329 (3.3); 10331 (3.3).
Contreras, J. L. 792 (4.2).
Converse, O. 33 (6.3).
Conzatti, C. 2116 (6.3).
Cook, M. T. (6.3).
Cook, O. F., & R. D. Martin 233 (6.2).
Cooley, G. R., & G. R. Proctor 10982 (6.3).
Cooper, G. P. 127 (6.3); 528 (6.3).
Cooper, N. 577 (9).
Cortés, M. E. 145 (6.3); 456 (3.2); 477 (3.2); 555 (3.2).
Cortés, L., & R. Torres C. 18 (3.1a); 660 (3.1a).
Cory, V. L. 45066 (10.2).
Cosminsky, S. 213 (6.3).
Coss, H. T. et al. 20369 (9).
Cowan, R. S. 2288 (9).
Crane, C. J. 289 (6.2).
Croat, T. B. 13953 (6.3); 21927 (6.3); 32542 (6.3); 39993 (6.3).
Croat, T. B., & D. P. Hannon 63385 (6.3).
Crosswhite, F. S. 848 (9).
Cuatrecasas, J. 6507 (6.3); 27492 (6.3).
Curran, H. M. 258 (6.3).
Curran, H. M., & M. Haman 182 (6.3).
Curtiss, A. H. 437 (6.3).
D'Arcy, W. G., & J. J. D'Arcy 6060 (3.1a).
McKelvey, S. D. 340 (9); 476 (9); 900C (9).
Daniel, J. 353 (6.3).
Daniel, T. F. 172 (9); 4293 (9).
Darrow, R. 1942 (9).
Darwin, S. P. 2455 (6.2).
Daubenmire, R. 469 (6.3).
Davidse, G., & A. E. Brant 31939 (6.2); 32585 (6.2).
Davidse, G., & R. W. Pohl 1434 (6.3); 2101 (4.2).
Davidson, C. et al. 6816 (6.3).
Davis, L. I., & E. Kincaid 55–11 (6.3).
Deam, C. C. 292 (6.3); 295 (6.3).
Deaver, C. F. 5959 (9); 6253 (10.1).
Delgado S., A., & J. García 1068 (4.2).
Delgado S., A. et al. 638 (4.2); 964 (6.3).
Demaree, D. 8039 (9); 44292 (9).
Díaz L., C. L. 3340 (3.1b).
Díaz P., M. 453 (2).
Dieckman, L. 140 (6.3); 140–79 (6.3); 158 (6.2).
Diego, N. 5183 (3.1a).
Diguet, L. (9).
Dodge, C. W. 17517 (6.3).
Dodge, C. W., & W. S. Thomas 6263a (6.3).
Dodson, C. H., & A. Gentry 6480 (6.3).
Dodson, C. H., & L. H. Thien 1788 (6.1).
Dodson, C. H. et al. 7462 (6.1).
Dorantes L., J. 34 (6.3); 479 (6.3).
Dorantes L., J. et al. 1118 (6.3).
Dorantes, B. 3555 (3.3).
Dressler, R. L., and Q. Jones 54 (3.2).
Dreyer, D. 217 (9).
Dudley, C. 112 (9).
Dugand, A. 282 (6.3); 5257 (6.3); 5611 (6.3).
Duke, J. A. 4873 (3.1a); 7323 (6.3); 9744 (6.3); 10179 (6.3); 10311 (3.1a); 10526 (6.3); 14086 (3.1a).
Durán, R., & I. Olmsted 798 (6.2); 814 (6.2).
Dwyer, J. D. S–69 (6.2); 2816 (6.3); 2872 (6.3); 4467 (6.3); 9575 (3.1a); 10701 (6.2); 10934 (6.2).
Dwyer, J. D. et al. 21 (6.2); 23 (6.2).
Dziekanowski, C. et al. 3408 (4.2).
Eastwood, A. 6166 (9); 16668 (9); 17719 (9).
Eastwood, A., & J. T. Howell 9422 (10.1).
Ebinger, J. E. 874 (3.1a).
Edwards, J. B. P557 (6.3).
Ekman, E. L. 4622 (2); 9846 (6.3); 12762 (6.3); 18712 (2).
Elias, T. 45912 (6.3).
Elmore, F. H. E7 (6.3).
Emrick, G. M. 99 (6.3).
Englesing, F. C. 134 (6.3).
Enriquez, O. G. 298 (6.2); 7612 (3.3).
Ernst, W. R. 2562 (6.3).
Ervendberg, L. C. 28 (3.2).
Espina, Mr. 14 (6.3).
Evangelista, V., & M. Mendoza 54 (3.2).
Evans, L. 54 (6.3).
Evans, M. 1524 (6.3).
Everett, P. C., & E. K. Balls 23116 (9).
Everman, B. W. 1924 (9).
Fénix, E. 166 (6.3).
Feddema, C. 2039 (6.3).
Felger, R. S. 75–40 (9); 7891 (9); 7903 (9); 12512 (9); 14260 (9).
Felger, R. S., & R. L. Bezy 14031 (9).
Felger, R. S., & F. Cooper 10380 (9).
Felger, R. S., & C. Lowe 2598 (9).
Felger, R. S., & A. Russell 8905 (9).
Felger, R. S. et al. 10329 (9); 10365 (9); 11118 (9); 20742 (9).
Ferrera B., A. 150 (4.2).
Ferris, R. S. 986 (9); 8623 (9); 8763 (9).
Ferris, R. S., & R. Bacigalupi 8241 (9).
Fischer, P. C. 5847 (9).
Fisher, G. L. 45171 (3.2).
Florence, J. 3826 (NY) (6.3).
Flores, G. 1 (6.3).
Flores, J. S. 9797 (6.2).

Forero, L. E., & N. Hernández 1609 (6.3).
Fosberg, F. R. 8310 (9); 35490 (6.3).
Fox, R. B. 2099–4 (6.3).
Francis, D., & D. Enos 73 (9).
Frankie, G. W. 327 (6.3); 353 (6.3).
Gómez L., J. 9977 (6.3).
Gammill, J. A. 294 (6.3).
Gander, F. F. 2532 (9).
García, E. 124 (10.2).
García et al. 2754 (3.1a).
Garth, J. C645 (9).
Gaumer, G. F. 448 (6.2); 23603 (6.2).
Gaviño, G. 16 (9).
Gee, E. 1902 (9).
Gentle, P. H. 401 (6.2); 1138 (6.2); 2529 (6.2); 2583 (3.2);
Gentry, A. 324 (6.3); 4900 (3.1a); 9657 (6.1); 10314 (6.3); 12306 (6.3).
Gentry, A., & E. Renteria 23688 (6.3).
Gentry, A. et al. 11054 (6.3); 18059 (6.3); 18156 (6.3); 18191 (6.3); 54072 (6.3).
Gentry, H. S. 2658 (10.4); 4475 (9); 4481 (9); 7589 (9); 7968 (9).
Gereau & Saynes V. 2139 (4.2).
Germán, M. T. et al. 218 (6.3).
Germán, M. T., & O. Téllez 956 (4.2).
Gierisch, R. K. 4668 (10.1).
Gillis, W. T. 9054 (6.3).
Gilly, C. L. 29 (6.3).
Gilly, C. L. et al. 33 (6.3); 104 (6.3).
Gilmartin, A. J. 830 (6.1).
Goldberg, D. E., & W. Cornell 77–109 (9).
Goldman, E. A. 395 (9).
Góngora, E. 591 (3.3).
Gonzales, A. 95 (2).
Goodding, L. N. 4183 (10.2).
Gordon, B. L. 78 (6.3).
Graham, E. H. 3798 (9); 3878 (9); 3907 (9).
Graham, H. W. 2.7.1904 (9); 27.6.1912 (9).
Greenman, J. M. 94 (6.3).
Grimes, J. et al. 2723 (3.2).
Guízar, E. 1924 (6.3).
Gutiérrez B., C. (6.3).
Gutiérrez, L. R. 95 (3.2).
Halse, R. R. 1934 (10.1).
Hamilton, S. H. 114 (6.3); 126 (6.3).
Hansen, B. et al. 1371 (9); 1376 (9).
Harbison, C. F. 1955 (9).
Harmon, W. E., & A. Fuentes 1844 (6.3).
Harriman, N. A. 15666 (3.2); 17668 (6.3).
Hartman, C. V. 241 (9).
Hastings & Thornber 9020 (9).
Haught, O. 3020 (6.1).
Havard, V. 51 (10.2); 81 (6.3); 154 (6.3).
Hayes, S. 159 (10.2); 203 (9).
Heithaus, E. R. 34 (6.3).
Henrickson, J. 1546 (9).
Henrickson, J., & R. Rutherford 1985 (9).
Hermann, H. A. van 525 (6.3).
Hernández A., S., & A. Figueroa N. 9 (3.1a).
Hernández G., H. 81 (3.3); 82 (3.3); 1348 (3.3).
Hernández M., R. 3917 (3.2).
Hernández M., R., & J. Dorantes 1803 (6.3); 1813 (3.1a).
Hernández, H. M., & A. Shilom Ton 551 (4.2).
Hernández M., R., & P. Tenorio L. 6957 (4.2).
Hernández M., R. et al. 6029 (4.2).
Hernández, S. O. 93 (3.2).
Hernández X., E., & A. J. Sharp X–268 (3.2).
Herrera, G. 1223 (3.1a).
Hess, W., & L. Byrne 4656 (4.2).
Hewit, M. 406 (6.3).
Heyder, H. M. 30 (6.2).
Higgins, L. C. 2575 (6.3).
Hill, J. H. 11 (6.3).
Hill, S. R. 17619 (6.3).
Hinckley, L. C. 423 (10.2); 2823 (10.2); 3025 (10.2); 10978 (10.1); 15574 (10.1); 18163 (10.2).
Hinton, G. B. 3331 (6.3); 5713 (6.3); 5869 (6.3); 5928 (4.2); 7983 (4.2); 10230 (6.3); 11246 (6.3); 12581 (6.3); 15741 (6.3); 15922 (6.3); 16234 (6.3); 16288 (6.3); 16289 (6.3); 23005 (11); 23109 (10.3).
Hioram, Bro. 1969 (2).
Hitchcock, C. L., & C. V. Muhlick 23211 (9).
Hodge, W. H. 1308 (6.3).
Hodge, W. H., & B. T. Hodge 1255 (6.3).
Hoehne, F. C. 994 (6.3); 43078 (6.3).
Holdridge, L. R. 1105 (6.3).
Holmgren, N. H. 3384 (10.1).
Holmgren, N. H. et al. 8785 (10.1).
Houk, 31 (6.3).
Howard, R. 4916 (2); 10388 (6.3).
Howard, R. et al. 11 (2).
Huft, M. 1899 (3.1a).
Hughes, C. E. 385 (4.2); 430 (6.3); 437 (6.3); 488 (6.2); 489 (6.2); 500 (6.2); 545 (6.3); 622 (6.3); 630 (6.3); 678 (6.2); 681 (6.3); 700 (6.3); 779 (6.3); 781 (6.3); 790 (6.3); 939 (6.2); 943 (6.2); 944 (6.2); 946 (6.2); 993 (6.1); 1009 (6.1); 1010 (6.1); 1012 (6.1); 1015 (6.1); 1186 (4.1); 1193 (6.1); 1194 (6.1); 1458 (4.2); 1459 (4.2); 1468 (4.2); 1475 (4.2); 1488 (4.1); 1498 (4.2); 1529 (4.2); 1678 (6.3); 1726 (4.2).
Hughes, C. E., & G. P. Lewis 1276 (6.2); 1281 (6.2); 1282 (6.2); 1291 (4.1).
Hughes, C. E. et al. 1588 (4.2); 1688 (4.2); 1692 (4.2).
Hunnewell, F. W. 11007 (6.3).
Hunter. J. R. 561 (6.3).
Huston, J. S. 0564 (6.2).
Ibarra M., G. 487 (6.3).

Iltis, H. H., & M. Nee 1549 (6.3).
Isely, D. 7836 (6.3).
Jack, J. G. 4242 (6.3); 4542 (2); 4957 (2); 5569 (2); 6039 (2); 6562 (2); 7797 (2); 8100 (2).
Jacques-Georges, A. 203 (6.3); 204 (6.3).
Jarquin, E., & A. Lot 53 (3.2).
Jepson, W. L. 5223 (9); 8790 (9); 17097 (9); 20-X-1912 (9).
Johnson, C. D. 23–73 (6.3); 81–68 (6.3); 118–68 (6.3); 147–68 (6.3); 148–68 (6.3); 206–68 (6.3); 275–68 (6.3); 387–93 (6.3); 397–73 (6.3); 465–73 (6.3); 639–79 (6.3); 835–79 (6.3).
Johnston, I. M. 3205 (9); 3277 (9); 3444 (9); 3780 (9); 3786 (9); 3969 (9); 4041 (9); 8203 (11).
Jones, M. E. 1926 (9); 1928 (9); 5286 (10.2); 23005 (6.3); 98913 (9).
Judziewicz, E. et al. 3481 (6.3).
Kamb, P. 1998 (9).
Keeler, K. H. 169 (6.3); 188 (6.3).
Keil, D. 14066 (9).
Kelly, I. 967 (6.3).
Kennedy, H. et al. 2017 (3.1a).
Kennedy, P. B. 7012 (9).
Kenoyer, L. A. A53 (3.2); A638 (3.2); 678 (4.2).
Killip, E. P. 43405 (6.3); 44537 (6.3).
Killip, E. P., & A. C. Smith 15437 (6.3).
Kimber, C. 1328 (6.3).
Kimnach, M. 718B (6.3).
King, R. M. 1283 (6.3); 2377 (6.3); 2458 (6.3); 2466 (6.3);
Kinloch, J. B. 56 (6.2).
Langenheim, J. H. 3278 (6.3).
Langlassé, E. 710 (6.3).
Lao, E. A. 88 (6.3).
Laughlin, R. M. 753 (4.2).
Lavin, M. 5611 (2); 6179a (10.1); 6182a (10.1); 6183a (10.1); 7147 (6.3).
Lavin, M., & A. Delgado 8201 (4.1); 8207 (4.1); 8210 (3.3); 8212 (3.3); 8217 (3.2); 8228 (3.1a).
Lavin, M., & H. Hedrick 6157 (10.1).
Lavin, M., & J. Lavin 8013 (6.3); 8016 (6.3).
Lavin, M., & K. Lavin 4642 (9); 4653 (9); 4654 (9); 4657 (9).
Lavin, M., & M. Luckow 5789 (3.2); 5819 (4.2); 7149 (6.3).
Lavin, M., & S. Sundberg 5037 (10.3).
Lavin, M., & T. Zanoni 8025 (6.3).
Lavin, M. et al. 4623 (6.3); 5355a (6.3); 7144 (2); 8029 (6.3).
Lawrance, A. E. 638 (6.3).
Leavenworth, Mr., & Mrs. W. C. 894 (3.2).
Lehto, E. 3836 (9); L23825 (10.1); 24252 (6.3).
Lems, K. 5056 (6.3).
León, Bro. 10698 (2); 12749 (2); 17090 (2); 18124 (2); 4501 (2).
León, Bro., & Bro. Alain 5026 (2).
León, Bro., & M. Victorin 17180 (2).
León, H. 24 (6.3).
Leonard, E. C. 2812 (6.3); 9694 (6.3).
Lepper, L. et al. 34358 (2).
Lewis, W. H. et al. 393 (6.3); 779 (6.3); 5367 (3.1a).
Liesner, R. 2311 (6.3); 4296 (6.3).
Liesner, R., & J. Dwyer 1546 (6.3).
Liogier, A. H. 21030 (6.3).
Little, E. L. 6444 (6.1); 6570 (6.1); 6683 (6.1).
Lloyd, R. M., & D. M. J. Mueller 3679 (6.3).
Loomis, H. F. 1953 (9).
López F., A. R. 179 (4.2).
Losure, M. B. 176 (10.2).
Lott, E. J. 190 (10.2).
Lott, E. J., & T. H. Atkinson 2446 (9).
Luna, A. 576
Lundell, C. L. 390 (3.2); 1272 (6.2); 1356 (6.2); 2179 (6.2); 2420 (6.2); 2852 (6.2); 2885 (6.2); 14765 (11); 15709 (6.2); 15710 (6.2).
Lundell, C. L., & A. A. Lundell 7266 (3.2).
Muller 85 (6.3).
Macbride, J. F. 2759 (6.3).
MacDougall, T. 550.S (4.2); 7942 (4.2).
Magallanes, J. A. S. 955 (6.3); 969 (3.1b); 1144 (3.1b); 1593 (3.1b); 1662 (3.1b); 1714 (9); 2044 (3.1b).
Magallanes, J. A. S., & P. Basurto 1830 (3.1b); 1835 (3.1b); 1885 (3.1b).
Magaña, M. A., & A. Guadarrama 1225 (3.3).
Magaña, M. A., & A. Suárez 640 (6.3).
Magaña, M. A., & S. Zamudio 774 (6.2).
Maguire, B., and R. J. Becraft 2664 (10.1); 2665 (10.1).
Maguire, B., & A. H. Holmgren 25127 (10.1).
Mahler, W. F., & J. W. Thieret 5881 (6.3); 5886 (6.3); 6006 (9).
Manning, W. E., & M. S. Manning 53585a (4.2).
Marie, T. 323 (9).
Marquez, W. et al. 619 (3.2).
Marshall, M. L. 81 (6.3).
Martínez C., G. 1867 (6.3).
Martínez S., E. 11343 (6.2); 18361 (6.2).
Martínez S., E., & B. Morales 3390 (6.3).
Martínez S., E., & C. H. Ramos 24032 (3.1a).
Martínez S., E., & A. Reyes 22023 (4.1).
Martínez S., E., & J. C. Soto N. 3606 (6.3).
Martínez S., E., & O. Téllez 204 (4.2); 2948 (3.2).
Martínez S., E. et al. 3006 (6.2); 20091 (4.1).
Mason, H. L. 13803 (6.3).
Mathias, M. 114 (6.3).
Matthew, K. M. 26861 (6.3).

Matuda, E. 15 (6.3); 15958 (4.2).
Matuda, E. et al. 27960 (6.3); 38314 (6.2).
Maxwell, R. H. 151 (3.1a).
McGregor, R. C. 42885 (6.3).
McKibben, M. 1904 (9).
McLaughlin, S. P. 3472 (9).
McMinn, H. E. 1447 (9).
McMurry, F. B 1398 (9).
McVaugh, R. 11855 (3.1b); 14900 (6.3); 22184 (6.3); 22487 (6.3); 22495 (6.3); 22594 (6.3); 22944 (6.3); 23568 (6.3); 25176 (6.3); 26186 (6.3).
McVaugh, R., & W. N. Koelz 1050 (6.3).
Mearns, E. A. 2700 (9).
Mears, J., & H. Mears 16786 (9).
Meave, J. et al. B–245 (6.3).
Medina, M. 821 (2).
Mejía, M., & T. Zanoni 9371 (6.3).
Mejía, M. et al. 10350 (6.3).
Mell, C. D. 2245 (6.3).
Mendoza, M., & V. Evangelista 41 (3.2).
Merrill, E. D. 289 (6.3); 613 (6.3); 894 (6.3).
Merritt, M. L. 61686–2 (6.3).
Mexia, Y. 931 (6.3); 8888 (6.3).
Meyer, R. 28667–1 (6.3).
Mille, L. 889 (6.1).
Millspaugh, C. F. 783 (6.3); 1055 (6.3); 1553 (6.2).
Miranda, F. 2670 (4.2); 3982 (4.2); 5077 (4.2); 7942 (4.2); 8856 (3.2); 9013 (3.2).
Mittleman, M. 4 (9).
Mittleman, M., & R. Engard M14 (10.2).
Moldenke, A. R., & H. N. Moldenke 25335 (9).
Moldenke, H. N. 5808 (6.3).
Molina R., A. 333 (4.2); 390 (4.2); 1765 (4.2); 3340 (4.2); 6649 (3.3); 7583 (4.2); 7829 (4.2); 8283 (4.2); 8520 (4.2); 10772 (4.2); 13626 (4.2); 13727 (4.2); 14230 (4.2); 14244 (4.2); 14518 (4.2); 18751 (4.2).
Molina R., A., & A. R. Molina 12355 (4.2); 14003 (4.2); 14131 (4.2); 24284 (4.2); 25825 (4.2); 31497 (4.2).
Mooers, B. H. 756 (10.1).
Moore, G. 425 (10.2).
Moore, H. E. 1665 (4.2); 2668 (4.2); 2929 (3.2); 5031 (4.2); 6077 (2).
Moore, H. E., & C. E. Wood 3640 (4.2).
Moore, R. H. 3672 (6.3).
Moore et al. 6794 (10.1).
Morales, J. B. 39 (3.1b).
Moran, R. 3609 (9); 3766 (9); 12971 (9).
Moreno, P. 130 (6.3); 493 (6.2); 2629 (6.3); 3504 (6.3); 5796 (6.3).
Mori, S. A. 10302 (6.3).
Munz, P. A. 9583 (9).
Munz, P. A., & D. Keck 9345 (9).
Nabhan, G., & A. Rea 286 (9).
Nee, M. 10124 (6.3); 29778 (3.3); 29986 (6.3).
Nee, M., & B. F. Hansen 18455 (3.2).
Nee, M., & K. Taylor 26474 (6.3); 26552 (6.3); 29353 (6.3); 29631 (6.3).
Neese, E. 5942 (10.1).
Neese, E., & S. L. Welsh 16880 (10.1).
Neese, E., & S. White 9049 (10.1).
Nelson, A. 10294 (9).
Nelson, A., & R. Nelson 3311 (9).
Nelson, E. W. 423 (3.2); 2570 (6.3).
Nevling, L., & A. Gómez-Pompa 883 & 884 (6.3).
Newberry, C. 353 (9).
Nichol, A. A. 28.4.1939 (9).
Niering, W. A., & R. H. Whitaker 8.5.1963 (9).
Novelo, A. 355 (6.3).
Opler, P. A. 510 (6.3).
Orcutt, C. R. 2014 (9); 4290 (6.3).
Orcutt, O. R. 3453 (6.3).
Ortega U., A. 168 (4.2).
Ortega, J. G. 6831 (6.3).
Ortega, R. V. O. 231 (6.3).
Ortiz, F. 1686 (6.3).
Ortiz, G. 737 (3.3).
Otero, J. I. M–83 (6.3); 381 (6.3).
Oviedo, R. 27992–27995 (2).
Pabst, G. 5412 (6.3).
Palma, J. G. 85–40 (6.2).
Palmer, E. 360 (6.3); 360? (6.3); 1279 (6.3).
Parish, S. B. 8123 (9); 10833 (9); 10857 (9).
Parish, S. B., & W. F. Parish 800 (9).
Parry, C. C. 1881 (9).
Parry, C. C., & E. Palmer 172 (10.3).
Pavlik, B. M. 153 (10.1).
Peabody, F., & Sears 1992 (10.2).
Peck, M. E. 402 (6.2); 422 (3.2).
Peebles, R. H. 7909 (9); 9311 (9); 8.7.1943 (9).
Peebles, R. H., & G. J. Harrison 4226 (9).
Peirson, F. W. 5885 (9).
Pennell, F. W., & H. H. Rusby 176 (6.3).
Pennington, C. W. 261 (9).
Pennington, T. D., & J. Sarukhán K. 9438 (6.3); 9464 (6.3).
Pérez, L. A. 114 (6.3); 1726 (3.1b).
Pérez, L. A., & C. H. Ramos 1577 (6.2).
Pérez, S. 4 (6.3).
Persaud, R. 1744 (6.3).
Peters, R. E. K. 115A (9).
Pierre, L. Jn. 72 (6.3).
Pilz, G. E. 1913 (6.3).
Piper, C. V. 5133 (6.3).
Pipoly, J. J. 5062 (3.1a).
Pittier, H. 5101 (3.1a); 6609 (3.1a); 6976 (3.1a); 7747 (6.3); 8710 (6.3); 13588 (6.3).
Plowman, T. C. 3012 (6.3).
Pohl, R. W. 12603 (4.2).
Ponce C., F. 102 (6.3); 131 (3.2); 256 (3.3).

Ponce C., F., & C. Alvarez del C. 294 (6.3).
Porsch, O. 742 (6.3).
Porter, D. M. et al. 4025 (3.1a); 5176 (6.3).
Powell, A. M. 3186 (11); 3339 (11); 3379 (10.2); 3454 (11).
Powell, A. M., & S. Powell 3639 (11); 3751 (11).
Prance, G. T. et al. 19303 (6.3).
Pringle, C. G. 99 (2); 1881 (9); 7285 (10.3); 7482 (10.3); 13,868 (4.2); 15138 (9); 15138a (9); 11.8.1884 (9).
Proctor, G. R. 27778 (6.3); 30292 (6.2); 35764 (6.2).
Purpus, C. A. 1188 (10.3); 2474 (10.3); 5891 (3.2); 8721 (3.2); 8836 (6.3).
Ramamoorthy, T. P. et al. 3640 (6.2).
Rea, A. M. 1190 (9).
Record, S. J. 3 (6.2).
Reichenbacher, F. W., & T. Wilson 1730 (10.1).
Renson, C. 72 (6.3).
Reznicek, A. A., & D. R. Gregory 299 (6.3).
Richardson, A. 1308 (4.2).
Rico, L., & E. Martínez S. 741 (3.3).
Ripley, H. D., & R. C. Barneby 3647 (10.1); 4932 (10.1); 6511 (10.1); 14158 (10.3).
Rixford, S. P. 1911 (9).
Rodríguez, D., & E. Martínez 38 (4.2).
Roig, J. T. 5728 (2); 6526 (2).
Romero R., H. 80 (6.3).
Rosas, M. R. 612 (3.1a).
Rose, J. N. 16521 (9); 16582 (9); 21843 (6.3).
Rose, J. N. et al. 13104 (9); 14291 (6.3); 15048 (9).
Rosentreter, R. 150 (10.1); 1328 (10.1).
Rowntree, L. 1937 (9).
Ruano, M. 868 (6.3).
Rugel, F. 265 (2).
Rzedowski, J. 7481 (6.3).
Saldanha, C. J. 12853 (6.3); 16326 (6.3).
Sanders, A. C., & K. R. Neisess 5707 (9).
Sandino, J. C. 296 (6.3); 1835 (6.3); 1950 (6.3).
Santana, E. J., & N. Cervantes 321 (3.1b).
Santiago & Ramona 27590 (2).
Sargent, C. S. 28 (6.3).
Sarukhán, J. et al. 2256 (3.2); 2930 (3.2).
Sauer, J. D. 7/5 (6.2).
Sauer, J., & D. Gade 3060 (6.3).
Sauer, J. D., & H. N. Sauer 4269 (6.3).
Saunders, J. et al. 770 (6.3).
Schaffner, J. G. 622 (10.3).
Schellenger, E. E. 12 (9).
Schmalzel, R. J. 481 (6.3).
Schnooberger, I. 8054 (3.2).
Schott, A. 1854 (9).
Schotts 13 (6.3).
Schubert, B. G., & C. Rojas 1863 (6.3).
Schuh, R. T., & B. M. Massie 37 (9).
Schultes, R. E., & B. P. Reko 573 (3.2); 578 (6.3).
Seler, C., & E. Seler 3883 (6.2).
Sessé et al. 3755 (3.2).
Seymour, F. C. 2231 (6.3).
Shafer, J. A. 546 (6.3); 1180 (6.3); 1397 (6.3); 1616 (2); 3271 (2); 10454 (2); 12287 (2); 12302 (2); 12123 (6.3); 13744 (2).
Shapiro, G. 216 (3.2); 288 (6.3).
Shatz, G. E. et al. 1193 (3.2).
Sherff, E. E. 9.8.1908 (9).
Shilom Ton, A. 2262 (3.3); 3382 (6.3); 3715 (6.3); 3899 (6.3); 4243 (3.3); 5863 (3.3); 5976 (3.3); 6221 (3.3); 6246 (3.3); 6289 (3.3); 6734 (3.3); 6776 (3.3); 8985 (3.3); 9069 (3.3).
Shreve, F. 6560 (9); 7386 (9).
Shultz, L. M., & J. S. Shultz 8334 (9).
Simon, C. 80–447 (6.3).
Simpson, D. R., & J. Schunke 516 (6.1).
Singleton, W. R. 50 (6.3); 67 (6.3).
Smith A. C. 10326 (6.3).
Smith, E. E. 782 (2).
Smith, F. 1032 (10.1).
Smith, H. H. 15 (6.3).
Smith, H. M. 132 (6.3).
Smith, J. D. 2314 (6.3); 2813 (6.3).
Smith, J. G. 109 (3.2).
Smith, L. B., & A. R. Hodgdon 3187 (2).
Smith, L. C. 453 (6.3).
Soejarto, D. D. 364 (6.3).
Sohns, E. R. 1712 (6.2).
Solomon, J. 346 (9).
Soto N., J. C., & L. Cortés A. 2599 (3.1b).
Soule, M. 5 (9); 20 (9).
Sousa Peña, M. 230 (9).
Sousa S., M. 721 (3.2); 768 (3.2); 780 (3.2); 2622 (3.2); 2888 (6.3); 8337 (4.2); 10504 (4.2).
Sousa S., M., & P. Basurto 9971 (4.2); 9973 (4.2).
Sousa S., M. et al. 5515 (6.3); 6300 (4.2); 6705 (4.2); 7076 (3.1a); 7954 (3.2); 8283 (4.2); 9150 (6.3); 10050 (4.2); 10475 (4.2); 11053 (6.2); 11382 (3.3); 12077 (3.2); 12331 (6.2); 12689 (6.3); 12729 (6.3); 13012 (6.3); 13046 (3.2); 13096 (3.1a); 13103 (3.1a); 13095 (3.1a); 13212 (4.1).
Spalding, V. M. 11.10.1906 (9).
Spellman, D. L., & W. W. Newey 2025 (3.2).
Spellman, D. L., & Stoddart 2439 (6.2).
Spencer, M. F. 165 (9); 1920 (9); 1925 (9).
Sponslu, O. J. 6968 (9).
Standley, P. C. 8450 (6.3); 11439 (6.3); 14226 (4.2); 15227 (4.2); 15467 (4.2); 15712 (6.3); 17685 (6.3); 19263 (6.3); 21897 (6.3); 22545 (4.2); 22601 (6.3); 23133 (6.3); 23750 (4.2);

53630 (6.3); 60498 (6.3); 64026 (6.3); 66617 (6.3); 75805 (6.3); 79274 (6.3); 87562 (6.3).
Standley, P. C., & J. Chacón P. 6373 (6.3).
Standley, P. C., & A. Molina R. 4481 (4.2).
Standley, P. C., & E. Padilla V. 3496 (6.3).
Stark, B. D. 1540 (9).
Steggerda, M. 30C (6.2).
Stehlé, H. 2942 (6.3).
Stern, W. L., & K. L. Chambers 157 (6.3).
Stern, W. L. et al. 289 (6.3); 1230 (6.3).
Stevens, W. D. 5306 (6.3); 5851 (6.3); 6195 (6.3); 6227 (6.3); 6622 (6.3); 6645 (6.3); 7240 (6.3); 7430 (6.3); 12019 (6.3); 12263 (6.3); 22872 (6.3).
Stewart 1116 (11); 2217 (11).
Steyermark, J. A. 29372 (6.3); 32275 (6.3); 33130 (6.3); 39219 (6.3); 39238 (3.3); 44432 (3.3); 45719 (3.3); 50926 (4.2); 50927 (4.2).
Stoffers, A. L. 1682 (6.3).
Stork, H. E. 4560 (6.3).
Studhalter, R. A. 1520 (9).
Sytsma, K., & W. G. D'Arcy 3466 (6.3).
Taylor, J., & C. Taylor 7352 (6.3); 30304 (9).
Téllez, O., & E. Cabrera 1320 (6.2); 1381 (6.2); 1452 (6.2); 1733 (6.2); 1857 (6.2); 1986 (6.2).
Téllez, O. et al. 400 (6.3); 649 (4.2); 650 (4.2).
Tenorio L., P., & C. Romero 12846 (9); 13058 (9).
Tenorio L., P. et al. 3430 (6.3).
Thackery, F. A. 454 (9).
Thompson, B., & B. Fields 329 (6.3).
Thompson, J. B. 644 (6.3); 1093 (6.3).
Thornber, J. J. 23.3.1902 (9); 24.3.1908 (9); 25.1.1905 (9); 26.9.1912 (9).
Thorne, K. H., & J. Chandler 5612 (10.2).
Thorne, K. H., & D. Atwood 5248 (10.1).
Thorne, K. H., & B. T. Welsh 924A (10.1).
Tidestrom, I. 2409 (10.1); 12620 (9).
Tiehm, A. 6351 (10.1); 6379 (10.1).
Tilforth, C. et al. 1318 (9).
Toro, R. A. 240 (6.3).
Torres C., R. 3176 (4.2).
Torres C., R., & L. Cortes 11519 (3.2).
Torres C., R., & J. Martínez 4822 (4.2); 10990 (3.2).
Torres C., R. et al. 433 (4.2); 9755 (6.3);
Troll, C. 456 (4.2).
Tucker, J. M. 862 (6.3).
Tún Ortiz, R. 790 (6.2); 888 (6.2); 976 (6.2); 1720 (3.2).
Turner, B. L., & A. M. Powell 6096 (10.2).
Turner, R. M. 61–2 (9).
Tyson, E. L. 2701 (6.3); 3151 (6.3); 3608 (6.3); 3626 (6.3); 3723 (6.3); 5340 (3.1a); 6625 (3.1a).
Tyson, E. L., & K. Blum 3781 (6.3).
Tyson, E. L., & A. Gentry 6409 (3.1a).
Tyson, E. L. et al. 2886 (6.3); 3164 (3.1a).
Ucán E., 708 (6.2); 725 (6.2); 777 (6.2).
Uitzil, H. et al. 01 (6.2).
Underwood, L. M., & F. S. Earle 97 (6.3).
Valdés, J. VR–546 (10.2).
Valerio R., J. 905 (4.2); 2113 (4.2); 2299 (6.3).
Valeur, E. J. 341 (6.3); 596 (6.3).
Valverde, T. 596 (6.1).
Van Devender, T. R. 88–209 (9).
Van Hermann, H. A. 525 (6.3); 748 (2); 749 (2); 5039 (2).
Vanoverbergh, M. 3404 (6.3).
Vázquez, C. 621 (6.3).
Vázquez, M. et al. V–369 (3.3); V–493 (3.3); V–567 (3.3); V–661 (3.2); V–1762 (3.3).
Velázquez L., C. 277 (6.3).
Ventur, P. 213 (6.2).
Ventura A., F. 3224 (6.3); 8178 (3.1a); 8229 (3.2); 10961 (3.1a); 11455 (3.2); 12596 (3.2); 12674 (3.1a); 14950 (3.2); 14966 (3.2); 15184 (3.1a); 19730 (3.2); 20337 (3.2); 20646 (3.3); 21226 (3.3).
Vera, J. S. 2833 (6.3).
Vestal, P. A., & M. R. Vestal 151 (10.2); 480 (10.2).
Vidal 2570 (6.3).
Villanueva G., R. 70 (6.3); 79 (6.3); 86 (6.3).
Vincelli, P. C. 109 (6.3).
Vovides, A. P. 548 (3.2).
Wagner, R. J. 514 (6.3).
Walker, S. 74H06 (6.3).
Warnock, B. H. T429 (10.2); 7585 (11); 10578 (10.2); 20837 (10.2); 20938 (11).
Warnock, B. H. et al. 1161 (11).
Warnock, M. J. 2086 (9).
Waterfall, U. T. 5240 (10.2); 5350 (10.2); 6717 (10.2).
Watson, S. 495 (6.3).
Webster, G. L. 3682 (2).
Webster, G. L. et al. 81 (2).
Webster, H. L. 1919 (9).
Wedel, H. von 505 (6.3); 1198 (6.3); 2747 (6.3).
Welsh, B. T. 414 (10.1).
Welsh, S. L. 5382 (10.1).
Welsh, S. L., & S. L. Welsh 12811 (10.1).
Welsh, S. L. et al. 16771 (10.1); 22908 (10.1).
Wendt, T. et al. 796 (11); 1951 (11).
Weston, A. S. et al. 4348 (6.3).
Wheelock, Mrs. C. M–47A(3.2).
White, D., & R. Mott 64 (6.2).
White, S. S. 5273 (6.3).
Whitefoord, C. 3160 (3.2).
Whiting, A. F., & H. S. Gentry 9011 (9); 9012 (9).
Wieslander, A. E. 168 (9).
Wiggins, I. L. 5463 (9). 6041 (9); 11632 (9); 17524 (9); 17724 (9); 19049 (9).
Wiggins, I. L., & D. B. Wiggins 14890-A (9).

Wight, A. E. 191 (6.3).
Willard, L. 7 (6.3).
Williams, L. O. 9083 (6.3); 9119 (6.3); 9184 (6.3); 9776 (6.3).
Williams, L. O., & A. Molina R. 10169 (4.2); 10320 (4.2); 10556 (4.2); 11283 (6.3); 11658 (4.2); 11965 (4.2); 12212 (6.3); 12556 (4.2); 12658 (4.2); 14578 (3.3); 15043 (4.2); 17813 (3.3).
Williams, L. O., & T. P. Williams 24510 (6.3).
Williams, M. P. 149 (10.1).
Williams, M. J. 781433 (10.1).
Williams, M. J., & A. Tiehm 8081 (10.1); 81196 (10.1); 81202 (10.2).
Williams, M. J., & M. Williams 78753 (10.1).
Williams, R. S. 341 (6.3); 3072 (6.3).
Wilson, P. 136 (6.3); 412 (6.3); 11346 (2).
Wing, E. S. 46 (6.3).
Wislizenus, A. 126 (10.2).
Wit, P. 36 (6.3).
Wolf, C. B. 2291 (9); 2295 (9); 3656 (9).
Wonderly, W. L. 6 (6.3).
Woodworth, R. H., & P. A. Vestal 477 (6.3).
Worthington, R. D. 12401 (10.2); 12415 (6.3).
Wright, C. 962 (10.2); 2350 (6.3).
Wuerthner, G. 83–27 (10.1).
Yale Forestry School 579762 (6.3).
Yuncker, T. G. 17775 (6.3).
Yuncker, T. G. et al. 5834 (4.2); 8814 (6.3).
Zárate, S. 318 (6.3).
Zanoni, T. et al. 18956 (6.3).
Zimmerman, A. D. 2440 (10.2).

INDEX TO SCIENTIFIC NAMES

Accepted names are in roman type; the main entry for each is in **boldface**. Synonyms are in *italics*.

Abronia 128
Acacia 47, 65, 68, 72, 96, 133
Acalypha 25
Acanthaceae 113
Acer 25
Aceraceae 25, 27
Acrodiclidium 25
Actinidia 25
Adolphia 133
Aeschynomene 68
  americana L. 11
Aeschynomeneae 3, 5, 7, 10, 11, 16, 24, 36, 42
Aganope Miq. 7
Agati Adans. 42, 43
  grandiflora (L.) Desv. 42
Agave 68, 132, 138
Alchornea 25
Alnus 25, 26
Alvaradoa 68
Ambrosia 128
Amorpheae 36
Amphiodon Huber 145
  effusus Huber 145
Andira 60
Aphelandra 113
Aporusa 25
Apurimacia Harms 144
  boliviana (Britton) Lavin 144
  dolichocarpa (Griseb.) Burkart 144
  michelii (Rusby) Harms 144
Artemisia 128
Astragalianae 2, 3
Astragalus L. 37, 126
Atriplex 128
Austrosteenisia Geesink 144
Balboa Liebm. ex Didr. 114
  diversifolia Liebm. 114
Bauhinia L. 96
Belairea A. Rich. 7
Bembecidium Rydb. 2, 4, 78, 79
  cubense Rydb. 79
Benthamantha Alef. 113
Betula 25, 26
Betulaceae 25
Bignoniaceae 68
Boraginaceae 25
Bouteloua 138
Brachysema R. Br. 70
Brickellia 132
Brittonamra Kuntze 113
Bromus 128
Brongniartieae 13, 145
Brosimum 96
Bursera 68, 87, 96
Burseraceae 25
Byrsonima 96
Caesalpinia 47, 87, 96
Caesalpinioideae 16, 33
Cajalbania Urb. 79
  immarginata (C. Wright) Urb. 79
Callerya Endl. 9, 20
Calliandra 60
Callistylon Pittier 112
  arboreum (Griseb.) Pittier 112
Canarium 25
Caragana Fabr. 4, 5, **9**, 10, 14, 15, 16, 17, 18, 19, 20, 21, 22, 24, 32
Carapa 25
Carpinus 25, 26
Carya 26

Castanea 25
Cedrela 25, 87, 96
Centris
pallida 119
Cephalocereus 96
Ceratozamia 25
Cercidiphyllaceae 25
Cercidiphyllum 25, 26
Cercidium
floridum 119
microphyllum 119
Chaetocalyx DC. 7
Chiovendaea Speg. 111
hypoleuca Speg. 111
Chloroleucon 87
Chrysophyllum 65
Chrysothamnus 128
Cinnamomum 25
Cladastris Raf. 33
Clethra 65
Cnidoscolis 68
Cochlospermum 96
Coleogyne 128
Combretaceae 25
Combretum 96
Comocladia 68, 96
Cordia 25, 87, 96
Cornaceae 25
Cornus 25
Corynella DC. 2, 4, 35, 78, 79
gracilis Griseb. 78
Corynellinae Rydb. 2, 3, 35
Corynitis Spreng. 79
Coursetia DC. 3, 4, 5, 6, 16, 17, 18, 19, 21, 23, 24, 28, 35, 36, 37, 39, 42, 77, 106, 107, **108–117,** 126, 132, 142, 144, 145, 152
sect. Coursetia **112,** 113
sect. Craccoides (DC.) Lavin 108, 112, **113–117**
sect. *Glandulosae* Rydb. 112
sect. Madrenses Rydb. ex Lavin & Sousa 56, **109–111,** 113
sect. Neocracca (Kuntze) Lavin 108, **112–113,** 114
sect. Poissonia (Baill.) Lavin **111,** 112, 113
sect. *Polyphyllae* Rydb. 109
ser. *Monophyllae* Harms 111
andina Lavin 117
apantensis M. Sousa 111
axillaris Coulter & Rose 112
barrancana Lavin 114
boliviana Britton 144
brachyrhachis Harms 112
cajamarca Lavin 112
caribaea (Jacq.) Lavin 113, 114
var. astragalina (H. B. K.) Lavin 114
var. caribaea 114, 116
var. chiapensis (Rydb.) Lavin 114
var. ochroleuca (Jacq.) Lavin 113, 114
var. pacifica (M. Sousa & Lavin) Lavin 114
var. sericea (A. Gray) Lavin 114, 126
var. tomentosa Lavin 114
var. trifoliolata (Rydb.) Lavin 114
chiapensis Lavin & M. Sousa 110, 111
dubia (H. B. K.) DC. 117
elliptica M. Sousa & V. Rudd 111
ferruginea (H. B. K.) Lavin 112
fruticosa (Cav.) Macbr 108, 112
glabella (A. Gray) Lavin 114
glandulosa A. Gray 112
gracilis Lavin 117
grandiflora Benth. 117
guatemalensis V. Rudd 114
hassleri Chodat 117
heterantha (Griseb.) Lavin 112, 113
hidalgoana Lavin 114
hintonii V. Rudd 114
hypoleuca (Speg.) Lavin 111
insomniifolia Lavin 112, 137
intermontana Lavin 117
madrensis Micheli 109, 111
maraniona Lavin 112
mollis Rob. & Greenm. 114, 115
oaxacensis M. Sousa & V. Rudd 111
orbicularis Benth. 111
paniculata M. Sousa & Lavin 111
paucifoliolata M. Sousa 111
planipetiolata Micheli 111
polyphylla Brandegee 109, 111
var. acutifolia M. Sousa & Lavin 111
var. breviloba M. Sousa & Lavin 111
var. polyphylla 111
pumila (Rose) Lavin 114
robinioides M. Sousa & Lavin 111
rostrata Benth. 112
tumbezensis Macbr. 117
vicioides (Nees & Mart.) Benth. 117
weberbaueri Harms 117
Cracca Benth. 3, 4, 113, 145
Cracca L. 2, 113, 138, 145
constricta (S. Watson) Tidestr. 142
Craccinae 2, 138, 145
Crataegus 25
Crataeva 96
Crescentia 87
Cryptocarya 25
Cyrtocarpa 96
Cycadaceae 25
Dalbergia L. f. 5, **7,** 11, 13, 18, 19, 21, 42, 96
Dalbergieae 1, 3, 5, 7, 10, 13, 16, 24, 34, 36, 42, 145
Dalea L. 125
Darwinia Raf. 43
exaltata Raf. 43

Darwinia Rudge 43
Daubentonia DC. 42, 44
Daubentoniopsis Rydb. 42, 43
Davilla 25
Dialium 65
Dillenia 25
Dilleniaceae 25
Dioon 25
Diphysa Jacq. 3, 54, 60
  americana (Mill.) M. Sousa 52
Diphysinae 3
Diplotropis Benth. 33
Drypetes 25
Ehretia 25
Ephedra 128
Eriogonum 128
Erythrina 65
Eugenia 65
Euphorbia 138
Euphorbiaceae 25
Eurotia 128
Fagaceae 25, 27
Fagus 25
Ficus 87
Fiebrigiella Harms 7
Fortunearia 26
Fothergilla 26
Fraxinus 25
Galega
  ochroleuca Jacq. 113
Galegeae 2, 3, 4, 5, 10, 11, 24, 35, 37, 42, 126, 138, 145
Genistidium I. M. Johnst. 3, 5, 7, 10, 15, 16, 17, 18, 19, 21, 23, 35, 39, 106, 107, 118, 119, 125, 126, 134, **135–137,**138, 142, 145
  dumosum I. M. Johnst. 135, 136, **137–140,** 152
Gingko 25
Gingkoaceae 25
Gliricidia H. B. K. 3, 5, 6, 9, 12, 13, 14, 16, 18, 19, 21, 22, 23, 24, 28, 36, 37, 38, 52, 54, 56, 60, 67, 70, 77, **81–82,** 86, 93, 109, 124, 126, 144, 145, 146
  brenningii (Harms) Lavin 12, 70, 71, 81, 82, **83–86,** 109, 152
  cubensis (H. B. K.) C. Wright 47
  ehrenbergii (Schltdl.) Rydb. 71, 82
  guatemalensis Micheli 50, 71
  lambii Fernald 94
  latifolia (A. Rich. ex R. de la Sagra) Griseb. 47
  lutea J. R. Johnst. 144
  maculata (H. B. K.) Steud. 81, 82, 83, **86–93,** 102, 152
    var. *multijuga* Micheli 94
  malacocarpa A. Molina 54
  meistophylla (J. D. Smith) Pittier 71
  michelii Rusby 144
  platycarpa Griseb. 47
  sepium (Jacq.) Steud. 13, 46, 50, 52, 56, 71, 81, 83, 86, 87, 88, 90, 91, 92, **93–102,** 152
  f. *maculata* (H. B. K.) Urb. 86
  spilophylla (Standl.) Rydb. 144
Glottidium Desv. 42, 44, 45
Glyptopleura 128
Guazuma 68, 84, 87, 96
Haematoxylum 87, 96
Hamamelidaceae 25, 27
Harpalyce DC. 145
Hebestigma Urb. 3, 5, 6, 15, 17, 18, 19, 21, 22, 23, 36, 37, 38, **45–46,** 54, 70, 102, 137, 145
  cubense (H. B. K.) Urb. 13, 45, **47–50,** 102, 152
    var. *latifolium* (A. Rich. ex R. de la Sagra) Urb. 47
Hechtia 132, 138
Heliconiaceae 27
Heliocarpus 65, 87
Hesperothamnus Brandegee 9
Homalobus Nutt. 2
Humboldtiella Harms 112
  ferruginea (H. B. K.) Harms 112
Hura 96
Hybosema Harms 3, 5, 6, 9, 12, 14, 16, 18, 19, 21, 23, 36, 38, 52, **66–67,** 77, 82, 86
  ehrenbergii (Schltdl.) Harms 50, 66, 67, 68, 70, **71–75,** 82, 152
  robustum M. Sousa & Lavin **68–71,** 75, 111, 152
Inga 65, 84
Indigofera
  coronilloides Martens & Galeotti 114
Indigofereae 36
Ipomoea 87, 96
Juglandaceae 25
Juglans 25
Juliania 96
Juniperus 72, 128, 132
Karwinskia 87
Kentrophyta Nutt. 2
Krameria 138
Kunstleria Prain 5, **8,** 13, 15, 18, 19, 21, 144
Larrea 128
Lauraceae 25
Laurus 25
Lennea Klotzsch 3, 5, 6, 8, 13, 17, 18, 19, 21, 22, 23, 36, 37, 38, **50–52,** 67, 70, 77, 82, 137
  brunnescens Standl. 54, 55
  melanocarpa (Schltdl.) Vatke ex Harms 13, 50, 53, 54, **60–64,** 66, 152
  modesta (Standl. & Steyerm.) Standl. & Steyerm. 13, 53, 54, 62, **64–66,** 152

robinioides Klotzsch 60, 64
salvadorensis Standl. 54
viridiflora Seem. 14, 52, **53–60,** 66
var. novogaliciensis Lavin & M. Sousa 52, 55, **56–60,** 70, 152
var. viridiflora 52, **53–56,** 57, 58, 59, 60, 152
Leucaena 68, 87, 132
Lindera 25
Liquidambar 26, 65, 72
Lonchocarpinae 9
Lonchocarpus H. B. K. 5, **8,** 9, 13, 14, 15, 18, 19, 21, 22, 56, 65, 68, 86, 96, 144, 145
sect. *Paniculati* Benth. 8
latifolius A. Rich. ex R. de la Sagra 47
maculatus (H. B. K.) DC. 86
meistophylla J. D. Smith 71
modestus Standl. & Steyerm. 64
roseus (Mill.) DC. 93
sepium (Jacq.) DC. 93
Lupinus L. 11
Lysiloma 68, 87, 96
Machaerium Pers. 33, 42
Machilus 25
Macroptilium (Benth.) Urb. 70
Mallotus 25
Manilkara 87
Margaritolobium Harms 144
luteum (J. R. Johnst.) Harms 144
Meliaceae 25
Meliosma 25
Millettia Wight & Arn. **9,** 14, 17, 18, 19, 20, 24, 37, 46, 50, 144
sect. Podocarpae Dunn 5, 9, 14, 18, 20, 21, 50, 144
atropurpurea (Wall.) Benth. 9
japonica (Sieb. & Zucc.) A. Gray 9
luzonensis A. Gray 50, 93
reticulata Benth. 9
sagraei Benth. 50
Millettieae 1, 2, 3, 5, 7, 8, 9, 10, 11, 15, 16, 17, 24, 28, 36, 37, 42, 46, 86, 102, 141, 144, 145
Mimosa 133
Mimosoideae 16, 33
Mirbelieae 70
Monimiaceae 25
Monoplectra Raf. 43
Moringa 68
Muellera L. 144
Muhlenbergia 133
Musaceae 25
Mutingia 84
Myrospermum Jacq. 11
Myroxylon L. f. 11
Myrtaceae 43
Nama 128
Nectandra 25
Nemacladus 128
Neocracca Kuntze 3, 4, 112
Neolitsea 25
Nipa 27
Nissolia Jacq. 7
Notodon Urb. 2, 4, 78
gracilis (Griseb.) Urb. 79
Ocotea 25
Oleaceae 25
Olneya A. Gray 3, 5, 6, 15, 16, 17, 18, 19, 21, 23, 36, 39, 86, 106, 107, 112, **117–119,** 121, 123, 124, 125, 132, 142
tesota A. Gray **119–124,** 145, 152
Opuntia 133, 138
Ormocarpinae 3, 5, **7,** 18, 19, 21
Ormocarpopsis R. Vig. 7
Ormocarpum P. Beauv. 7
Ormosia 33
Ostryocarpus Hook. f. 5, **7,** 18, 19, 21
Ostryoderris Dunn 7
Pachira 87
Papilionoideae 1, 2, 3, 13, 22, 32, 33, 36, 103
Paraderris (Miq.) Geesink 8, 15
Penstemon 128
Persea 25
Peteria A. Gray 2, 3, 5, 6, 15, 16, 17, 18, 19, 21, 23, 36, 37, 39, 106, 107, 114, 118, 119, **124–127,** 134, 135, 138, 145
glandulosa (A. Gray ex S. Watson) Rydb. 121, 127, **133–134,** 135, 152
nevadensis Tidestrom 126, 128
pinetorum C. L. Porter 121, 126, 127, **134–135,** 152
scoparia A. Gray 121, 124, 126, 127, **131–133,** 134, 135, 152
var. *glandulosa* A. Gray ex S. Watson 133
thompsonae S. Watson 103, 121, 126, **127–131,** 135, 152
Peteria Raf. 124
Phaseoleae 70
Philenoptera A. Rich. 5, **8,** 18, 19, 21
Pictetia DC. 7
Pinaceae 25
Pinus 25, 65, 72, 128, 132
Piscidia L. 87, 144
Platanus 25
Platycarya 26, 27
Platycyamus Benth. 5, 6, **8,** 11, 13, 15, 18, 19, 21, 22, 24
Poa 128
Poaceae 128, 132
Poecilanthe Benth. 3, 5, 6, **7,** 11, 13, 18, 19, 21, 24, 145
grandiflora Benth. 145
Poissonia Baill. 3, 4, 111
solanacea Baill. 111

Poitea Vent. 2, 3, 4, 5, 6, 9, 10, 11, 12, 13, 14, 15, 16, 17, 18, 19, 21, 22, 23, 35, 36, 38, 42, 46, 51, 52, 67, **75–81,** 82, 86, 112, 132, 142, 152
sect. *Eupoitea* Urb. 75
sect. Corynella (DC.) Lavin **79, 81**
sect. Poitea **78–79**
sect. Sabinea (DC.) Lavin **79**
sect. Sauvallella (Rydb.) Lavin **79**
sect. *Vilmorinia* (DC.) Urb. 78
campanilla DC. 70, 79
carinalis (Griseb.) Lavin 79
dubia (Poir.) Lavin 79, 81
florida (Vahl) Lavin 11, 15, 78, 79, 80
galegoides Vent. 15, 71, 75, 78, 79
var. galegoides 79
var. stenophylla Ekman ex Lavin 79
glyciphylla (Poir.) Urb. 79
gracilis (Griseb.) Lavin 79, 81
immarginata (C. Wright) Lavin 78, 79
longiflora Urb. 79
multiflora (Sw.) Urb. 78, 79
paucifolia (DC.) Lavin 67, 81
punicea (Urb.) Lavin 79
Populus 25
Porophyllum 138
Poulsenia 96
Prosopis 128, 132
Pseudosmodingium 96
Pterocarpus Jacq. 33, 34, 96
Pterocarya 26
Pterosesbania 42
Quercus 25, 65, 72, 132
Rehsonia Stritch 9, 10
Resupinaria Raf. 43
grandiflora (L.)Raf. 43
Rhacoma 96
Robinia L. 3, 5, 6, 7, 9, 10, 13, 14, 15, 16, 17, 18, 19, 21, 22, 23, 33, 35, 36, 37, 39, 52, **102–108,** 109, 118, 124, 125, 126, 127, 137, 141, 142, 152
cubensis H. B. K. 47
ehrenbergii Schltdl. 71
hispida L. 105, 106, 107, 108
maculata H. B. K. 86
melanocarpa Schltdl.60
neomexicana A. Gray 104, 105, 106, 107
pseudoacacia L. 102, 105, 106, 107, 108, 146
rosea Mill. 93
sepium Jacq. 93
spilophylla Standl. 144
variegata Schltdl. 93
viscosa Vent. 105, 106, 107, 108
Robinieae **35–37**
Robiniinae 2, 3, 35
Rosaceae 25
Sabiaceae 25
Sabinea DC. 2, 4, 78, 79
sect. *Eusabinea* Urb. 79
sect. *Sabineopsis* Urb. 79
punicea Urb. 79
Salicaceae 25, 27
Salix 25
Salvia 128
Sapium 25
Sarcobatus 128
Sauvallella Rydb. 4, 78, 79
Schefflerodendron Harms 5, **9,** 12, 14, 17, 18, 19, 21
Selaginella 138
Senna 68, 87, 96
Sequoia 25
Sesban Adans. 39
Sesbania Adans. 2, 3, 5, 6, 7, 9, 14, 16, 17, 18, 19, 21, 22, 23, 24, 28, 29, 30, 31, 35, 36, 37, 38, **39–45,** 70, 82, 109, 112, 124, 132, 137, 152
subg. *Agati* (Adans.) Baker 43
subg. *Daubentonia* (DC.) Baker 44
subg. *Pterosesbania* Gillett 44
sect. Daubentonia (DC.) Benth. 40, 42, 43, **44**
sect. Daubentoniopsis Lavin 40, 42, **43–44**
sect. Glottidium Lavin 13, 40, 42, 43, **44–45**
sect. Sesbania 13, 14, 40, 41, 42, **43,** 44
brenningii Harms 82, 83
emerus (Aubl.) Urb. 43
grandiflora (L.) Poir. 40, 42, 43
herbacea 41
longifolia (Cav.) DC. 43, 44
macrantha Phill. & Hutch. 44
pachycarpa DC. 44
punicea (Cav.) Benth. 44
rogersii Phill. & Hutch. 44
sesban (L.) Merr. 39
tetraptera Hochst. ex Baker 44
tripettii Aust. 44
vesicaria (Jacq.) Ell. 13, 44
virgata Poir. 44
Sesbanieae 35
Sesbaniinae Benth. 3, 35, 42
Siparuna 25
Solanum 96
Sophoreae 11, 13
Sphinctospermum Rose 3, 5, 7, 10, 15, 16, 17, 18, 19, 21, 23, 35, 39, 106, 107, 118, 119, 125, 126, 134, 137, **140–142,** 152
constrictum (S. Watson) Rose 137, 140, **142–143**
Strelitziaceae 27
Styphnolobium Schott ex Endl. 33, 34
Swartzia Schreb. 32
Tabebuia 87, 96
Talisia 87
Taraxacum 106

Taxodiaceae 25
Taxodium 25
Templetonia 13
Tephrosia Pers. 2, 3, 5, **8–9,** 14, 18, 19, 21, 22, 125, 138, 141, 145
  subg. Macronyx (Dalz.) Hook. 3, 141
  sect. *Craccoides* DC. 113
  constricta S. Watson 142
Tephrosieae 3
Terminalia 25
Tetracentron 26
Tetracera 25
Thamnosma 128
Tilia 25
Tiliaceae 25
Tipuana (Benth.) Benth. 33, 34
Townsendia 128
Toxotropis Turcz. 79
  jaegeri Turcz. 79
Trichilia 96
Ulmaceae 25
Ulmus 25
Vauquelinia 132
Vilmorinia DC. 78
  multiflora (Sw.) DC. 78
Wisteria Nutt. 4, 5, **9–10,** 14, 16, 17, 18, 19, 20, 21, 22, 24, 32, 37
  frutescens (L.) Poir. 10, 16
Xeroderris Roberty 7
Xylocopa 47
Yucaratonia Burkart 3, 81
  brenningii (Harms) Burkart 81, 82, 83
Yucca 128, 132
Zelkova 25